T0181020

Texts and Readings in Physical Sciences

Volume 20

The *Texts and Readings in Physical Sciences* series publishes high-quality textbooks, research-level monographs, lecture notes and contributed volumes. Undergraduate and graduate students of physical sciences and applied mathematics, research scholars and teachers would find this book series useful. The volumes are carefully written as teaching aids and highlight characteristic features of the theory. The books in this series are co-published with Hindustan Book Agency, New Delhi, India.

More information about this series at http://www.springer.com/series/15139

Subhashish Banerjee

Open Quantum Systems

Dynamics of Nonclassical Evolution

HINDUSTAN
BOOK AGENCY

 Springer

Subhashish Banerjee
Department of Physics
Indian Institute of Technology Jodhpur
Jodhpur, Rajasthan, India

ISSN 2366-8849   ISSN 2366-8857   (electronic)
Texts and Readings in Physical Sciences
ISBN 978-981-13-4815-0   ISBN 978-981-13-3182-4   (eBook)
https://doi.org/10.1007/978-981-13-3182-4

Contents

List of Figures

About the Author

Subhashish Banerjee is teaching at the Department of Physics, Indian Institute of Technology Jodhpur, India. He obtained his PhD degree from the School of Physical Sciences, Jawaharlal Nehru University, New Delhi, India, in 2003. He has been involved in the study of the theory of open quantum systems since his Ph.D. A major theme of his work has been to see how the theory of open quantum systems provides a common umbrella to understand various facets of quantum optics, quantum information processing, quantum computing, quantum cryptography, the foundations of quantum mechanics, relativistic quantum mechanics and eld theory. He has published two books and more than 70 articles in international journals of repute.

Preface

Open quantum systems is the study of quantum dynamics of the system of interest, taking into account the effects of the ambient environment. It is ubiquitous in the sense that any system could be envisaged to be surrounded by its environment which could naturally exert its influence on it. It traces its roots from Quantum Optics and has found applications in diverse areas, ranging from condensed matter to quantum cosmology. Open Quantum Systems allows for a systematic understanding of irreversible processes such as decoherence and dissipation, of essence in order to have a correct understanding of realistic quantum dynamics and also for possible implementations. This would be essential for a possible development of quantum technologies. Interest has been revived in recent times due to the upsurge of theoretical and experimental progress.

We try to put down in this book, in a comprehensive manner, the basic ideas of open quantum systems and the tools needed for the same. Emphasis is given to both the traditional master equation as well as the functional (path) integral approaches. In fact, this book can be used as a beginning guide for understanding and use of path integrals. The basic paradigm of open systems, the harmonic oscillator and the two-level system are discussed in detail. The traditional topics of dissipation and tunneling as well as the modern field of quantum information find a prominent place in the text.

Despite its importance, the subject of Open Quantum Systems is not present in the curriculum of Indian Universities and Institutes; it is treated, at best, as an abstruse subject. One of the main goals, and *hopes*, of this book would be to bring a change in this scheme of things. Assuming a basic background of quantum and statistical mechanics, this book will help to familiarize the reader with the basic tools of open quantum systems.

This book is aimed at taking a reader with a basic background of quantum and statistical mechanics to the level where he/she can start appreciating research problems of current interest. A good background of undergraduate physics should suffice to begin with the present book. In any case, an introductory chapter on quantum statistical mechanics and path integrals are included, with references to more advanced literature. The book aims to highlight the ubiquity of Open Quantum Systems based on simple models and calculations.

As I reflect back, I find that there are a number of people to whom I owe the development of this book. The first person who comes to mind is Prof. R.

Ramaswamy who has been a teacher and friend to me for a long time. It was his suggestion that started this project. His constant help and advice, along with that of Prof. Debashis Ghoshal and Mr. D. K. Jain of Hindustan Book Agency made this book possible. If I were to trace the roots of my involvement with this subject, then I would say that it started with the works of Profs. A. O. Caldeira and A. J. Leggett. I also immensely benefited from the works of Prof. Vinay Ambegaokar, with whom I was fortunate to have a brief interaction, and Profs. H. Grabert, G. Ingold, P. Hanggi, G. S. Agarwal and B. L. Hu. My sincere thanks to all of them. I have also benefitted very much from the classic book on the subject by Profs. H-P. Breuer and F. Petruccione, with both of whom I have had the opportunity of some interaction. Over the years there have been a number of people in the scientific community to whom I have looked up to for inspiration and this would be an appropriate juncture to thank them. They are Profs. R. Ghosh, R. Rajaraman, (late) D. Kumar, R. Simon, J. Kupsch, G. Rajasekaran, H. S. Mani and C. S. Seshadri. My journey in this field would not have been possible without discussions and collaborations with a number of colleagues: Richard Mackenzie, Andreas Buchleitner, Christophe Couteau, Sibasish Ghosh, C. M. Chandrashekar, V. V. Sreedhar, R. Jagannathan, R. Parthasarathy, Abhishek Dhar, R. Srikanth, (late) N. Kumar, Hema Ramachandran, A. R. Usha Devi, A. K. Rajagopal, Subhash Chaturvedi, Subhasish Dutta Gupta, Arun Jayanavar, Pankaj Agrawal, Arun Pati, Debasis Sarkar, Archan S. Majumdar, Guruprasad Kar, Somshubhro Bandyopadhyay, Prasanta K. Panigrahi, Anirban Pathak, V. Ravishankar, Sankalpa Ghosh, Sandeep Goyal, George Thomas and S. Uma Sankar. In particular, I need to thank S. Omkar, Pradeep Kumar, N. Siddharth, Javid A. Naikoo and Supriyo Dutta for their extensive help with the numerical and formatting related issues associated with the book. I also want to express my gratitude to my family and friends for all their help and support. I conclude by thanking my wife Pallavi and son Shubhonkar for making me realize there is more to life than open quantum systems.

Chapter 1

Introduction

Quantum theory of open systems represents a very important problem in quantum-statistical mechanics as it attempts to provide a natural route for reconciliation of non-unitary processes such as damping and dephasing or decoherence with the process of quantization [1, 2]. One starts with the conservative composite closed system consisting of the system of interest and its environment to which the standard rules of quantization are applied. The total system, comprising of the system of interest and its ambient environment, are evolved via a unitary evolution and then the environmental coordinates are eliminated to give a closed equation (reduced dynamics) for the dissipative system alone. In this picture, friction comes about by the transfer of energy from the "small" system (the system of interest) to the "large" environment. The energy, once transferred, disappears into the environment and is not given back within any time of physical relevance (but only in the so-called Poincaré recurrence time).

There are various approaches to the quantum theory of open systems. The traditional approaches include the master equation and Langevin equation approach [3, 4], but in a number of scenarios, it is seen that the functional integral (path integral) approach [5] provides a practical method of description. Quantum optics provided one of the first testing grounds for the application of the formalism of open quantum systems [6, 7]. Application to other areas was intensified by the works of [8, 9, 10] and [11], among others. The recent upsurge of interest in the problem of open quantum systems is because of the spectacular progress in manipulation of quantum states of matter, encoding, transmission and processing of quantum information, for all of which understanding and control of the environmental impact are essential [12, 13, 14].

There are many scenarios in nature which can be described by a system with one or few degrees of freedom in contact with a complex environment whose number of degrees of freedom is very large (tending to infinity). The coupling of a system, with few quantum degrees of freedom, to a thermal reservoir results in fluctuating forces reflecting the characteristics of the environment (reservoir or bath) and the coupling. In the classical regime, the dynamics of such systems is described by a Langevin equation which is a phenomenolog-

© Hindustan Book Agency 2018 and Springer Nature Singapore Pte Ltd. 2018
S. Banerjee, *Open Quantum Systems*, Texts and Readings in Physical Sciences 20,
https://doi.org/10.1007/978-981-13-3182-4_1

ical equation with a frictional force proportional to the velocity and driven by a fluctuating force. A prototype example of this is the theory of Brownian motion. One of the earlier reviews of this is [15].

The approaches to quantum open systems can be broadly classified into two categories. They either modify the procedure of quantization or use the system-plus-reservoir approach. In the context of the former approach, Kostin [16] introduced a theory with a nonlinear Schrödinger equation. The same equation was found later in [17] using Nelson's stochastic quantization procedure [18]. However, besides violating the superposition principle, this theory shows some highly controversial results such as stationary damped states. In [19] a canonical quantization procedure was developed using complex variables. Despite reproducing some interesting results, such as the Fokker-Planck equation for the Wigner transform of the density operator, the theory appears obscure in some points such as the unphysical noise source for the equation of motion and another one for the momentum equation. The more natural and successful approach is the system plus reservoir method of quantum open systems [1, 2] and will be used consistently in this book.

Another approach to open systems has been developed over the last few decades, i.e., the Stochastic Schrödinger Equations (SSEs) approach that evolves the wave function of the system as a vector, in the Hilbert space of the system, following a stochastic trajectory. The resultant reduced density matrix can be recovered as a sum of the projectors of stochastic trajectories. Depending on the method used in the derivation, there are many different SSEs that recover the reduced density matrix of the open system and are called unravelings of the reduced density matrix [20, 21]. In this book, we will not dwell on this approach.

As stated above, the formalism of open quantum systems allows for a natural description of processes such as decoherence and dissipation, due to the influence of the environment on the system of interest. It is ubiquitous in the sense that any system could be envisaged to be surrounded by its environment which could naturally exert its influence on it. This becomes clearer when we try to gauge the range of applications of the ideas of open quantum systems, from quantum optics to condensed matter physics and issues related to quantum cosmology and quantum gravity to the recent developments in quantum information processing; some of which will be covered, in due course, here.

We try to put down in this book, in a comprehensive manner, the basic ideas of open quantum systems and the tools needed for the same. Emphasis is given to both the traditional master equation as well as the functional integral approaches. The basic paradigm of open systems, and perhaps of all of physics, the harmonic oscillator is studied in detail from a number of different perspectives. The other paradigm model of open quantum systems is the two-level system, also called, in the parlance of quantum information processing, the qubit. This is studied using path integral methods. As a matter of fact, this book can be used as a beginning guide for understanding and use of path integrals, needless to say, appropriately supplemented by a number of great books on the subject, some of which find their place in the bibliography.

This book has been written keeping in mind an advanced undergraduate or a graduate student wanting to learn about open quantum systems. For this reason, an introduction is made to the various tools of quantum statistical mechanics and path integrals. This is followed by a discussion of master equations and the influence functional approach, a path integral method used here. These form the core required for a coherent understanding of the rest of the material presented. The tools developed are next applied to the two paradigm models of open systems, *viz.* the dissipative harmonic oscillator and the dissipative two-level system. The dissipative harmonic oscillator is studied using both master equation and path integral techniques. Discussion of the dissipative two-level system follows naturally into the problem of quantum tunneling, which is studied here using path integral techniques. As we have had occasion to remark earlier, the field of quantum information provides a natural breeding ground for ideas of open quantum systems. This endeavor is thus undertaken next. The book culminates with a brief discussion of some modern trends in the use of open system ideas. A number of results obtained with the help of different colleagues appear in this book. They follow naturally from the formalism developed. Problems are given intermittently and serve the purpose of further sharpening the arguments presented.

In the process of elucidating various aspects of open quantum systems, well known models from different fields of study, such as the Lindblad evolution and dissipative Jaynes-Cummings model, of importance in quantum optics, Caldeira-Leggett and the spin-Boson models are also discussed. Applications of open quantum systems to quantum optics, quantum information, condensed matter and high energy physics can easily be gleaned from the material. This should, hopefully, serve to highlight the ubiquity of open quantum systems. So without further ado, let us begin our journey.

Chapter 2

A Primer on Quantum Statistical Mechanics and Path Integrals

2.1 Introduction

In this chapter, we will focus on some of the basic tools of quantum statistical mechanics as well as get introduced to the subject of path integration. These are vast enterprises in themselves and are needed to have an understanding of the subject of open quantum systems. Here we introduce some of the basic concepts in quantum statistical mechanics and path integration. The harmonic oscillator, which has a ubiquitous presence in all realms of physics, will be discussed in detail, both from the perspective of quantum statistical mechanics as well as path integration. One of the themes pursued in this chapter is to discuss concepts from dual point of views, that is, using ideas of (conventional) quantum statistical mechanics as well as by using path integration. For example, we discuss the partition function as well as the evolution of the density matrix from this perspective. This, we believe, would encourage the reader to develop a global viewpoint on the issues studied. Reference is made to advanced literature on the subject in the end.

2.2 Quantum Statistical Mechanics

Statistical mechanics is about trying to find how macrosystems emerge from their microscopic origins. Often, this requires also the application of the laws of quantum physics. The amalgamation of statistical and quantum mechanical ideas is quantum statistical mechanics. As the title of this chapter suggests, what we will discuss here will be the bare rudiments of this vast subject. However, we feel that the topics discussed are suffice to give the reader an appreciation of the subject. Thus, we talk about the basic setup of quantum mechanics; how states and operators are defined, their transformations, the various pictures used in different contexts. We then discuss the Baker Camp-

© Hindustan Book Agency 2018 and Springer Nature Singapore Pte Ltd. 2018
S. Banerjee, *Open Quantum Systems*, Texts and Readings in Physical Sciences 20,
https://doi.org/10.1007/978-981-13-3182-4_2

bell Hausdorff theorem, which is very useful in computations with operators; we come across operators repeatedly in quantum statistical mechanics. The study of open quantum systems, the principal object of this book, is basically a theory of quantum statistical mechanics. Here we encounter mixed states frequently. They are handled conveniently by density matrices. The ubiquitous harmonic oscillator, without a discussion of which no study of quantum statistical mechanics would be appropriate, is then discussed using the method of annihilation and creation operators. We then briefly discuss the notions of the partition function and entropy. A guide is provided to more complete literature.

2.2.1   States, Operators, Evolutions and Transformations

A quantum mechanical system lives, *mathematically*, in a Hilbert space $\mathcal{H}$. This is a complex, complete, linear vector space equipped with a positive semi-definite inner product [22, 23]. A very convenient representation of quantum *states*, and one which we will follow consistently in this book, is the Dirac bra and ket formalism [24]. In this, a state is represented by $|\psi\rangle$, called *ket psi* and is a column vector whose entries are, in general, complex numbers. Its Hermitian conjugate is $\langle\psi|$, called *bra psi*. As an example, the states of a two-level system, also called *qubit* in the parlance of quantum information, are

$$|0\rangle = \begin{pmatrix} 1 \\ 0 \end{pmatrix}, \quad |1\rangle = \begin{pmatrix} 0 \\ 1 \end{pmatrix}. \tag{2.1}$$

A complex, linear vector space is a set of elements (vectors), which are closed under addition and admit multiplication with complex scalars, which is linear and associative [23, 25]. Further, there exists a unique zero and identity. An important concept in these issues is the notion of an inner product. Given two states $|\psi\rangle$ and $|\phi\rangle$, the inner product between them can be defined by $\langle\psi|\phi\rangle$.

A vector space which is equipped with an inner product is called an inner product space. The inner product obeys the properties of Hermiticity, that is, it is equal to its transposed conjugate counterpart and linearity. Further, the notion of an inner product allows for the definition of a distance function on the state space. Thus, we have that $\langle\psi|\psi\rangle \geq 0$, with equality if $|\psi\rangle = 0$. This is the property of positive semi-definiteness of the inner product. The length or the norm of a vector $|\psi\rangle$ is

$$\left\| |\psi\rangle \right\| = \sqrt{\langle\psi|\psi\rangle}. \tag{2.2}$$

Thus we see that the length of a vector is related to its inner product. Another important concept is that of a *basis*. A basis is a set of vectors which is linearly independent and complete. This implies that if a vector space $\mathcal{V}$ is spanned by a basis, say for example $\{|\phi_i\rangle\}$, $i = 1, \cdots n$, then any vector $|\psi\rangle$ in $\mathcal{V}$ can be expressed as a linear combination of the basis, that is, $|\psi\rangle = \sum_i^n |\phi_i\rangle$, where n is the number of elements or vectors in $\mathcal{V}$. Another crucial concept, related to the basis, is its completeness, which implies that $\sum_i^n |\phi_i\rangle\langle\phi_i| = 1$. When the basis spanning the vector space is both linearly independent and complete, it follows that the number of elements of the basis is equal to the dimension of the vector space.

An *operator* can be represented conveniently by an *outer product*, for example, given two vectors $|\psi\rangle$ and $|\phi\rangle$, an operator can be constructed as $|\psi\rangle\langle\phi|$, which from simple matrix multiplication can be seen to have the form of a matrix, in contrast to an inner product, which would be a number. The general definition of an operator is that it acts on a vector to produce another vector

$$\left(|\psi\rangle\langle\phi|\right)|\chi\rangle = \langle\phi|\chi\rangle|\psi\rangle = c|\psi\rangle, \tag{2.3}$$

where c is a complex number. The form of the matrix representing an operator depends upon the basis chosen to represent the matrix elements. The basis in which the matrix representation of the operator is diagonal is called the diagonal basis. In another basis, the matrix may not be diagonal. The sum of diagonal elements of a matrix representation of an operator is called its *trace*. The trace operation is independent of the basis chosen to represent the operator and is cyclic. Given two operators A and B,

$$\mathrm{Tr}(AB) = \mathrm{Tr}(BA). \tag{2.4}$$

This can be proved as follows:

$$
\begin{aligned}
\mathrm{Tr}(AB) &= \sum_i \langle i|AB|i\rangle = \sum_{i,j} \langle i|A|j\rangle\langle j|B|i\rangle \\
&= \sum_{i,j} \langle j|B|i\rangle\langle i|A|j\rangle = \sum_j \langle j|BA|j\rangle \\
&= \mathrm{Tr}(BA).
\end{aligned}
\tag{2.5}
$$

Here we have made use of the completeness of basis $\sum_i |i\rangle\langle i| = 1$ and the fact that an inner product is an ordinary c number and can be moved in any order. Trace is used to compute the average of an operator in a given state $\langle A\rangle_{|\psi\rangle}$ as

$$\langle A\rangle_{|\psi\rangle} = \langle\psi|A|\psi\rangle = \mathrm{Tr}(A|\psi\rangle\langle\psi|). \tag{2.6}$$

Problem 1: Prove that for operators A, B and C, the following holds: $\mathrm{Tr}(ABC) = \mathrm{Tr}(BCA) = \mathrm{Tr}(CAB)$.

We have defined in Eq. (2.3), the basic action of an operator. A special case of this is the eigenvalue equation of operators,

$$A|\psi\rangle = \lambda|\psi\rangle. \tag{2.7}$$

Here, $|\psi\rangle$ is the eigenvector and λ is the eigenvalue of the operator A. Two kinds of operators find prominent use in quantum mechanics, viz. the Hermitian and Unitary operators. An operator is Hermitian if it is equal to its transpose conjugate, also known as the adjoint of the operator and represented by the symbol †, while the adjoint of a unitary operator yields its inverse. It can be easily shown that the eigenvalues of a Hermitian operator are real, while those of a unitary operator are complex, with unit modulus [23, 25]. Since the eigenvalues of a Hermitian operator is real, it can be used to represent physical observables, such as energy and momentum. On the other hand, the unitary operators are used for the evolution of the state

$$|\psi(t)\rangle = U(t,0)|\psi(0)\rangle. \tag{2.8}$$

This implies that the unitary operator $U(t,0)$ evolves the state $|\psi\rangle$ from time $t = 0$ to time t. From this equation, it can be seen that the unitary evolution preserves the state norm, that is,

$$\langle\psi(t)|\psi(t)\rangle = \langle\psi(0)|U^\dagger U|\psi(0)\rangle = \langle\psi(0)|\psi(0)\rangle, \tag{2.9}$$

because $U^\dagger U = \mathcal{I}$.

Problem 2: A well known matrix, for a two level system, is the Pauli matrix:

$$\sigma_Z = \begin{pmatrix} 1 & 0 \\ 0 & -1 \end{pmatrix}.$$

Find the eigenvalues and eigenvectors of this matrix. This is the matrix representation of the Pauli-Z operator in the computational basis $\{|0\rangle, |1\rangle\}$, given in Eq. (2.1).

Another important transformation, effected on operators, is the similarity transformation. In quantum mechanics, unitary operators are used to effect this transformation. Thus, for example, given an operator R and a unitary matrix U, built from the eigenvectors of R, the similarity transformation would be

$$D = U^\dagger RU, \tag{2.10}$$

where D is the diagonalized form of the matrix R. We could say that D and R are similar. Commutators and anti-commutators are very important operations associated with operators. Given two operators A and B, their commutator would be $\left[A, B\right] = AB - BA$, while their anti-commutator would be $\left[A, B\right]_+ = AB + BA$. An interesting aspect of this is that if $\left[A, B\right] = 0$ and we assume non-degeneracy, that is, different eigenvectors correspond to different eigenvalues, then A and B can be measured simultaneously. We conclude this discussion by stating a useful representation of an operator, its spectral representation. Given an operator A, its spectral representation is given by

$$A = \sum_{i=1}^{n} \lambda_i |i\rangle\langle i|. \tag{2.11}$$

Here λ_i, $|i\rangle$ are the eigenvalues and eigenvectors of the operator A, respectively.

Problem 3: Prove the relation in Eq. (2.11).

2.2.2   Various Pictures

In applications of quantum mechanics, we often make use of various pictures. The most prominent among them are the Heisenberg, Schrödinger and the interaction pictures [23, 26]. In the Heisenberg picture, the time dependence is carried by the operators while the state vector is time independent. This is reversed in the Schrödinger picture while the interaction picture is intermediate between the two.

Heisenberg picture: Here the evolution of an operator O^H in time t is given as

$$O^{\mathcal{H}}(t) = e^{\frac{i}{\hbar}Ht}O^{\mathcal{H}}(0)e^{-\frac{i}{\hbar}Ht}. \tag{2.12}$$

Here the superscript $\mathcal{H}$, on the operator $O^{\mathcal{H}}$, denotes Heisenberg. Differentiating the equation with respect to time t yields the equation of motion

$$i\hbar\frac{\partial}{\partial t}O^{\mathcal{H}}(t) = \left[O^{\mathcal{H}}(t), H\right], \tag{2.13}$$

where H is the Hamiltonian. Now $O^{\mathcal{H}}(0)$ in Eq. (2.12) is the operator at time $t = 0$. Also, the state vectors in this picture do not evolve with time, that is, $|\alpha, t\rangle^{\mathcal{H}} = |\alpha, 0\rangle^{\mathcal{H}} = |\alpha\rangle^{\mathcal{H}}$.

Schrödinger picture: Here $O^S = O^{\mathcal{H}}(0)$, while

$$|\alpha, t\rangle^S = e^{-\frac{i}{\hbar}Ht}|\alpha, 0\rangle^S, \tag{2.14}$$

where $|\alpha, 0\rangle^S = |\alpha\rangle^{\mathcal{H}}$. Here the superscript S, on the operator O^S, denotes Schrödinger. Differentiating this with respect to time t yields

$$i\hbar\frac{\partial}{\partial t}|\alpha, t\rangle^S = H|\alpha, t\rangle^S. \tag{2.15}$$

Note that operator averages are unchanged, irrespective of the picture used for computation. Thus, we have

$$\langle\beta, t|O^S|\alpha, t\rangle^S = \langle\beta|e^{\frac{i}{\hbar}Ht}O^Se^{-\frac{i}{\hbar}Ht}|\alpha\rangle^{\mathcal{H}} = \langle\beta|O^{\mathcal{H}}(t)|\alpha\rangle^{\mathcal{H}}. \tag{2.16}$$

The transformations between the two pictures is mediated by canonical transformations, that is, transformations preserving commutation relations. Thus, if $\left[A^{\mathcal{H}}, B^{\mathcal{H}}\right] = C^{\mathcal{H}}$, then $\left[A^S, B^S\right] = C^S$.

Interaction picture: This is intermediate between the above two pictures. It is very suited to discuss scenarios where the total Hamiltonian can be split up into the free Hamiltonian and an interaction Hamiltonian $H = H_0 + H_I$.

Here the free term is H_0, while the interacting part is H_I. In this picture, the operators evolve as

$$O^I(t) = e^{\frac{i}{\hbar}H_0 t} O^S e^{-\frac{i}{\hbar}H_0 t}, \qquad (2.17)$$

while the state vector evolves as

$$|\alpha, t\rangle^I = e^{\frac{i}{\hbar}H_0 t}|\alpha, t\rangle^S. \qquad (2.18)$$

Here the superscript I denotes the interaction picture. For $H_I = 0$, $O^I(t) = O^\mathcal{H}(t)$ and $|\alpha, t\rangle^I = |\alpha\rangle^\mathcal{H}$; while for $t = 0$, $O^I(0) = O^S = O^\mathcal{H}(0)$ and $|\alpha, 0\rangle^I = |\alpha\rangle^\mathcal{H} = |\alpha, 0\rangle^S$. Differentiating Eq. (2.18) we get

$$i\hbar\frac{\partial}{\partial t}|\alpha, t\rangle^I = H_I^I|\alpha, t\rangle^I, \qquad (2.19)$$

while differentiating Eq. (2.17) we get

$$i\hbar\frac{\partial}{\partial t}O^I(t) = \left[O^I(t), H_0\right]. \qquad (2.20)$$

Here H_I^I is the interaction Hamiltonian in the interaction picture.

Problem 4: Derive Eqs. (2.19) and (2.20).

2.2.3 Baker Campbell Hausdorff (BCH) Theorem

A crucial thing with operators, is that functions of operators cannot be factorized like functions of ordinary (c) numbers. This is captured by the BCH theorem [27], and comes in handy in many applications. Consider the function $f(\xi)$ of the parameter ξ:

$$f(\xi) = e^{\xi\hat{A}}e^{\xi\hat{B}}. \qquad (2.21)$$

Differentiation with respect to the parameter ξ gives

$$\frac{\partial f}{\partial \xi} = (\hat{A} + e^{\xi\hat{A}}\hat{B}e^{-\xi\hat{A}})f(\xi). \qquad (2.22)$$

In this sub-section, we will denote operators with a hat on top. The nontrivial term in the above equation is $e^{\xi\hat{A}}\hat{B}e^{-\xi\hat{A}}$. If $\hat{A}$ and $\hat{B}$ had been c numbers, then this would become B. However, as we will see below, this term will have a non-trivial expansion, due to the operator nature of $\hat{A}$ and $\hat{B}$. In fact, it turns out that

$$e^{\xi\hat{A}}\hat{B}e^{-\xi\hat{A}} = \hat{B} + \xi[\hat{A}, \hat{B}] + \frac{\xi^2}{2}[\hat{A}, [\hat{A}, \hat{B}]] + \cdots. \qquad (2.23)$$

This can be proved as follows, using the technique of parametric differentiation. Let

$$\hat{g}(\xi) = e^{\xi\hat{A}}\hat{B}e^{-\xi\hat{A}}, \cdots \hat{g}(0) = \hat{B}. \qquad (2.24)$$

Then,

$$\frac{\partial \hat{g}}{\partial \xi} = [\hat{A}, \hat{g}(\xi)], \cdots \frac{\partial \hat{g}}{\partial \xi}\Big|_{\xi=0} = [\hat{A}, \hat{B}],$$

$$\frac{\partial^2 \hat{g}}{\partial \xi^2} = [\hat{A}, [\hat{A}, \hat{g}(\xi)]], \cdots \frac{\partial^2 \hat{g}}{\partial \xi^2}\Big|_{\xi=0} = [\hat{A}, [\hat{A}, \hat{B}]]. \tag{2.25}$$

$$\tag{2.26}$$

Using Taylor's expansion, we have

$$\hat{g}(\xi) = \hat{g}(0) + \xi \frac{\partial \hat{g}}{\partial \xi}\Big|_{\xi=0} + \frac{\xi^2}{2} \frac{\partial^2 \hat{g}}{\partial \xi^2}\Big|_{\xi=0} + \dots. \tag{2.27}$$

This proves Eq. (2.23). Setting $\xi = 1$ in Eq. (2.23), we get

$$e^{\hat{A}}\hat{B}e^{-\hat{A}} = \hat{B} + [\hat{A}, \hat{B}] + \frac{1}{2}[\hat{A}, [\hat{A}, \hat{B}]] + \dots. \tag{2.28}$$

Problem 5: Let $\hat{A} = \frac{i}{\hbar}\hat{p}, \hat{B} = \hat{q}$ and the parameter $\xi \in \mathbb{R}$, such that $[\hat{q}, \hat{p}] = i\hbar$. Show that

$$e^{\frac{i}{\hbar}\hat{p}\xi}\hat{q}e^{\frac{-i}{\hbar}\hat{p}\xi} = \hat{q} + \xi.$$

Thus, momentum $\hat{p}$ is the generator of displacement $\hat{q}$.

If $[\hat{A}, [\hat{A}, \hat{B}]] = 0 = [\hat{B}, [\hat{A}, \hat{B}]]$, this is so when $[\hat{A}, \hat{B}]$ is a c number, then, from Eq. (2.23), we see that $e^{\xi\hat{A}}\hat{B}e^{-\xi\hat{A}} = \hat{B} + \xi[\hat{A}, \hat{B}]$. Parametric differentiation of the function $f(\xi)$, Eq. (2.21), gives

$$\frac{\partial f}{\partial \xi} = (\hat{A} + \hat{B} + \xi[\hat{A}, \hat{B}])f(\xi). \tag{2.29}$$

It can be seen that, in the above equation, $[\hat{A} + \hat{B}, [\hat{A}, \hat{B}]] = [\hat{A}, [\hat{A}, \hat{B}]] + [\hat{B}, [\hat{A}, \hat{B}]] = 0$. This allows for the factorization of $(\hat{A} + \hat{B}) + \xi[\hat{A}, \hat{B}]$ as ordinary commuting variables. Using $f(0) = 1$, the solution of Eq. (2.29) is given by

$$f(\xi) = e^{((\hat{A}+\hat{B})\xi + \frac{\xi^2}{2}[\hat{A}, \hat{B}])}$$

$$= e^{\xi(\hat{A}+\hat{B})}e^{\frac{\xi^2}{2}[\hat{A}, \hat{B}]}. \tag{2.30}$$

Setting $\xi = 1$ we get

$$e^{\hat{A}}e^{\hat{B}} = e^{(\hat{A}+\hat{B})}e^{\frac{1}{2}[\hat{A}, \hat{B}]}, \tag{2.31}$$

$$e^{(\hat{A}+\hat{B})} = e^{\hat{A}}e^{\hat{B}}e^{\frac{-1}{2}[\hat{A}, \hat{B}]}. \tag{2.32}$$

These are two well-known forms of the BCH theorem.

Problem 6: Let there be two $n \times n$ matrices A and B such that $[A, [A, B]] = [B, [A, B]] = O$. Show that:

$$
\begin{aligned}
e^{A+B} &= e^A e^B e^{-\frac{1}{2}[A,B]} \\
&= e^B e^A e^{\frac{1}{2}[A,B]}.
\end{aligned}
$$

2.2.4　Density Matrices

The usage of density operator is ubiquitous in studies related to Open Quantum Systems. Here we recapitulate some common properties of density operators.

Consider the average of an operator M in Schrödinger picture (SP) (where the state vector evolves with time, but not the operator), in the state $|\psi_S(t)\rangle$ as

$$
\begin{aligned}
\langle M \rangle &= \langle \psi_S(t)|M_S|\psi_S(t)\rangle \\
&= \text{Tr} M_S |\psi_S(t)\rangle\langle\psi_S(t)|.
\end{aligned} \tag{2.33}
$$

In many cases, it is not possible to determine exactly the state $|\psi_S(t)\rangle$ to which the system belongs. The best one can have is the probability p_ψ of the system being in the state $|\psi_S(t)\rangle$. Then the above expression for the operator average becomes modified to

$$
\begin{aligned}
\langle\langle M \rangle\rangle &= \Sigma_\psi p_\psi \text{Tr} M_S |\psi_S(t)\rangle\langle\psi_S(t)| \\
&= \text{Tr} M_S \rho_S(t),
\end{aligned} \tag{2.34}
$$

where the density matrix (or operator) $\rho_S(t)$ is

$$
\rho_S(t) = \Sigma_\psi p_\psi |\psi_S(t)\rangle\langle\psi_S(t)|, \tag{2.35}
$$

and $\Sigma_\psi p_\psi = 1$. The density matrix satisfies two properties:

$$
\begin{aligned}
\text{Tr}\rho &= 1, \\
\text{Tr}\rho^2 &\leq 1,
\end{aligned} \tag{2.36}
$$

with equality for pure and inequality for mixed states.

Problem 7: Prove the above identities for density matrices.

Using the Schrödinger equation

$$
i\hbar \frac{\partial}{\partial t}|\psi\rangle = H|\psi\rangle,
$$

it can be seen easily that Eq. (2.35) yields the following equation of motion

$$i\hbar\frac{\partial}{\partial t}\rho = \left[H, \rho\right].\tag{2.37}$$

This is the well known Schrödinger-vonNeumann equation [27].

Let us consider an example of density matrices, which as we will see in the sequel, is a rather important one, a two-level system representing a qubit. Consider the Hamiltonian

$$H_S = \frac{\hbar\omega}{2}\sigma_z,\tag{2.38}$$

where $\sigma_z = \begin{pmatrix} 1 & 0 \\ 0 & -1 \end{pmatrix}$ is the usual Pauli matrix. Now consider as the system eigenbasis $|j, m\rangle$, these are the well-known Wigner-Dicke states [28]. The eigenvalue equation of H_S is

$$\begin{aligned} H_S|j, m\rangle &= \hbar\omega m|j, m\rangle \\ &= E_{j,m}|j, m\rangle, \end{aligned}\tag{2.39}$$

where $-j \le m \le j$. Consider the initial state to be

$$|\psi(0)\rangle = \cos\left(\frac{\theta_0}{2}\right)|0\rangle + e^{i\phi_0}\sin\left(\frac{\theta_0}{2}\right)|1\rangle.\tag{2.40}$$

By the way, this represents the state of a qubit with θ_0 and ϕ_0 being the polar and azimuthal angles, respectively. The time-evolved density matrix is

$$\rho^s_{m,n}(t) = \begin{pmatrix} \cos^2(\frac{\theta_0}{2}) & \frac{1}{2}\sin(\theta_0)e^{-i(\omega t+\phi_0)} \\ \frac{1}{2}\sin(\theta_0)e^{i(\omega t+\phi_0)} & \sin^2(\frac{\theta_0}{2}) \end{pmatrix}.\tag{2.41}$$

Problem 8: Check for yourself that the state, Eq. (2.41), is attained by evolving the initial state, Eq. (2.40), by the Hamiltonian, Eq. (2.38). Does Eq. (2.41) represent a pure or a mixed state?

We will now derive a very convenient representation of a single field mode, $H = \hbar\omega(a^\dagger a + 1/2)$, basically a harmonic oscillator , in thermal equilibrium, whose state can be written as

$$\rho = \frac{e^{-\beta H}}{\mathcal{Z}}.\tag{2.42}$$

Here

$$\mathcal{Z} = \text{Tr}\left(e^{-\beta\hbar\omega(a^\dagger a+1/2)}\right) = \frac{1}{2\sinh(\frac{\beta\hbar\omega}{2})}.\tag{2.43}$$

Also, using the completeness of basis composed of the eigenstates $|n\rangle$ of the harmonic oscillator, that is, $\sum_n |n\rangle\langle n| = 1$, we have

$$e^{-\beta H} = \sum_n e^{-\beta\hbar\omega(a^\dagger a+1/2)}|n\rangle\langle n| = e^{-\frac{\beta\hbar\omega}{2}}\sum_n e^{-n\beta\hbar\omega}|n\rangle\langle n|.\tag{2.44}$$

Thus, Eq. (2.42) can be written as

$$\rho = \sum_n (1 - e^{-\beta\hbar\omega})e^{-n\beta\hbar\omega}|n\rangle\langle n|. \tag{2.45}$$

We will make use of this representation later. Next, we make use of the state ρ, Eq. (2.45), to present a very useful identity

$$\mathrm{Tr}\left[\rho\exp(x\hat{a}^\dagger + y\hat{a})\right] = \exp\left[\frac{xy}{2}\coth\left(\frac{\beta\hbar\omega}{2}\right)\right]. \tag{2.46}$$

Here x, y are ordinary numbers and $\beta = 1/(k_B T)$, where k_B is the Boltzmann constant and T is the temperature. The proof is sketched below. Consider *LHS*: using Eq. (2.45), the LHS of Eq. (2.46) is

$$(1 - e^{-\beta\hbar\omega})\mathrm{Tr}\left[\sum_n e^{-n\beta\hbar\omega}|n\rangle\langle n|\exp(x\hat{a}^\dagger + y\hat{a})\right]. \tag{2.47}$$

The trace term can be written, using completeness of basis, as

$$\sum_m \sum_n \langle m|e^{-n\beta\hbar\omega}|n\rangle\langle n|\exp(x\hat{a}^\dagger + y\hat{a})|m\rangle. \tag{2.48}$$

Now we make use of the BCH identity

$$e^{(x\hat{a}^\dagger + y\hat{a})} = e^{-xy/2}e^{y\hat{a}}e^{x\hat{a}^\dagger}. \tag{2.49}$$

Note that

$$e^{x\hat{a}^\dagger}|m\rangle = \sum_{\alpha=0}^{\infty} \frac{x^\alpha}{\alpha!}\left(\frac{(m+\alpha)!}{m!}\right)^{1/2}|m+\alpha\rangle. \tag{2.50}$$

Using Eqs. (2.50) and (2.49), Eq. (2.48) can be written as

$$\sum_n e^{-n\beta\hbar\omega}\sum_{\alpha=0}^{\infty} e^{-\frac{xy}{2}}\frac{(xy)^\alpha}{(\alpha!)^2}\frac{(n+\alpha)!}{n!}. \tag{2.51}$$

Hence, the *LHS* of Eq. (2.46) becomes

$$(1 - \zeta)e^{-u/2}\sum_{n,\alpha}\zeta^n\frac{(u)^\alpha}{(\alpha!)^2}\frac{(n+\alpha)!}{n!}. \tag{2.52}$$

Here $xy = u$, $e^{-\beta\hbar\omega} = \zeta$. Using

$$\frac{1}{(1-\zeta)^{\alpha+1}} = \sum_{n=0}^{\infty}\frac{(n+\alpha)!}{\alpha!}\frac{\zeta^n}{n!}, \tag{2.53}$$

Eq. (2.52) can be easily seen to be $\exp\{\frac{u}{2}\frac{1+\zeta}{1-\zeta}\}$, which is equal to $\exp\left[\frac{xy}{2}\coth\left(\frac{\beta\hbar\omega}{2}\right)\right]$, the *RHS* of Eq. (2.46).

Next, we present another useful identity involving a density matrix ρ_{sqth}, representing a squeezed thermal state

$$\rho_{sqth} = \hat{S}(r,\phi)\rho_{th}\hat{S}^\dagger(r,\phi), \tag{2.54}$$

where ρ_{th} represents the thermal bath and is given in Eq. (2.45). Also, $\hat{S}(r,\phi)$ is the squeezing operator [29] and is given by

$$\hat{S}(r,\phi) = \exp\left[r\left(\frac{\hat{a}^2}{2}e^{-i2\phi} - \frac{\hat{a}^{\dagger^2}}{2}e^{i2\phi}\right)\right]. \tag{2.55}$$

Here r and ϕ are the amplitude and phase of squeezing, respectively, while $\hat{a}$ and $\hat{a}^\dagger$ are the harmonic annihilation and creation operators, respectively. The identity we are interested in, and which we will make use of later, is

$$\text{Tr}\left[\rho_{sqth}\hat{D}(x-x')\right] = \exp\left[-\frac{1}{2}\coth\left(\frac{\beta\hbar\omega}{2}\right)\left|(x-x')\cosh(r) + (x-x')^*\sinh(r)e^{i2\phi}\right|^2\right]. \tag{2.56}$$

Here

$$\hat{D}(x) = \exp\left[x\hat{a} - x^*\hat{a}^\dagger\right], \tag{2.57}$$

is the displacement operator [29]. The proof of Eq. (2.56) is sketched below. The *LHS* can be re-written, using the cyclicity of trace, as

$$\text{Tr}\left[\rho_{th}\hat{S}^\dagger(r,\phi)\hat{D}(x-x')\hat{S}(r,\phi)\right]. \tag{2.58}$$

Using the identity

$$\hat{S}^\dagger(r,\phi)\hat{D}(x)\hat{S}(r,\phi) = \hat{D}(x\cosh(r) + x^*\sinh(r)e^{i2\phi}), \tag{2.59}$$

and Eqs. (2.57) and (2.46), the identity, Eq. (2.56), is proved.

Problem 9: Prove the identity Eq. (2.59).

2.2.5 Harmonic Oscillator

Consider the Hamiltonian

$$H = \frac{1}{2}(p^2 + \omega^2 q^2). \tag{2.60}$$

This is the Hamiltonian of the harmonic oscillator with mass $m = 1$, frequency ω and q, p being the standard position and momentum operators, respectively, satisfying $[q, p] = i\hbar$. We will use the harmonic oscillator as an opportunity to introduce the concept of annihilation a and creation $a^\dagger$ operators, a very useful concept that has ramifications on many aspects of quantization [24]. Let

$$
\begin{aligned}
a &= \frac{1}{\sqrt{2\hbar\omega}}(\omega q + ip), \\
a^\dagger &= \frac{1}{\sqrt{2\hbar\omega}}(\omega q - ip).
\end{aligned}
\tag{2.61}
$$

It is easy to see that

$$
\left[a, a^\dagger\right] = 1.
$$

Using Eqs. (2.61) in Eq. (2.60), the Hamiltonian takes the form

$$
H = \hbar\omega\left(a^\dagger a + \frac{1}{2}\right).
\tag{2.62}
$$

Problem 10: Derive Eq. (2.62).

Thus, the problem of finding the eigenspectrum of the harmonic oscillator is equivalent to finding the spectrum of the operator $N = a^\dagger a$. This problem is treated in all the standard textbooks of quantum mechanics, see for example [22, 23, 25], and leads to the following useful relations

$$
\begin{aligned}
a|n\rangle &= \sqrt{n}|n-1\rangle, \\
a^\dagger|n\rangle &= \sqrt{n+1}|n+1\rangle.
\end{aligned}
\tag{2.63}
$$

Since, the annihilation operator reduces n by one, while the creation operator raises it by one; they are called the ladder operators. Here n is a positive semidefinite integer, $n = 0, 1, 2, \cdots$. From Eq. (2.63), it can be seen that

$$
a^\dagger a|n\rangle = n|n\rangle.
$$

This justifies the terminology, number states for $|n\rangle$. Using this, the eigenvalue equation of the harmonic oscillator becomes

$$
H|n\rangle = \hbar\omega\left(n + \frac{1}{2}\right)|n\rangle.
\tag{2.64}
$$

Eigenfunctions of harmonic oscillator

We will next generate the eigenfunctions of the harmonic oscillator using the method of creation and annihilation operators [30]. Let us first construct the wavefunction of the vacuum ψ_0. We know that $a\psi_0 = 0$. Using Eq. (2.61), we have

$$
\frac{1}{(2\hbar w)^{\frac{1}{2}}}(wq + \hbar\frac{\partial}{\partial q})\psi_0(q) = 0.
\tag{2.65}
$$

This can be solved to give $\psi_0(q) = ce^{-\frac{wq^2}{2\hbar}}$, where c is a constant. It can be determined by the normalization of the wave function

$$\int_{-\infty}^{\infty} dq |\psi_0(q)|^2 = 1, \tag{2.66}$$

giving $c = \left(\frac{w}{\hbar\pi}\right)^{\frac{1}{4}}$. Thus the wave function of the vacuum, in the coordinate q representation, can be written as

$$\psi_0(q) = \left(\frac{w}{\hbar\pi}\right)^{\frac{1}{4}} e^{-\frac{wq^2}{2\hbar}}. \tag{2.67}$$

From this, the other eigenfunctions can be obtained by repeated application of the creation operator. For example,

$$\psi_n(q) = \frac{1}{\sqrt{n}} (a^\dagger)^n \psi_0(q)$$

$$= \left(\frac{w}{\hbar\pi}\right)^{\frac{1}{4}} \frac{1}{\sqrt{n!}} \frac{1}{(2\hbar w)^{\frac{n}{2}}} \left(wq - \hbar\frac{d}{dq}\right)^n e^{-\frac{w}{2\hbar}q^2}$$

$$= P_n(q) e^{-\frac{w}{2\hbar}q^2}.$$

Here $P_n(q) = H_n(\sqrt{w}q)$, where H_n is the Hermite polynomial of order n.

2.2.6   Partition Function

Partition function is a very useful tool in statistical mechanics, classical as well as quantum. Before we get into this, it would be appropriate to briefly discuss the notion of an ensemble . In statistical mechanics, we are interested in the macroscopic dynamics of systems, characterized by macroscopic parameters such as temperature T, pressure P and volume V. Now a typical macroscopic system is composed of a large number of microscopic subsystems, also called microstates. An ensemble is the assembly of all possible microstates, consistent with the constraints with which the system is characterized macroscopically [31, 32, 33]. Thus

- *Microcanonical Ensemble*: is an assembly of all states with fixed energy E, fixed size; number of particles or systems N and volume V. This is appropriate for describing a close, isolated system.

- *Canonical Ensemble*: is an assembly of all states with fixed size, but energy E can vary. This is appropriate for describing a system in contact with a heat bath.

- *Grand-Canonical Ensemble*: is an assembly of all states with size and energy fluctuating.

Now consider the Gibbs or the canonical distribution, characterized by the probability, of finding a sub-system in the nth state by

$$p_n = Ce^{-\beta H(p,q)}, \tag{2.68}$$

where $\beta = 1/(k_B T)$, k_B being the Boltzmann constant and T being the equilibrium temperature. Also, $H(p, q)$ is the system Hamiltonian as a function of the phase space variables p and q. C is a constant, determined by the requirement that the probability p_n is normalized and is the inverse of the partition function. Explicitly, the partition function for the canonical distribution would be

$$\mathcal{Z} = \int \frac{d^{3N} p\, d^{3N} q}{N! (2\pi\hbar)^{3N}} e^{-\beta H(p,q)}. \tag{2.69}$$

Here N denotes the number of sub-systems, $d^{3N} p\, d^{3N} q$ is the phase space volume, $\hbar = h/(2\pi)$ and the factor $(2\pi\hbar)^{3N}$ in the denominator is due to the quantum-classical correspondence between the number of quantum states in a given energy interval and the phase space occupied, classically. This has an important ramification in the definition of entropy, a point to which we will return to shortly. In the context of quantum statistical mechanics, the partition function has a similar form and can be represented by

$$\mathcal{Z} = \text{Tr}\left(e^{-\beta H}\right) = \sum_n e^{-\beta E_n}. \tag{2.70}$$

The partition function allows expressing a number of important *bulk* thermodynamic quantities in a compact form, thereby facilitating their calculation. Thus, for example, the mean energy can be expressed in terms of the partition function of the canonical distribution as

$$\langle H \rangle = -\frac{\partial}{\partial \beta} \ln(\mathcal{Z}). \tag{2.71}$$

Further, the partition function allows for the definition of another important quantity called the free energy F as

$$\mathcal{Z} = e^{-\beta F}. \tag{2.72}$$

The free energy, an extensive quantity, is related to the mean energy U by $F = U - TS$, where S is the entropy. Also, other thermodynamic quantities can be expressed in terms of F as

$$
\begin{aligned}
P &= \left. -\frac{\partial}{\partial V} F \right|_{T=\text{constant}}, \\
S &= \left. -\frac{\partial}{\partial V} F \right|_{V=\text{constant}}.
\end{aligned}
\tag{2.73}
$$

As an illustration of these concepts, we will compute the thermodynamic properties of N independent harmonic oscillators in equilibrium at temperature T. This constitutes Einstein's model of heat capacity . The Hamiltonian of a single harmonic oscillator is

$$H_1 = \hbar\omega(a^\dagger a + 1/2) = \hbar\omega(n + 1/2), \cdots n = 0, 1, .. \tag{2.74}$$

The corresponding partition function is

$$\mathcal{Z}_1(\beta) = \Sigma_{n=0}^{\infty} e^{-\beta E_n} = \frac{1}{2\sinh\left(\frac{\beta\hbar\omega}{2}\right)}. \tag{2.75}$$

The total partition function is $\mathcal{Z}(\beta) = \left[\mathcal{Z}_1(\beta)\right]^N = \frac{1}{2^N \sinh^N\left(\frac{\beta\hbar\omega}{2}\right)}$. Using Eq. (2.71), the mean energy $U = \langle H \rangle$ is

$$U = \frac{N\hbar\omega}{2}\coth\left(\frac{\beta\hbar\omega}{2}\right) = N\left[\frac{\hbar\omega}{2} + \frac{\hbar\omega}{e^{\beta\hbar\omega} - 1}\right]. \tag{2.76}$$

Also, using the partition function, the free energy F, Eq. (2.72), is

$$F = Nk_B T \ln(e^{\frac{\beta\hbar\omega}{2}} - e^{-\frac{\beta\hbar\omega}{2}}), \tag{2.77}$$

entropy S is

$$S = -Nk_B \ln(e^{\frac{\beta\hbar\omega}{2}} - e^{-\frac{\beta\hbar\omega}{2}}) + \frac{N\hbar\omega}{2}\coth\left(\frac{\beta\hbar\omega}{2}\right). \tag{2.78}$$

Using Eq. (2.76), the specific heat capacity, at constant volume, is

$$C = \left.\frac{\partial U}{\partial T}\right|_{V=\text{constant}} = Nk_B(\beta\hbar\omega)^2 \frac{e^{\beta\hbar\omega}}{(e^{\beta\hbar\omega} - 1)^2}. \tag{2.79}$$

Thus, the heat capacity at high T is $C \approx Nk_B$, independent of T.

2.2.7   Entropy

Entropy S is a central concept in statistical mechanics [32, 31, 34]. It is the logarithm of the statistical weight of the sub-system $\Delta\Gamma$

$$S = k_B \ln \Delta\Gamma = -k_B \langle \ln p_n \rangle = -k_B \sum_n p_n \ln p_n. \tag{2.80}$$

$\Delta\Gamma$ is the number of quantum states in a macroscopic energy interval ΔE, equal in order of magnitude to the mean fluctuation of energy of the subsystem and is the degree of broadening of the macroscopic state of the subsystem *w.r.t* its microscopic states. Here, p_n is the probability distribution of energy levels of the sub-system and the angular brackets in the *RHS* of Eq. (2.80) denotes average with respect to the distribution p_n. In classical statistical mechanics, this is connected to the phase space volume as $\Delta\Gamma = \frac{\Delta p \Delta q}{(2\pi\hbar)^n}$, where n is the number of degrees of freedom of the subsystem considered. This relation is obtained by taking a quasi classical correspondence between the volume of a region of phase space and the *corresponding* number of quantum states. Also, n quantifies the quantum, classical phase space correspondence. Thus, the classical entropy is

$$S = k_B \ln \left(\frac{\Delta p \Delta q}{(2\pi\hbar)^n} \right). \tag{2.81}$$

In the definitions of entropy, the proportionality constant is k_B, the Boltzmann constant. In purely classical statistics, the entropy can also be defined by $S = k_B \ln(\Delta p \Delta q)$. This is defined only to within an additive constant which depends on the choice of units, and only differences of entropy, are definite quantities independent of the choice of units. Using the definition in Eq. (2.81), allows for a dimensionless definition of entropy. Consider two bodies in thermal equilibrium with each other, forming a closed system. Then the entropy of the system is maximum. From this, it can be shown that the derivative of the entropy with respect to the energy is a constant. This constant defines the inverse of the absolute temperature of the body. From this it follows that the temperatures of bodies in thermal equilibrium with each other are equal.

Entropy plays myriad roles not only in its traditional arena of classical (quantum) statistical mechanics, but also in information theory. There, entropy tells us about the average uncertainty in a random variable and is measured in bits [35]. It is also the basis for the Shannon data compression theorem [36], where it sets the limit to which data can be safely compressed. Entropy forms the basis of the definition of the mutual information, which is the entropy of a random variable conditioned on the knowledge of another random variable. The mutual information is in turn used to define the channel capacity of a communication channel. In the quantum regime, von-Neumann entropy is used, which is similar to the Shannon entropy [27, 37], with the density matrix used to represent the probability distribution. The concept of entropy also plays an important role in quantum information [38], where the von-Neumann entropy is used to quantify entanglement between two systems. Also, quantum mutual information is a measure of total quantum correlations. We will have occasion to discuss these, in some detail, later on in this book when we discuss aspects of open quantum systems in quantum information.

We will now discuss a novel use of entropy, that is, the use of the principle of maximum entropy, initiated by Jaynes [39]. We will use this to obtain the canonical or Gibbs distribution. Consider the entropy

$$S = -k_B \text{Tr}(\rho \ln \text{p}). \tag{2.82}$$

This is subject to the constraint

$$\text{Tr}\rho = 1, \tag{2.83}$$

which is the statement of normalization of the state ρ. Further, we know the average energy of the system

$$\langle E \rangle = \text{Tr}\rho \text{H}, \tag{2.84}$$

where H is the system Hamiltonian. Now we maximize the entropy, Eq. (2.82) keeping into account the constraints of Eqs. (2.83) and (2.84). We then have

$$
\begin{aligned}
\mathrm{Tr}(1 + \ln \rho)\delta\rho &= 0, \\
\mathrm{Tr}\delta\rho &= 0, \\
\mathrm{Tr}H\delta\rho &= 0. \quad\quad (2.85)
\end{aligned}
$$

We now apply the method of undetermined multipliers by multiplying the first constraint, of normalization, by α and the second one, of average energy, by β to get

$$
\mathrm{Tr}(1 + \alpha + \ln \rho + \beta H)\delta\rho = 0. \quad\quad (2.86)
$$

Since $\delta\rho$ is arbitrary, all variations are independent. Eq. (2.86) will be satisfied for

$$
\ln \rho = -1 - \alpha - \beta H, \quad\quad (2.87)
$$

which implies that

$$
\rho = e^{-1-\alpha}e^{-\beta H}. \qu\quad\quad (2.88)
$$

The first term on the *RHS* of Eq. (2.88) is nothing but the partition function $\mathcal{Z} = \mathrm{Tr}e^{-\beta H}$, as can be checked by demanding the normalization of the state ρ. The average energy $\langle E \rangle = -\frac{\partial}{\partial \beta} \ln \mathcal{Z}$. From this, the constraint β can be obtained. It can be shown that $\beta = 1/(k_B T)$. Using this in Eq. (2.88), we get

$$
\rho = \frac{e^{-\frac{H}{k_B T}}}{\mathrm{Tr}e^{-\frac{H}{k_B T}}}, \quad\quad (2.89)
$$

the canonical distribution.

2.3 Path Integrals

The path integral approach provides an alternative to the operator approach to problems involving quantum mechanics. In contrast to the usual evolution equations, such as the Schrödinger equation where the properties of a state at a given time are determined from their knowledge at an infinitesimally earlier time, the path integral allows for a global understanding of the state evolution. Further, in the path integral formalism, the formalism of operators is replaced by ordinary c numbers, though the price to be paid is the involvement of infinite products of integrals. The path integral approach has become very important in the quest for understanding gauge field theories. Also, in the context of open quantum systems, the subject of the present book, path integrals play a major role. Here we will provide an introduction to the method of path integrals and illustrate them by working out the propagators for the free particle as well as the harmonic oscillator. Then we develop the path integral formalism for the partition function, thereby providing a glimpse into the notion of imaginary time path integrals. This approach is very useful in studies related to quantum statistical mechanics. We then provide a path integral description of the evolution of the density matrix. The construction made here will be directly carried over to the development of the influence functional approach to open quantum systems, in Chapter IV.

2.3.1   Introduction to the Path Integral

Let us consider the amplitude

$$\psi(x,t) = \langle x,t|\psi\rangle. \tag{2.90}$$

Here $|x,t\rangle = e^{\frac{i}{\hbar}Ht}|x\rangle$, where H is the Hamiltonian and $|\psi\rangle$ is the state at time $t = 0$. We know from our discussions about various pictures that this is also the state in the Heisenberg picture. This satisfies (a). *Orthonormality*: $\langle x',t|x,t\rangle = \delta(x'-x)$, and (b). *Completeness*: $\int dx|x,t\rangle\langle x,t| = 1$. Hence, $|x,t\rangle$ can be used as a basis. Using these, we can express the amplitude, in Eq. (2.90), as

$$
\begin{aligned}
\langle x',t'|\psi\rangle &= \int dx\langle x',t'|x,t\rangle\langle x,t|\psi\rangle \\
&= \int dx\langle x',t'|x,t\rangle\psi(x,t).
\end{aligned}
\tag{2.91}
$$

This is the starting point of the construction of the path integral. The transition amplitude $\langle x',t'|x,t\rangle$ in Eq. (2.91) is the so called Feynman kernel or propagator [40, 41, 42, 43, 44, 1]. The path integral provides a means to construct the transition amplitude (Feynman kernel) from the classical Hamiltonian of the system, without any explicit reference to non-commuting operators or Hilbert space vectors [45]. As evident from Eq. (2.91), knowledge of the Feynman kernel $\langle x',t'|x,t\rangle$ allows for the determination of the state's evolution at any time. Thus, the path integral provides a global approach to the time evolution problem.

The next step is to split the time interval (t,t'), in Eq. (2.91) into n slices, such that $t_n = t + n\epsilon$, $n = 1,2,\cdots,N-1$. This implies that $t'-t = N\epsilon$. At each of the n lattice (grid) points, a complete set of basis states $\{|x_n,t_n\rangle\}$ is inserted to give

$$\langle x',t'|x,t\rangle = \int dx_{N-1}\cdots\int dx_2\int dx_1\langle x',t'|x_{N-1},t_{N-1}\rangle\cdots\langle x_2,t_2|x_1,t_1\rangle\langle x_1,t_1|x,t\rangle. \tag{2.92}$$

In Eq. (2.92), $\langle x_{n+1},t_{n+1}|x_n,t_n\rangle$ is the transfer T matrix and is, using $|x,t\rangle = e^{\frac{i}{\hbar}Ht}|x\rangle$,

$$\langle x_{n+1},t_{n+1}|x_n,t_n\rangle = \langle x_{n+1}|e^{-\frac{i}{\hbar}H(p,x)\epsilon}|x\rangle, \tag{2.93}$$

where $\epsilon = t_{n+1} - t_n$. It should be noted that in the above equation, $H(p,q)$ in the *RHS* is an operator, as in quantum mechanics, observables, such as position x and momentum p are operators. The construction in Eq. (2.92) can be interpreted as a coherent superposition, interference, of all paths from x,t

to x', t'. Now we know from our previous discussions on operators that it is not possible to simply factorize functions, such as exponentials, of operators as for the corresponding c number cases. However, for infinitesimal ϵ, we can express Eq. (2.93) as

$$\langle x_{n+1}, t_{n+1} | x_n, t_n \rangle = \langle x_{n+1} | 1 - \frac{i\epsilon}{\hbar} H(p, x) | x_n \rangle + o(\epsilon^2). \qquad (2.94)$$

Using completeness of the momentum basis $\{|p\rangle\}$,

$$\langle x_{n+1} | H(p, x) | x_n \rangle = \int \frac{dp_n}{2\pi\hbar} \langle x_{n+1} | p_n \rangle \langle p_n | H(p, x) | x_n \rangle. \qquad (2.95)$$

Now, $\langle p_n | H(p, x) | x_n \rangle = \langle p_n | x_n \rangle H(p_n, \bar{x}_n)$. Here $\bar{x}_n = x_n$ or $\frac{1}{2}(x_{n+1} + x_n)$ (the so called *mid-point prescription*), depending upon whether the momentum operators stand to the left of the position operators or they are symmetrically distributed, obeying a Weyl ordering, repectively [43, 44]. This allows us to write Eq. (2.94) as

$$\langle x_{n+1}, t_{n+1} | x_n, t_n \rangle = \int \frac{dp_n}{2\pi\hbar} \exp\left[\frac{i}{\hbar} p_n (x_{n+1} - x_n)\right] \left(1 - \frac{i\epsilon}{\hbar} H(p_n, \bar{x}_n)\right) + o(\epsilon^2). \qquad (2.96)$$

It should be noted that the p_n, x_n in the *RHS* of the above equation are now ordinary c numbers. Substituting Eq. (2.96) in Eq. (2.92) and invoking the limits $\epsilon \to 0$ or $N \to \infty$, we get

$$\langle x', t' | x, t \rangle = \lim_{N \to \infty} \int \prod_{n=1}^{N-1} dx_n \prod_{n=0}^{N-1} \frac{dp_n}{2\pi\hbar} \exp\left[\frac{i\epsilon}{\hbar} \sum_{n=0}^{N-1} p_n \frac{(x_{n+1} - x_n)}{\epsilon}\right] \qquad (2.97)$$

$$\times \prod_{n=0}^{N-1} \left(1 - \frac{i\epsilon}{\hbar} H(p_n, \bar{x}_n)\right)$$

$$= \lim_{N \to \infty} \int \prod_{n=1}^{N-1} dx_n \prod_{n=0}^{N-1} \frac{dp_n}{2\pi\hbar} \exp\left[\frac{i\epsilon}{\hbar} \sum_{n=0}^{N-1} \left(p_n \frac{(x_{n+1} - x_n)}{\epsilon} - H(p_n, \bar{x}_n)\right)\right]. \qquad (2.98)$$

Here use is made of the identity

$$\lim_{N \to \infty} \prod_{n=0}^{N-1} \left(1 + \frac{x_n}{N}\right) = \exp\left(\lim_{N \to \infty} \frac{1}{N} \sum_{n=0}^{N-1} x_n\right). \qquad (2.99)$$

In the limit of $N \to \infty$, the lattice (grid) points x_n and p_n come arbitrary close and can be viewed as sampled values of continuously defined functions $x(t)$ and $p(t)$. The object constructed in Eq. (2.98) is called a *functional integral* or a *path integral*. Using compact notations $\int \prod_{n=1}^{N-1} dx_n = \int Dx$, $\int \prod_{n=0}^{N-1} \frac{dp_n}{2\pi\hbar} = \int Dp$, $\frac{(x_{n+1} - x_n)}{\epsilon} = \dot{x}(t_n)$, $\epsilon \sum_{n=0}^{N-1} f(t_n) = \int_t^{t'} d\tau f(\tau)$, the Feynman kernel can be written as a *path integral in phase space*

$$\langle x', t' | x, t \rangle = \int Dx \int Dp \exp \left[\frac{i}{\hbar} \int\limits_t^{t'} d\tau \left(p\dot{x} - H(p, \bar{x}) \right) \right]. \tag{2.100}$$

This path integral is over all functions $p(t)$ in momentum space and in coordinate space satisfies the boundary conditions $x(t) = x$, $x(t') = x'$. The path integral, thus provides a means of computing the transition amplitude as a sum over all paths, between the given boundary conditions. The modulus square of the amplitude then gives the probability, which is the Born rule .

Now consider the usual Hamiltonian

$$H(p, x) = \frac{p^2}{2m} + V(x), \tag{2.101}$$

where $V(x)$ is the potential. Using Eq. (2.101) in Eq. (2.96), we see that

$$\langle x_{n+1}, t_{n+1} | x_n, t_n \rangle \approx \int \frac{dp_n}{2\pi\hbar} \exp \left[\frac{i\epsilon}{\hbar} \left(p_n \dot{x}_n - \frac{p_n^2}{2m} - V(\bar{x}) \right) \right]. \tag{2.102}$$

The exponent in the *RHS* of Eq. (2.102) is quadratic in p_n and hence forms a Gaussian integral which can be easily done to yield

$$\langle x_{n+1}, t_{n+1} | x_n, t_n \rangle \approx \left(\frac{2\pi i \hbar \epsilon}{m} \right)^{-1/2} \exp \left[\frac{i\epsilon}{\hbar} \left(\frac{1}{2} m \dot{x}_n^2 - V(\bar{x}_n) \right) \right]. \tag{2.103}$$

Here use is made of $p = \frac{1}{2} m \dot{x}^2$. Inserting Eq. (2.103) in Eq. (2.98), we have

$$
\begin{aligned}
\langle x', t' | x, t \rangle &= \lim_{N \to \infty} \left(\frac{2\pi i \hbar \epsilon}{m} \right)^{-N/2} \int \prod_{n=1}^{N-1} dx_n \exp \left[\frac{i\epsilon}{\hbar} \sum_{n=0}^{N-1} \left(\frac{1}{2} m \dot{x}_n^2 - V(\bar{x}_n) \right) \right] \\
&= \mathcal{N} \int Dx \exp \left[\frac{i}{\hbar} S(x, \dot{x}) \right]. \tag{2.104}
\end{aligned}
$$

Here $S(x, \dot{x})$ is the action and is

$$S(x, \dot{x}) = \int\limits_t^{t'} d\tau \left(\frac{1}{2} m \dot{x}^2 - V(\bar{x}) \right) = \int\limits_t^{t'} d\tau L(x, \dot{x}). \tag{2.105}$$

$L(x, \dot{x})$ is the Lagrangian , $\mathcal{N}$ is a normalization constant and we have made use of the abbreviated notations defined above. Thus, we see that the Feynman kernel is defined in terms of the action, a property which makes it very useful in many scenarios, including open quantum systems and quantum field theory.

If $S \gg \hbar$, then the terms in the argument of the exponential, in the *RHS* of Eq. (2.104) become highly oscillating and cancel each other. The only paths where this cancellation is avoided are where the action remains stationary, that is, where $\delta \int_t^{t'} d\tau L(x, \dot{x}) = 0$. Hence, the classical description of the system can be derived from the corresponding quantum theory using the stationary phase approximation to the path integral. For a more mathematical introduction to the subject, the reader is encouraged to look into [46].

Problem 11: Do the Gaussian integral to arrive at Eq. (2.103).

2.3.2   Illustrative examples

Now we illustrate the formalism developed above by working out examples of the Feynman kernel for a free particle as well as that of a particle in a harmonic oscillator .

Free Particle Path Integral: For the case of a free particle, $V(x) = 0$. The Eq. (2.104) gives

$$\langle x', t' | x, t \rangle = \lim_{N \to \infty} \left(\frac{2\pi i \hbar \epsilon}{m} \right)^{-N/2} \int \prod_{n=1}^{N-1} dx_n \exp \left[\frac{im}{2\hbar\epsilon} \sum_{n=0}^{N-1} \left(x_{n+1} - x_n \right)^2 \right].$$

(2.106)

Repeatedly using the Gaussian integral identity

$$\int_{-\infty}^{\infty} dx_n \exp \left[-\alpha(q_{n+1} - q_n)^2 \right] \exp \left[-\beta(x_n - x_{n-1})^2 \right]$$

$$= \sqrt{\frac{\pi}{\alpha + \beta}} \exp \left[-\frac{\alpha\beta}{\alpha + \beta}(x_{n+1} - x_{n-1})^2 \right],$$

(2.107)

$N - 1$ times, Eq. (2.106) becomes

$$\langle x', t' | x, t \rangle = \lim_{N \to \infty} \sqrt{\frac{m}{2\pi i \hbar N \epsilon}} \exp \left[\frac{1}{N\epsilon} \frac{im}{2\hbar} (x_N - x_0)^2 \right]$$

$$= \sqrt{\frac{m}{2\pi i \hbar (t' - t)}} \exp \left[\frac{im}{2\hbar} \frac{(x' - x)^2}{(t' - t)} \right].$$

(2.108)

It can be easily seen that the Feynman kernel, Eq. (2.108), satisfies the free particle Schrödinger equation $i\hbar \frac{\partial \psi}{\partial t} = H\psi$, where $H = -\frac{\hbar^2}{2m} \frac{\partial^2 \psi}{\partial x^2}$. This motivates the adjective *propagator* for the Feynman kernel.

Problem 12: Derive Eq. (2.107).

Harmonic Oscillator Path Integral: The harmonic oscillator serves as a paradigm for many systems in physics. Keeping in mind its importance, we now work out in detail the Feynman kernel or the propagator for the harmonic oscillator. This will also serve to provide a firm insight into the working of the path integral method. Here the potential $V(x) = \frac{1}{2}m\omega^2 x^2$. Note that $\bar{x}$ in Eq. (2.105) is, here, equal to x. We start with Eq. (2.104)

$$\langle x', t|x, 0\rangle = \mathcal{N} \int Dx \exp\left[\frac{i}{\hbar}S(x, \dot{x})\right]$$

$$= \mathcal{N} \int Dx \exp\left\{\frac{i}{\hbar}\int_0^t ds\left(\frac{1}{2}m_n\left[\dot{x}^2(s) - \omega^2 x^2(s)\right]\right)\right\}. \quad (2.109)$$

Here, for simplicity of notation, we have replaced the initial time t by zero and final time t' by t. We now need to evaluate the functional integral in Eq. (2.109) where we have to sum over all paths $x(s)$ of the oscillator with $x(0) = x$ and $x(t) = x'$. Since the functional integral is Gaussian, its dependence on the boundary values x, x', can be obtained by expanding about the path $\tilde{x}(s)$ minimizing the action in the exponent of Eq. (2.109), the classical path. Let us express

$$x(s) = \tilde{x}(s) + \alpha(s), \quad (2.110)$$

where $\tilde{x}(s)$ is the solution of the stationary value of the action, and is the classical path. $\alpha(s)$ is the quantum fluctuation about this path, and $\tilde{x}(0) = x$ and $\tilde{x}(t) = x'$. We have by Taylor expansion of the action about the classical path

$$S[X] = S[\tilde{x}] + \int ds \frac{\delta S[x]}{\delta x(s)}\Big|_{x=\tilde{x}} + \frac{1}{2!}\int ds_1 ds_2 \alpha(s_1)\alpha(s_2)\frac{\delta^2 S[x]}{\delta x(s_1)\delta x(s_2)}\Big|_{x=\tilde{x}}. \quad (2.111)$$

It should be noted that as the classical path $\tilde{x}(s)$ is stationary for the action, the first derivative of the action in the *RHS* of Eq. (2.111) vanishes. As a result, Eq. (2.109) becomes

$$\langle x', t|x, 0\rangle = \mathcal{N} \exp\left\{\frac{i}{\hbar}\int_0^t ds\left(\frac{1}{2}m\left[\dot{\tilde{x}}^2 - \omega^2 \tilde{x}^2\right]\right)\right\}$$

$$\times \int D\alpha \exp\left\{\frac{i}{\hbar}\int_0^t ds\left(\frac{1}{2}m\dot{\alpha}^2 - \omega^2 \alpha^2\right)\right\}. \quad (2.112)$$

The functional integral over $\alpha(s)$ sums over all paths $\alpha(s)$ with $\alpha(0) = \alpha(t) = 0$ so that the dependence on $x(0)$ and $x(t)$ is completely included in the first exponential. To calculate the functional integral in Eq. (2.112) we made use of the fact that $\tilde{x}(s)$ is a trajectory that minimizes the action. Consider the first term in Eq. (2.112). The Euler-Lagrangian equation of motion is

$$\frac{\delta S[x]}{\delta x(s)}\Big|_{x=\tilde{x}} = 0, \quad (2.113)$$

which implies that

$$m\ddot{\tilde{x}}(s) + m\omega^2\tilde{x}(s) = 0. \qquad (2.114)$$

The solution of Eq. (2.114) satisfying the boundary conditions is

$$\tilde{x}(s) = \frac{\beta(s)}{\beta(t)}x' + \frac{\beta(t-s)}{\beta(t)}x, \qquad (2.115)$$

where $\beta(s) = \frac{1}{m\omega}\sin(\omega s)$. To compute the action of the trajectory $\tilde{x}(s)$ we perform an integration by parts and use the equation of motion (2.114). This yields

$$\int_0^t ds\left(\frac{1}{2}m\left[\dot{\tilde{x}}^2 - \omega^2\tilde{x}^2\right]\right) = \frac{1}{2}m(\tilde{x}(t)\dot{\tilde{x}}(t) - \tilde{x}(0)\dot{\tilde{x}}(0)). \qquad (2.116)$$

The *RHS* of Eq. (2.115) is then inserted in Eq. (2.116), to yield the classical contribution to the action. We now want to evaluate the contribution to the functional integral, Eq. (2.116), from the quantum fluctuations $\alpha(s)$, see Eq. (2.110). This can be obtained by expanding the α dependent part of the functional integral, Eq. (2.112), into a Fourier series, using $\alpha(0) = \alpha(t) = 0$

$$\alpha(s) = \sum_{n=1}^{\infty} c_n \sin(\nu_n s); \quad \nu_n = \pi n/t. \qquad (2.117)$$

Due to the orthogonality of the sine functions, the integrand in the second term in the *RHS* of Eq. (2.112) becomes

$$\exp\left\{\frac{i}{\hbar}\int_0^t ds\left(\frac{1}{2}m\dot{\alpha}^2 - \omega^2\alpha^2\right)\right\} = \exp\left\{i\frac{mt}{4\hbar}\sum_{n=1}^{\infty}\left((c_n)^2(\nu_n^2 - \omega^2)\right)\right\}, \qquad (2.118)$$

while the integration measure becomes

$$\int D\alpha \cdots = \prod_{n=1}^{\infty}\left(\mathcal{N}_1^{-1}\int_{-\infty}^{\infty} dc_n \cdots\right). \qquad (2.119)$$

Here $\mathcal{N}_1$ is a constant independent of ω which arises from the Jacobian of the transformation in Eq. (2.117). From the Eqs. (2.118) and (2.119), it can be seen that the functional integral factorizes into regular Gaussian integrals over the Fourier components c_n which can be done separately. We then have

$$f(t) = \int D\alpha \exp\left\{\frac{i}{\hbar}\int_0^t ds\left(\frac{1}{2}m\dot{\alpha}^2 - \omega^2\alpha^2\right)\right\} = C\prod_{n=1}^{\infty}\left(1 - \frac{\omega^2}{\nu_n^2}\right)^{-1/2}. \qquad (2.120)$$

The constant C has all the factors independent of ω, including $\mathcal{N}$ and $\mathcal{N}_1$. Using the mathematical identity

$$\prod_{n=1}^{\infty}\left(1 - \frac{\omega^2}{\nu_n^2}\right) = \frac{\sin(\omega t)}{\omega t}, \tag{2.121}$$

we get

$$f(t) = C\left(\frac{\omega t}{\sin(\omega t)}\right)^{1/2}. \tag{2.122}$$

The constant C can be determined by evaluating the Jacobian of the transformation, Eq. (2.117). Another way of doing this is by comparing Eq. (2.120) with the corresponding result for the case of a free particle, Eq. (2.108). Thus we see that

$$f(t, \omega = 0) = C = \left(\frac{m}{2\pi i\hbar t}\right)^{1/2}. \tag{2.123}$$

Using Eqs. (2.123), (2.122) and the result obtained by substituting the *RHS* of Eq. (2.115) in the *RHS* of Eq. (2.116), we get

$$\langle x', t | x, 0\rangle = \left(\frac{m\omega}{2\pi i\hbar \sin(\omega t)}\right)^{1/2} \exp\left\{\frac{i}{\hbar}\Phi(x', x)\right\}, \tag{2.124}$$

where

$$\Phi(x', x) = \frac{m\omega}{2\sin(\omega t)}\left\{(x^2 + x'^2)\cos(\omega t) - 2xx'\right\}. \tag{2.125}$$

This is the Feynman kernel or the propagator for the harmonic oscillator.

2.3.3 Partition Function as a Path Integral

Keeping in mind the ubiquity of the partition function in quantum statistical mechanics, we now provide a path integral description of it. The starting point of a path integral derivation of the partition function is the quantum mechanical partition function, introduced earlier in Eq. (2.70). This can be put in contact with the evolution operator via analytic continuation to imaginary time as

$$\mathcal{Z} = \text{Tr}\left(e^{-\beta H}\right) = \sum_n e^{-\beta E_n} = \text{Tr} e^{-i(t'-t)H/\hbar}, \tag{2.126}$$

where $t' - t = -\frac{i\hbar}{k_B T} \equiv -i\hbar\beta$. The trace can be computed using any convenient basis. For example, in the position basis $\{|x\rangle\}$, this amounts to integrating the above amplitude over $x' = x$, at the analytically continued time $t' - t = -i\hbar\beta$

$$\mathcal{Z} = \int_{-\infty}^{\infty} dx\zeta(x) = \int_{-\infty}^{\infty} dx\langle x|e^{-\beta H}|x\rangle = \int_{-\infty}^{\infty} dx\langle x, t'|x, t\rangle\bigg|_{t'-t=-i\hbar\beta}. \tag{2.127}$$

Here the object $\zeta(x) = \langle x|e^{-\beta H}|x\rangle = \langle x, t'|x, t\rangle|_{t'-t=-i\hbar\beta}$ plays the role of the partition function density. This has the form of the Feynman kernel or propagator with the added proviso that here the position of the end point is the same as that of the initial point. From the Eqs. (2.124) and (2.125) for the harmonic oscillator propagator, and using the imaginary time relation, we see that

$$\zeta(x) = \left(\frac{m\omega}{2\pi\hbar\sinh(\beta\hbar\omega)}\right)^{1/2} \exp\left\{-\frac{m\omega}{\hbar}\tanh\left(\frac{1}{2}\beta\hbar\omega\right)x^2\right\}, \qquad (2.128)$$

is the harmonic oscillator partition function density. In the general case, we proceed as in Eqs. (2.92) and (2.93) keeping in mind that now we are dealing with imaginary time. Thus, we basically split the factor $e^{-\beta H}$ into a product of $N+1$ factors $e^{-\epsilon H/\hbar}$ with $\epsilon = \hbar\beta/(N+1)$ and insert partitions of unity, using the $\{|x\rangle\}$ basis, between each factor, set the beginning and end positions equal and then integrate over it. The path integral would thus be composed of the product of factors like Eq. (2.96)

$$\langle x_{n+1}|e^{-\epsilon H/\hbar}|x_n\rangle \approx \int \frac{dp_n}{2\pi\hbar} \exp\left[\frac{i}{\hbar}p_n(x_{n+1}-x_n) - \frac{\epsilon}{\hbar}H(p_n,\bar{x}_n)\right]. \qquad (2.129)$$

The difference with the earlier, real time case, is that here there is no imaginary term i in front of the Hamiltonian H. The product of the terms like Eq. (2.129), in the manner indicated above, gives

$$\mathcal{Z} \approx \prod_{n=1}^{N+1}\left[\int_{-\infty}^{\infty} dx_n \int_{-\infty}^{\infty} \frac{dp_n}{2\pi\hbar}\right] \exp\left[-\frac{1}{\hbar}\xi_E^N\right], \qquad (2.130)$$

where ξ_E^N, the superscript E denoting Euclidean, is

$$\xi_E^N = \sum_{n=1}^{N+1}\left[-ip_n(x_{n+1}-x_n) + \epsilon H(p_n,\bar{x}_n)\right]. \qquad (2.131)$$

In the continuum limit, the sum goes over to the integral and we have the partition function

$$\mathcal{Z} = \int Dx \int \frac{Dp}{2\pi\hbar} e^{-\xi_E(p,x)/\hbar}, \qquad (2.132)$$

where we have used the compact notations introduced in our introduction of the path integral. Also,

$$\xi_E(p,x) = \int_0^{\hbar\beta} d\tau\left[-ip(\tau)\dot{x}(\tau) + H(p(\tau),x(\tau))\right]. \qquad (2.133)$$

This is the imaginary time or *Euclidean* action, and is a functional of $p(\tau)$, $x(\tau)$, which could be thought of as paths along an imaginary time axis $\tau = it$. The partition function, Eq. (2.132), is symmetrical with respect to x and p.

Problem 13: Derive Eq. (2.128). Use it to show that the partition function $\mathcal{Z} = \int_{-\infty}^{\infty} dx\zeta(x)$ is equal to the result obtained in Eq. (2.75) by another method.

Gaussian integrals play an important role in many practical calculations. Try out the following problems involving Gaussian integration.

Problem 14: For $\psi(x) = \left(\frac{2a}{\pi}\right)^{-1/4} e^{-ax^2/2}$, with a real find $\langle x^n \rangle$ for an interger $n > 0$. The average is to be taken w.r.t the state $\psi(x)$.

Problem 15: Let $\psi(x) = A(ax - x^2)$, $0 \leq x \leq a$. Normalize $\psi(x)$ and find $\langle x \rangle$, $\langle x^2 \rangle$. The average is to be taken w.r.t the state $\psi(x)$.

2.3.4   Density Matrix Evolution as a Path Integral

We have seen before in this chapter that when we deal with open quantum systems, more often than not we have to work with mixed states that entail working with density matrices . Here we present a path integral approach to studying the evolution of the density matrix. This would be of immense importance to us when we study the influence functional approach to open quantum systems later. We start with the evolution of the state $\rho(0)$ at time $t = 0$, evolving under the Hamiltonian H as

$$\rho(t) = e^{-\frac{i}{\hbar}Ht}\rho(0)e^{\frac{i}{\hbar}Ht}. \tag{2.134}$$

In the coordinate representation this can be expressed as

$$\langle x|\rho(t)|y\rangle = \int dx' dy' \langle x|e^{-\frac{i}{\hbar}Ht}|x'\rangle \langle x'|\rho(0)|y'\rangle \langle y'|e^{\frac{i}{\hbar}Ht}|y\rangle. \tag{2.135}$$

Here we have introduced partitions of unity, for example, $\int dx'|x'\rangle\langle x'| = 1$. We know from Eq. (2.104) that

$$\langle x|e^{-\frac{i}{\hbar}Ht}|x'\rangle = \int Dx \exp\left[\frac{i}{\hbar}S(x,\dot{x})\right] = K(x,t;x',0), \tag{2.136}$$

and

$$\langle y'|e^{\frac{i}{\hbar}Ht}|y\rangle = \int Dy \exp\left[-\frac{i}{\hbar}S(x,\dot{x})\right] = K^*(y,t;y',0). \tag{2.137}$$

Here we have used the compact path integral notation and have absorbed the normalization constant in the path integral measure. The Eqs. (2.136) and (2.137) could be thought of as the forward $K(x,t;x',0)$ and backward $K^*(y,t;y',0)$ Feynman propagators , respectively [47, 48]. They are functional integrals over paths x and y with $x(0) = x'$, $x(t) = x$, $y(0) = y'$ and $y(t) = y$.

Using Eqs. (2.136) and (2.137) in Eq. (2.135), the evolution of the density matrix is seen to be

$$\rho(x, y, t) = \langle x|\rho(t)|y\rangle = \int dx'dy'K(x, t; x', 0)K^*(y, t; y', 0)\langle x'|\rho(0)|y'\rangle.$$

$$(2.138)$$

This gives us the intuitive picture of the evolution of the density matrix involving a product of two propagators, one evolving forward in time and the other evolving backward in time. This construction would be very handy when we discuss the influence functional formalism of open quantum systems. We will then see that the interaction of the system with its environment results in coupling of the forward and backward propagators and is expressed as the influence functional.

2.4   Guide to advanced literature

In this chapter we have provided a glimpse to the vast subjects of quantum statistical mechanics and path integrals. This is required as the study of open quantum systems draws heavily from both these subjects. As has been mentioned in the text, there are a number of standard text books on the subject of (quantum) statistical mechanics, such as [27, 31, 32, 33, 34, 49, 50], to which the reader is encouraged to look into for persuing a topic discussed here in more detail. The subject of path integrals is an equally vast one. The first textbook on it was [40]. At present there are a number of very good textbooks on the subject, a small sample of which would be [41, 42, 51, 52, 53, 54]. Also, there is a preponderance of the use of path integrals in the modern treatments of gauge field theory, see for example, the books [43, 44]. Further, in [1] the study of open quantum systems is presented in considerable detail.

Chapter 3

Master Equations: A Prolegomenon to Open Quantum Systems

3.1 Introduction

In this chapter, we will discuss the foundations of open quantum systems whose roots are in (non) equilibrium (quantum) statistical mechanics. The approach to this problem is to begin with the Liouville equation, which is the equation of motion in (quantum) statistical mechanics, and then develop a scheme that ensures irreversible dynamics emerging from it [55, 56]. Irreversible dynamics is the observed macroscopic behavior and hence the need to understand how it emerges from the underlying reversible microscopic dynamics. A number of tools have been developed to achieve an understanding of emergent irreversibility. The Langevin equation is an important technique employed in this context. It attempts to provide a contracted description of the system, on the macroscopic scale and invokes a probabilistic description. We will discuss Langevin equations and illustrate the theory by working out a model of Brownian motion of harmonic oscillators, a paradigm model in these studies. The Fokker-Planck equation is another approach used in the study of irreversible behavior. It is an equation of the probability distribution and is connected to the Langevin equation, though under certain constraints. The Fokker-Planck equations will be elucidated by examples of the ubiquitous Brownian motion and the Smoluchowski equation modeling strong friction. This is followed by a brief discussion of the Boltzmann equation, of great utility in describing kinetic processes. The conditions under which this can be approximated by a Fokker-Planck equation are spelled out. We will then move on to the master equations by spelling out their stochastic background. The attempt would be to bring out the inherent framework of (non) equilibrium statistical mechanics in all these discussions. After this will be discussed a few classes of well known master equations employed in a large number of studies related to open quantum systems. We will start with a detailed discussion of the dynamical semigroup and its related Lindblad-Gorini-Kosakowoski-Sudarshan master equation [57, 58]. This equation, which

© Hindustan Book Agency 2018 and Springer Nature Singapore Pte Ltd. 2018
S. Banerjee, *Open Quantum Systems*, Texts and Readings in Physical Sciences 20,
https://doi.org/10.1007/978-981-13-3182-4_3

describes memoryless or Markovian evolution, has found prolific use in quantum optics and quantum information, a topic we shall return to later in this book. We then study a unique quantum mechanical process called the quantum non-demolition. This will be followed by a discussion of the Nakajima-Zwanzig master equation [59, 60, 61] and the time-convolutionless projection operator technique (TCL) [62, 63, 64], both of which are general master equations for dealing with non-Markovian scenarios.

3.2   Liouville Equation

The Liouville equation is one of the cornerstones of nonequilibrium statistical mechanics [55, 56, 34]. In the classical context, it is a statement of the fact that due to motion in phase space, $\{q, p\}$, the number of points in a given volume, of phase space, remain unchanged. Thus, if the ensemble is specified by a distribution function $\rho(q, p, t)$, which is proportional to the probability density of the ensemble in phase space, then invoking the Hamiltonian equations of motion [65]

$$\frac{\partial H}{\partial p} = \dot{q}, \quad \frac{\partial H}{\partial q} = -\dot{p}, \tag{3.1}$$

we have

$$\frac{\partial \rho}{\partial t} = -i\mathcal{L}\rho, \tag{3.2}$$

which is the Liouville equation. Here $\mathcal{L}$ is the Liouville operator which acts on a function $f(q, p)$ of phase space as

$$i\mathcal{L}f = \{f, H\}, \tag{3.3}$$

where $\{f, H\}$ is the standard Poisson bracket given by

$$\{f, H\} = \left(\frac{\partial f}{\partial q} \frac{\partial H}{\partial p} - \frac{\partial f}{\partial p} \frac{\partial H}{\partial q} \right). \tag{3.4}$$

When $\mathcal{L}$ is time independent, Eq. (3.2) can be solved to give

$$\rho(q, p, t) = e^{-i\mathcal{L}(t)} \rho(q, p, 0). \tag{3.5}$$

$e^{-i\mathcal{L}(t)}$ could be called the evolution operator. The Liouville operator can be used to generate the dynamical equations of motion of observables. Thus, for example, for a dynamical observable $O(q, p, t)$, using Eqs. (3.1), it can be shown that

$$\frac{dO}{dt} = \frac{\partial O}{\partial t} + \{O, H\}. \tag{3.6}$$

Problem 1: Derive Eq. (3.6).

The Liouville equation for the quantum case can be analogously written as

$$\frac{\partial \rho(t)}{\partial t} = -i\mathcal{L}_{qu}\rho(t) = -\frac{i}{\hbar}\big[H, \rho(t)\big]. \qquad (3.7)$$

Here $\big[H, \rho(t)\big]$ is the usual Dirac commutator . The classical and quantum Liouville equations have a similar form. However, in the classical case, the Liouvillean operator is basically a function while in the quantum mechanical case, it acts in the space of operators and is a superoperator. Further, $\rho(t)$ has different connotations in the classical and quantum cases. In the classical scenario, it is the distribution function of the ensemble, while in the quantum case, it denotes the density matrix. Also, the Poisson bracket is replaced by the Dirac commutation relation. The Liouville equation, both classical as well as quantum, possess time-reversal invariance. This property is closely related to the fact that the Liouville equation is a reflection of Hamiltonian dynamics. Thus, it is unable to, in its present form, explain the approach to equilibrium. Hence, the way forward to understand the statistical evolution of macroscopic systems is to take into account interaction with the ambient environment. This introduces violation of time-reversal invariance in the evolution, leading to the appropriate macroscopic dynamics. We shall try to bring out this point in various guises, in this book. This would be the crux for understanding open (quantum) systems.

3.3 Langevin Equation

One of the main aims of statistical mechanics is to achieve an understanding of the emergence of the macroscopic behaviour from the underlying microscopic one. The Liouville equation, discussed above, is the fundamental approach to this problem. However, implementing it in practice could be a daunting task. Thus, a number of approximate methods have been developed to address these issues. The methodology of Langevin equations [56] is a prominent example of this. This usually results in a contracted description of the system, on the macroscopic scale and needs a stochastic (probabilistic) prescription. We will begin our discussion of Langevin equations by using the backdrop of Brownian motion , a ubiquitous model of (quantum) statistical mechanics, which has its roots in the attempt to understand the random motions of a small particle immersed in a dense fluid [66, 67]. The notion of Brownian motion has been used to address random motions associated with some collective property of a macroscopic system, for e.g., SQUID (Superconducting Quantum Interference Device) ring threaded by an external flux near half-a-flux quantum. Such a superconducting device could be appropriate for the observation of macroscopic quantum coherence effects [68].

We begin our discussion of the Langevin equation with the random motion of a Brownian particle. The equation of motion, for the Brownian particle position x, would be

$$m\frac{d^2x}{dt^2} = F(t), \tag{3.8}$$

where $F(t)$ would be the force acting on the particle. In a typical, classical, Brownian motion scenario, the force is dominated by a frictional force $-\zeta v$, that is, a velocity v dependent force with the frictional constant ζ. If we insert this into the *RHS* of Eq. (3.8) and solve for the velocity, we get $v(t) = e^{-\zeta t/m}v(0)$. This is however in contradiction with our notions of thermodynamic equilibrium according to which the mean squared velocity of the particle should approach k_BT/m in equilibrium [32]. Here k_B is the Boltzmann constant and T is the equilibrium temperature. In order to achieve this consistency, the *RHS* of Eq. (3.8) must be modified. Along with the dissipative, frictional force, there should be a random (fluctuating force) $\mathcal{F}_{ran}(t)$ such that the equation of motion becomes

$$m\frac{d^2x}{dt^2} = -\zeta\frac{dx}{dt} + \mathcal{F}_{ran}(t). \tag{3.9}$$

This is the Langevin equation in its basic form and could be called the equation of motion approach to modelling the (quantum) stochastic evolution of the system. It strives to achieve a balance between dissipation and fluctuations. Both the friction, source of dissipation and noise, causing fluctuations, have their origin in the interaction of the Brownian particle with its environment, also called the reservoir or the heat bath.

The effect of the random, fluctuating force is given by its first and second moments

$$\langle\mathcal{F}_{ran}(t)\rangle = 0, \quad \langle\mathcal{F}_{ran}(t)\mathcal{F}_{ran}(t')\rangle = 2A\delta(t - t'). \tag{3.10}$$

Here A is a measure of the strength of the fluctuating force and the angular brackets indicate an average over an infinitesimal time interval. The delta function indicates no correlation between the impacts on the Brownian particle, due to $\mathcal{F}_{ran}$, at different time intervals. Due to the large number of impacts caused by the fluctuating force, it is a reasonable assumption to model the fluctuating force by a Gaussian distribution determined by the above two moments. The Langevin equation (3.9) can be easily solved for the velocity $v(t) = \frac{dx}{dt}$ to give

$$v(t) = e^{-\zeta t/m}v(0) + \int_0^t dt' e^{-\zeta(t-t')/m}\frac{\mathcal{F}_{ran}(t')}{m}. \tag{3.11}$$

From this, the mean squared velocity $\langle v^2(t)\rangle$ can be obtained using Eq. (3.10) as

$$\langle v^2(t)\rangle = e^{-2\zeta t/m}v^2(0) + \frac{A}{\zeta m}(1 - e^{-2\zeta t/m}). \tag{3.12}$$

It can be seen that in the long time limit, the mean squared velocity tends to $\frac{A}{\zeta m}$. Further, this should be equal to the equilibrium value of k_BT/m [32]. This implies that

$$A = k_BT\zeta, \tag{3.13}$$

which is a simple form of the famous fluctuation-dissipation (F-D) theorem [49, 50], connecting the fluctuating parameter A to the dissipative parameter ζ.

In a similar fashion, the mean squared displacement $\langle \Delta x^2(t) \rangle$ can be obtained from Eq. (3.9). Using $\Delta x(t) = \int_0^t ds v(s)$ and Eq. (3.11), we have

$$\langle \Delta x^2(t) \rangle = \frac{2k_B T}{\zeta}\left(t - \frac{m}{\zeta} + \frac{m}{\zeta}e^{-\zeta t/m}\right). \qquad (3.14)$$

In the long time limit, the mean squared displacement is dominated by the term $\frac{2k_B T}{\zeta}t$. Einstein's expression relating mean squared displacement to diffusion D of the Brownian particle is $\langle \Delta x^2(t) \rangle = 2Dt$ [67]. This implies that $D = \frac{k_B T}{\zeta}$.

Problem 2: Work out the details leading to Eqs. (3.12) and (3.14).

Upto now our discussion of Langevin equations deals with Markovian evolution, that is, the case where memory effects are not accounted for. In particular, this entails that the friction parameter ζ at a particular time depends upon the velocity v at the same time. In general, the friction will have memory, that is, the friction at time t will depend on the velocity at a time $t' < t$. Thus,

$$\zeta v(t) \rightarrow \int_{-\infty}^{t} ds\mu(t-s)v(s) = \int_0^\infty ds\mu(s)v(t-s). \qquad (3.15)$$

Here the friction parameter ζ is replaced by the memory function $\mu(t)$. In the above equation, the history is assumed to have begun at time $t = -\infty$. The corresponding Langevin equation is

$$\frac{dx(t)}{dt} + \int_0^\infty ds\mu(s)x(t-s) = \mathcal{F}_{ranx}(t). \qquad (3.16)$$

Here $\mu(s)$ is the memory and $\mathcal{F}_{ranx}(t)$ is the random force. The above equation is a prototype of a non-Markovian Langevin equation.

Langevin Equation for Brownian Motion of a Harmonic Oscillator:

Now we consider the Langevin equation for the Brownian motion of a harmonic oscillator. This is a prototype model for a number of studies in Open (Quantum) Systems. The model we consider is written in such a manner that it has relevance to both classical as well as quantum mechanical systems. We start with the total Hamiltonian H

$$H = H_S + H_R + H_{SR}, \qquad (3.17)$$

where

$$H_S = \frac{p^2}{2m} + V(x), \qquad (3.18)$$

is the system Hamiltonian. For the case of (quantum) Brownian motion of a harmonic oscillator, $V(x) = \frac{1}{2}m\omega^2 x^2$, else it could be an arbitrary potential.

Note that in the quantum case, the x, q and p's are operators satisfying appropriate commutation relations. Further,

$$H_R = \sum_i \left[\frac{p_i^2}{2} + \frac{1}{2}\omega_i^2 q_i^2 + \frac{c_i^2}{2\omega_i^2}x^2 \right], \tag{3.19}$$

is the reservoir of harmonic oscillators of, for convenience, unit mass, also called interchangeably the heat bath or the system's environment. The third term in the RHS of the above equation could be absorbed into the potential $V(x)$ in H_S. Further,

$$H_{SR} = -\sum_i c_i x q_i, \tag{3.20}$$

is the interaction part of the Hamiltonian with c_i being the coupling constant. This bilinear form of the interaction allows for easy tractability of the resultant dynamics. The equations of motion are

$$\frac{dx}{dt} = \frac{p}{m}, \qquad \frac{dp}{dt} = -V'(x) + \sum_i c_i \left(q_i - \frac{c_i}{\omega_i^2}x \right)^2, \tag{3.21}$$

$$\frac{dq_i}{dt} = p_i, \qquad \frac{dp_i}{dt} = -\omega_i^2 q_i + c_i x. \tag{3.22}$$

If the motion of the system variable $x(t)$ is known, then from the above equations, it is easy to solve for the reservoir oscillator $q_i(t)$. Substituting this into the equation for the system momentum $p(t)$, we get

$$\frac{dp}{dt} = -V'(x(t)) - \int_0^t ds\mu(s)\frac{p(t-s)}{m} + \mathcal{F}_{ranp}(t). \tag{3.23}$$

This has the form of a Langevin equation with the memory

$$\mu(t) = \sum_i \frac{c_i^2}{\omega_i^2}\cos(\omega_i t), \tag{3.24}$$

and the noise term

$$\mathcal{F}_{ranp}(t) = \sum_i c_i p_i(0)\frac{\sin(\omega_i t)}{\omega_i} + \sum_i c_i\left(q_i(0) - \frac{c_i}{\omega_i^2}x(0)\right)\cos(\omega_i t). \tag{3.25}$$

By appropriately choosing the reservoir spectrum $\{\omega_i\}$ and coupling constants $\{c_i\}$, the memory function $\mu(t)$ can be designed. Thus, for example, if the spectrum is continuous, then the sum over the frequencies can be replaced by an integral as $\sum_i \to \int d\omega\rho(\omega)$, where $\rho(\omega)$ is the reservoir density of states. Then the memory function becomes

$$\mu(t) = \int_0^\infty d\omega\rho(\omega)\frac{c^2(\omega)}{\omega^2}\cos(\omega t). \tag{3.26}$$

If $\rho(\omega) \propto \omega^2$, $c(\omega) = c$ (a constant), then $\mu(t) \propto \delta(t)$ and the Langevin equation (3.23) becomes Markovian. If the number of bath oscillators is very large, then the noise generated would have, invoking the central limit theorem, a simple Gaussian distribution. Assuming the bath oscillator initial conditions to be $e^{-\beta H_R}$, where $\beta = 1/(k_B T)$, then the average over the reservoir variables becomes easy. The linear terms average to zero, while the quadratic terms go as

$$\left\langle \left(q_i(0) - \frac{c_i}{\omega_i^2} x(0) \right)^2 \right\rangle = \frac{k_B T}{\omega_i^2}, \quad \langle p_i^2(0) \rangle = k_B T. \tag{3.27}$$

Using the above equation, we have

$$\left\langle \mathcal{F}_{ranp}(t)\mathcal{F}_{ranp}(t') \right\rangle = k_B T \mu(t - t'). \tag{3.28}$$

This is the fluctuation-dissipation theorem for (quantum) Brownian motion. In passing we remark that in the quantum mechanical scenario, Eq. (3.28) would be replaced by

$$\frac{1}{2} \left\langle \mathcal{F}_{ranp}(t)\mathcal{F}_{ranp}(t') + \mathcal{F}_{ranp}(t')\mathcal{F}_{ranp}(t) \right\rangle = \hbar K(t - t'), \tag{3.29}$$

where

$$K(t - t') = \sum_i \frac{c_i^2}{2\omega_i^2} \cos(\omega_i t) \coth\left(\frac{\hbar \omega_i}{2 k_B T} \right). \tag{3.30}$$

On taking the classical limit of Eq. (3.29), we recover Eq. (3.28).

3.4 Fokker-Planck Equation

The Fokker-Planck (FP) equation is another prominent technique used in the study of nonequilibrium processes. It is a kind of a Liouville equation and is connected to the Langevin equation for memoryless (Markovian) friction and Gaussian white noise. Let us consider the Langevin equation for the collective variable $\{x\}$, denoting a set $\{x_1, x_2, \cdots\}$. Assuming a Markovian dynamics, for example, memoryless friction and a white noise distributed according to a Gaussian distribution, the corresponding Langevin equation is

$$\frac{d\mathbf{x}}{dt} = v(\mathbf{x}) + \mathcal{F}(t). \tag{3.31}$$

Here $\mathcal{F}(t)$ is the Gaussian noise, with zero average, and second moment

$$\langle \mathcal{F}(t)\mathcal{F}(s) \rangle = 2K\delta(t - s). \tag{3.32}$$

The strategy behind developing the FP equation is that instead of concentrating on the solution of the equation (3.31), we concentrate on the probability distribution $p(\mathbf{x}, t)$ of the values of $\mathbf{x}$ at time t. Also, what we are really interested in is the behaviour of this distribution averaged over the noise $\mathcal{F}$. Since, the probability distribution is conserved, we have a continuity equation

$$\frac{\partial p}{\partial t} = -\frac{\partial}{\partial \mathbf{x}}\left(\frac{\partial \mathbf{x}}{\partial t}p\right). \tag{3.33}$$

This step is also encountered when we deal with the Liouville equation. Now we make contact of the FP equation with the Langevin equation. Substituting Eq. (3.31) into the *RHS* of Eq. (3.33) we get

$$\frac{\partial p(\mathbf{x}, t)}{\partial t} = -\frac{\partial}{\partial \mathbf{x}}\cdot\left(v(\mathbf{x})p(\mathbf{x}, t) + \mathcal{F}(t)p(\mathbf{x}, t)\right). \tag{3.34}$$

This can be rearranged as

$$\begin{aligned}\frac{\partial p(\mathbf{x}, t)}{\partial t} &= -\frac{\partial}{\partial \mathbf{x}}\cdot\left(v(\mathbf{x})p(\mathbf{x}, t)\right) - \frac{\partial}{\partial \mathbf{x}}.\mathcal{F}(t)p(\mathbf{x}, 0) \\ &+ \frac{\partial}{\partial \mathbf{x}}.\mathcal{F}(t)\int_0^t dse^{-(t-s)L}\frac{\partial}{\partial \mathbf{x}}.\mathcal{F}(s)p(\mathbf{x}, s).\end{aligned} \tag{3.35}$$

Here L is the symbol representing the following operation

$$Lf \equiv \frac{\partial}{\partial \mathbf{x}}\cdot\left(v(\mathbf{x})f\right). \tag{3.36}$$

Next, we average Eq. (3.35) over the noise $\mathcal{F}(t)$, using Eq. (3.32), to get

$$\frac{\partial}{\partial t}\langle p(\mathbf{x}, t)\rangle = -\frac{\partial}{\partial \mathbf{x}}.v(\mathbf{x})\langle p(\mathbf{x}, t)\rangle + \frac{\partial}{\partial \mathbf{x}}.K.\frac{\partial}{\partial \mathbf{x}}\langle p(\mathbf{x}, t)\rangle. \tag{3.37}$$

This is the FP equation. The first term on the *RHS* is independent of noise and is the *drift* term. The second term is responsible for *diffusion* and corresponds to the averaged effect of noise. The FP equation (3.37) is thus a stochastic differential equation. It should be noted that the equivalence between the Langevin and FP equation established here depended crucially on the noise being white and Gaussian distributed and may not exist for more complicated scenarios [69]. The FP equation has the general form of the probability conservation equation

$$\frac{\partial\langle p(\mathbf{x}, t)\rangle}{\partial t} = -\frac{\partial J(\mathbf{x}, t)}{\partial x}. \tag{3.38}$$

Here $J(\mathbf{x}, t)$ is the probability current and is a reflection of the conservation of probability which holds under FP evolution. The analogous treatment in a quantum setup would follow the same lines with the probability distribution $p(\mathbf{x}, t)$ being replaced by the density matrix $\rho(t)$.

   Examples:
   (a). *Brownian motion:*

We consider the Brownian motion of a particle moving in a potential $V(x)$. Let us restrict to the case of memoryless friction ζ. The corresponding Langevin equations are

$$\frac{dx}{dt} = \frac{p}{m}, \quad \frac{dp}{dt} = -V'(x) - \zeta\frac{p}{m} + \mathcal{F}_{ranp}(t). \tag{3.39}$$

Here $V'(x)$ is the differential of the potential $V(x)$ with respect to x. The noise has the second moment

$$\langle \mathcal{F}_{ranp}(t)\mathcal{F}_{ranp}(s)\rangle = 2\zeta k_B T\delta(t-s). \tag{3.40}$$

Making correspondence with the nomenclature used in discussing the F-P equation, that is,

$$\mathbf{x} = \begin{pmatrix} x \\ p \end{pmatrix}, v(\mathbf{x}) = \begin{pmatrix} p/m \\ -V'(x) - \zeta p/m \end{pmatrix},$$

$$\mathcal{F}(t) = \begin{pmatrix} 0 \\ \mathcal{F}_{ranp}(t) \end{pmatrix}, K = \begin{pmatrix} 0 & 0 \\ 0 & \zeta k_B T \end{pmatrix}, \tag{3.41}$$

we get, comparing with Eq. (3.37), the following FP equation

$$\frac{\partial p(x,t)}{\partial t} = -\frac{\partial}{\partial x}\left(vp(x,t)\right) - \frac{\partial}{\partial p}\left(-V'(x) - \zeta v\right)p(x,t) + \zeta k_B T\frac{\partial^2}{\partial p^2}p(x,t). \tag{3.42}$$

This is a parabolic differential equation and $v = p/m$ is the velocity. Also, $\langle p(x,t)\rangle$ has been depicted in the above equation, as well as in the next example, for notational simplicity, as $p(x,t)$.

(b). *Smoluchowski equation:*

This is a special case of the evolution governed by Eq. (3.39) and holds for the case of strong friction, that is, when the time scale m/ζ is smaller than all the other time scales in the problem, including that coming from motion due the potential term $V(x)$. Thus, the Brownian particle velocity relaxes to its stationary value very quickly. Hence, the $\frac{d^2x}{dt^2}$ term, in the Langevin equation, can be set to zero. The Langevin equation we deal with here is thus simplified to

$$\frac{dx}{dt} = -\frac{1}{\zeta}V'(x) + \frac{1}{\zeta}\mathcal{F}(t). \tag{3.43}$$

Following the discussions above, the corresponding FP equation is

$$\frac{\partial p(x,t)}{\partial t} = \frac{1}{\zeta}\frac{\partial}{\partial x}V'(x)p(x,t) + D\frac{d^2}{dx^2}p(x,t). \tag{3.44}$$

Here $D = \frac{k_B T}{\zeta}$ is the diffusion constant. The corresponding probability current, Eq. (3.38), is

$$J(x,t) = -\frac{1}{\zeta}\left(V'(x) + k_B T\frac{\partial}{\partial x}\right)p(x,t). \tag{3.45}$$

3.5   Boltzmann Equation

The Boltzmann equation was developed with the aim to provide a correct description of dynamical processes in a dilute gas [31]. It is associated with the famous H-theorem, providing a prescription for the approach to equilibrium of the gas asymptotically. The equation could be envisaged as an artifact of the Liouville equation and is

$$\frac{\partial f}{dt} + v.\Delta_r f + \frac{1}{m}F(x).\Delta_v f = \left(\frac{\partial f}{\partial t}\right)_{collision}, \tag{3.46}$$

where the *LHS* is basically the Liouville equation, (3.2), for single particle motion in a potential. $F(x)$ is the external force acting on the particle of mass m and $f(x, p, t)$ is the density function. The *RHS* is called the collision integral and accounts for changes in f due to collisions [56]. Since the *RHS* involves collisions and at least two particles are required for a collision, it makes the Boltzmann equation nonlinear. Assuming $F = 0$ and a uniform distribution, the collision terms can be assumed to be position independent. Define

$$f(p, t) = \frac{1}{V}\int d^3x f(x, p, t), \tag{3.47}$$

which is the normalized probability distribution in momentum space. Then the Boltzmann equation takes the form

$$\frac{\partial f(p, t)}{dt} = \left(\frac{\partial f}{\partial t}\right)_{collision}. \tag{3.48}$$

In the absence of external force, all the variation in the probability distribution can be ascribed to collisions. Defining $\zeta(p, k)$ to be the rate of collisions which change the momentum from p to $p - k$, the *RHS* of Eq. (3.48) becomes

$$\left(\frac{\partial f}{\partial t}\right)_{collision} = \int d^3k[\zeta(p + k, k)f(p + k) - \zeta(p, k)f(p)]. \tag{3.49}$$

The integral operators involved in the collision processes, contained in $\zeta(p, k)$, makes the dynamics of Eq. (3.48) complicated. At this stage if it can be assumed that the collisions are dominated by soft processes, that is, low energy collisions, then the integrand on the *RHS* of Eq. (3.49) can be converted into a differential operator [70, 71], that is,

$$\zeta(p + k, k)f(p + k) \approx \zeta(p, k)f(p) + k.\frac{\partial}{\partial p}(\zeta f) + \frac{1}{2}k_i k_j \frac{\partial^2}{\partial p_i \partial p_j}(\zeta f). \tag{3.50}$$

Using this in Eq. (3.49), we see that Eq. (3.48) becomes

$$\frac{\partial f}{\partial t} = \frac{\partial}{\partial p_i}\left[A_i(p)f + \frac{\partial}{\partial p_j}[B_{ij}(p)f]\right]. \tag{3.51}$$

This has the form of a Fokker-Planck equation with

$$A_i = \int d^3k \zeta(p,k)k_i, \tag{3.52}$$

$$B_{ij} = \frac{1}{2} \int d^3k \zeta(p,k)k_i k_j, \tag{3.53}$$

being the drift and drag coefficients, respectively.

3.6 Master Equation

In the endeavour to understand nonequilibrium statistical mechanics, master equations are a very convenient and well known tool. The master equation can be developed, classically, from the perspective of stochastic processes [72, 73, 74, 75]. We recall some standard definitions: $P_1(x_1, t_1)$ is the probability density that the stochastic variable X takes the value x_1 at time t_1. The joint probability density of n stochastic variables is written, in short, as P_n, which is positive semi-definite, can be reduced and is normalized. Another important term is the conditional probability $P_{1|1}(x_1, t_1|x_2, t_2)$, that is, the probability that the stochastic variable X takes the value x_2 at time t_2, given that it was x_1 at t_1. A standard relation between these quantities is

$$\int dx_1 P_1(x_1, t_1) P_{1|1}(x_1, t_1|x_2, t_2) = \int dx_1 P_2(x_1, t_1; x_2, t_2) = P_1(x_2, t_2). \tag{3.54}$$

Here $P_2(x_1, t_1; x_2, t_2)$ is the joint probability that the stochastic variable X is x_1 at t_1 and x_2 at t_2. Expanding the conditional probability in

$$P_1(i, t + \Delta t) = \sum_{j=1}^{J} P_1(j, t) P_{1|1}(j, t|i, t + \Delta t), \tag{3.55}$$

which is basically the discretized form of Eq. (3.54), in powers of Δt, and taking the limit of $\Delta t \to 0$, we get

$$\frac{\partial P_1(i, t)}{\partial t} = \sum_{j=1}^{J} \left[P_1(j, t) w_{j,i}(t) - P_1(i, t) w_{i,j}(t) \right], \tag{3.56}$$

the *master equation*. The master equation is inherently linear. Here $w_{j,i}(t)$ is the transition probability rate and is obtained by expanding the conditional probability term in Eq. (3.55). The transition probability rates satisfy the detailed balance condition when

$$P_S(i) w_{i,j} = P_S(j) w_{j,i}, \tag{3.57}$$

where $P_S(i)$ is the long time stationary limit of the probability $P_1(i, t)$. This conveys the information that at equilibrium, the flow of probability into a level i from a level j is balanced by the flow from i to j.

Now we turn to the quantum aspects of master equations. This can be done at various levels, as we will see in this chapter. We start with the *Pauli master equation*, for the Hamiltonian $H = H_0 + \lambda V$, where H_0 is the basic unperturbed Hamiltonian, while V is the perturbing term causing transitions between the unperturbed energy levels and λ is the term gauging the strength of the perturbation. The Pauli master equation is a gain-loss equation for the probability of occupation of a given state i, P_i,

$$\frac{dP_i(t)}{dt} = \sum_j \left(w_{ji} P_j(t) - w_{ij} P_i(t) \right). \tag{3.58}$$

Note that this has the form of Eq. (3.56). Eq. (3.58) tells us that the rate of change of probability of a state i at a time t is equal to the balance of flow from states j to state i and the flow from i to states j. Here, w_{ji} is the transition probability from states j to i. Its form is given by the standard Golden Rule [23], which using first order of perturbation theory is,

$$w_{ij} = \frac{2\pi}{\hbar} \lambda^2 |V_{ij}|^2 \delta(E_i - E_j). \tag{3.59}$$

Here E_i's are the eigenvalues of the unperturbed Hamiltonian H_0. The Pauli master equation allows for transitions between states with approximately equal total, unperturbed energy, that is, it is microcanonical in nature. The transition rates are symmetrical in states, $w_{ij} = w_{ji}$, called *microscopic reversibility*. A note is in order here. The Golden Rule formula is valid as long as λ is small, time t is large and $\lambda^2 t$ is of order 1, called the van Hove limit [76].

Now we turn to a master equation having a canonical nature, that is, exchange of energy is also taken into account. This is usually discussed under the guise of the *heat bath (reservoir) master equation* . We will now discuss this in some detail, as this is a protoype of open (quantum) systems. Consider the total Hamiltonian H

$$H = H_S + H_R + H_{SR}, \tag{3.60}$$

where H_S is the Hamiltonian of the system of interest, for which the master equation will be constructed. H_R is the reservoir Hamiltonian, while H_{SR} is the interaction part which connects the system to the reservoir and would be like the λV term, discussed above. The system and reservoir Hamiltonian's have the eigenvalue equations

$$H_S|i\rangle = E_i|i\rangle, \quad H_R|\alpha\rangle = \epsilon_\alpha|\alpha\rangle. \tag{3.61}$$

Using the Pauli master equation, (3.58), in the basis of the unperturbed system and reservoir Hamiltonians we have

$$\frac{dP_{i\alpha}}{dt} = \sum_{j\beta} \left(w_{j\beta,i\alpha} P_{j\beta} - w_{i\alpha,j\beta} P_{i\alpha} \right). \tag{3.62}$$

Here $w_{j\beta,i\alpha} = \frac{2\pi}{\hbar}\lambda^2|\langle j\beta|V|i\alpha\rangle|^2\delta(E_i+\epsilon_\alpha-E_j-\epsilon_\beta)$. Assuming that the reservoir remains in its state of thermal equilibrium,

$$P_{i\alpha}(t) \approx P_i(t)\rho_\alpha. \tag{3.63}$$

Here $P_i(t)$ is the nonequilibrium probability of the state i and ρ_α is the thermal equilibrium state of the bath. Substituting Eq. (3.63) in Eq. (3.62) and summing over the reservoir states $|\alpha\rangle$, that is, tracing out the reservoir degrees of freedom, we have the master equation of the system of interest S

$$\frac{dP_i}{dt} = \sum_j w_{ji}P_j - \sum_j w_{ij}P_i. \tag{3.64}$$

Here

$$w_{ji} = \sum_\alpha \sum_\beta w_{j\beta,i\alpha}\rho_\beta,$$

$$w_{ij} = \sum_\alpha \sum_\beta w_{i\alpha,j\beta}\rho_\alpha. \tag{3.65}$$

The master equation, (3.64), allows the determination of the relaxation of the system probability $P_i(t)$ (given by, say, the diagonal elements of the system density matrix and denoting, for example, the population) to its thermal equilibrium $P_{eql}(T)$, where the temperature T is determined by that of the reservoir. The principle of micocanonical reversibility of the Pauli master equation is now replaced by the principle of detailed balanced

$$w_{ji}e^{-\beta E_j} = w_{ij}e^{-\beta E_i}. \tag{3.66}$$

Here $\beta = 1/(k_B T)$.

We now couch the above example in terms of the standard lexicon of open quantum systems. The interaction Hamiltonian is expressed as a product of system and reservoir operators

$$H_{SR} = \lambda V = M\Theta, \tag{3.67}$$

where M is a system and Θ a reservoir operator. Thus $\lambda V_{j\beta,i\alpha} = M_{ji}\Theta_{\beta\alpha}$. The thermally averaged transition rate becomes

$$w_{ij} = \frac{2\pi}{\hbar}|M_{ij}|^2 \sum_\alpha \sum_\beta \delta(E_i - E_j + \epsilon_\alpha - \epsilon_\beta)|\Theta_{\alpha\beta}|^2\rho_\beta. \tag{3.68}$$

Using the integral representation of the delta function, the transition rate can be brought to the convenient form

$$w_{ij} = \frac{1}{\hbar^2}|M_{ij}|^2 \int_{-\infty}^{\infty} dt e^{i\omega_{ij}t}\langle\Theta(0)\Theta(t)\rangle_{eql}. \tag{3.69}$$

Here $\langle\Theta(0)\Theta(t)\rangle_{eql}$ is the thermal average, obtained using ρ_β, of the reservoir time correlation function. Thus, the transition rate is proportional to the spectral density of the reservoir time correlation function evaluated at the frequency $\omega_{ij} = (E_i - E_j)/\hbar$ of the transition.

Let us revisit the model of Brownian motion of a harmonic oscillator, Eqs. (3.17) to (3.20). By comparison, $M = x$ and $\Theta = -\sum_j c_j q_j$. Since the reservoir is composed of harmonic oscillators, the time correlation function of the reservoir operator is

$$\langle\Theta(0)\Theta(t)\rangle_{eql} = \sum_j c_j^2 \cos(\omega_j t)\langle q_j^2\rangle_{eql} + \sum_j \frac{c_j^2}{\omega_j} \sin(\omega_j t)\langle p_j q_j\rangle_{eql}. \qquad (3.70)$$

The equilibrium averages can be done in a straightforward manner, see for example, Eq. (3.27).

Problem 3: Sketch the steps leading to the Eq. (3.69).

Upto this point we have outlined the general framework of non-equilibrium (quantum) statistical mechanics. We now are in a position to make contact with some well known classes of master equations.

3.7 Quantum Dynamical Semigroups and Markovian Master Equation

We will now discuss an important class of master equations, the *Lindblad* equations [57, 58, 77]. Apart from being aesthetically appealing, they find use in a large number of studies in quantum optics and quantum information [29, 2]. Suppose that the system S and reservoir R are initially uncorrelated, that is,

$$\rho(0) = \rho^S(0) \otimes \rho^R. \qquad (3.71)$$

Let us define a dynamical map $V(t)$ describing the transformation of the reduced system at $t = 0$ to some $t > 0$ as

$$\rho^S(t) = V(t)\rho^S(0) = \mathrm{Tr}_R(U(t,0)[\rho^S(0) \otimes \rho^R]U^\dagger(t,0)), \qquad (3.72)$$

where $U(t,0)$ is the unitary operator giving the evolution of the full $S+R$ complex. If t is allowed to vary, it leads to a one-parameter family of dynamical maps with $V(0)$ being the identity map. The map $V(t)$ represents a convex-linear, completely positive and trace-preserving quantum operation, the meaning of this will become clear in Chapter 8. If the characteristic time scales over which the reservoir correlation function decays are much smaller than the characteristic time scales of the system, memory effects in the reduced dynamics can be neglected. This leads to a Markovian-type behaviour and may be formalized with the help of the semigroup property:

$$V(t_1)V(t_2) = V(t_1 + t_2), \quad t_1, t_2 \geq 0. \qquad (3.73)$$

Thus a quantum dynamical semigroup is a continuous, one-parameter family of dynamical maps satisfying the semigroup property. By its very definition, the semigroup property breaks the time reversal invariance in the dynamics and is a suitable starting point for obtaining irreversible dynamics from the Liouville equation, as discussed earlier in this Chapter. If the quantum dynamical semigroup $V(t)$ is contracting, that is,

$$||V^{\dagger}(t)A||_1 \leq ||A||_1, \tag{3.74}$$

for an operator A in the Hilbert space of the open system and where $||A||_1 = \mathrm{Tr_S}|A|$ is the trace norm, then there exists a linear map $\mathcal{L}$ called the generator of the semigroup

$$V(t) = \exp(\mathcal{L}t), \tag{3.75}$$

which leads to

$$\frac{d}{dt}\rho^S(t) = \mathcal{L}\rho^S(t), \tag{3.76}$$

the Quantum Markovian Master equation. The construction of the most general form of the generator $\mathcal{L}$ leads to the Lindblad equation.

3.7.1 Derivation of the Lindblad-Gorini-Kosakowoski-Sudarshan Master Equation

Here we discuss the details of the derivation of the Lindblad-Gorini-Kosakowoski-Sudarshan (LGKS) master equation and spell out the physical criteria behind its construction [2]. The derivation is along the lines, sketched earlier, of attempting to obtain irreversible behavior starting from the Liouville equation, which here would imply starting with the interaction picture von Neumann equation

$$\frac{d}{dt}\rho(t) = -i[H_{SR}(t), \rho(t)], \tag{3.77}$$

for the total density matrix $\rho(t)$. This gives

$$\rho(t) = \rho(0) - i\int_0^t ds[H_{SR}(s), \rho(s)]. \tag{3.78}$$

Inserting the integral into the von Neumann equation and tracing over the bath, in order to obtain the equation of motion of the system of interest S,

$$\frac{d}{dt}\rho^S(t) = -\int_0^t ds\mathrm{Tr_R}[H_{SR}(t), [H_{SR}(s), \rho(s)]], \tag{3.79}$$

where $\mathrm{Tr_R}[H_{SR}(s), \rho(0)] = 0$ is assumed. The *RHS* of the equation depends on the full density matrix $\rho(s)$. This is where the first approximation is made, the *Born approximation*. This assumes that the coupling between S and R is weak,

ρ^R is negligibly affected by the interaction and the total system after time t is $\rho(t) \equiv \rho^S(t) \otimes \rho^R$. This gives

$$\frac{d}{dt}\rho^S(t) = -\int_0^t ds\,\mathrm{Tr_R}[\mathrm{H_{SR}}(t), [\mathrm{H_{SR}}(s), \rho^S(s) \otimes \rho^R]]. \qquad (3.80)$$

A further simplification: $\rho^S(s) \longrightarrow \rho^S(t)$. Thus the evolution equation of the system at t depends only on the present state. This is the *Redfield* equation [78]. The Redfield equation is local in time, but depends on the choice of the initial preparation at $t = 0$, and hence is not Markovian. To make it Markovian we replace s by $t - s$ in the integrand and let the upper limit go to infinity. This results in

$$\frac{d}{dt}\rho^S(t) = -\int_0^\infty ds\,\mathrm{Tr_R}[\mathrm{H_{SR}}(t), [\mathrm{H_{SR}}(t-s), \rho^S(t) \otimes \rho^R]]. \qquad (3.81)$$

This is a Markovian equation and the approximation is called the *Markovian approximation*. It is justified when the time scale associated with the reservoir correlations τ_R is much smaller than the time scale τ_{rel} over which the state varies appreciably. Thus the Markovian evolution is defined on a *coarse-grained time scale*, where the dynamical behaviour over times of the order of τ_R are not resolved. Since τ_R depends on the reservoir temperature and τ_{rel} on the $S - R$ coupling strength, the Markovian approximation is easily justified for weak $S - R$ coupling and high T.

The approximations made till now would be collectively called the *Born-Markov* approximation. However, they do not guarantee a quantum dynamical semigroup evolution [79, 80]. A further approximation involving averaging over the rapidly oscillating terms in the master equation is performed, the *Rotating Wave Approximation*. For accomplishing this, the interaction Hamiltonian H_{SR} is decomposed into eigenoperators of the system Hamiltonian H_S. A generic interaction Hamiltonian in the interaction picture can be written as

$$H_{SR}(t) = \sum_{\alpha,\omega} e^{-i\omega t} A_\alpha(\omega) \otimes B_\alpha(t), \qquad (3.82)$$

where A, B denote operators belonging to the system and reservoir, respectively. Also,

$$A_\alpha(\omega) = \sum_{\epsilon'-\epsilon=\omega} \Pi(\epsilon) A_\alpha \Pi(\epsilon'), \qquad (3.83)$$

where $\Pi(\epsilon)$ projects the operator onto the eigenspace of H_S belonging to the eigenvalue ϵ. Thus $[H_S, A_\alpha(\omega)] = -\omega A_\alpha(\omega)$, i.e., $A_\alpha(\omega)$ lowers the energy of H_S by ω while $A_\omega^\dagger$ raises it by ω. Invoking the Heisenberg picture,

$$e^{iH_S t} A_\alpha(\omega) e^{-iH_S t} = e^{-i\omega t} A_\alpha(\omega), \qquad (3.84)$$

and

$$e^{iH_R t} B_\alpha e^{-iH_R t} = B_\alpha(t). \qquad (3.85)$$

The earlier condition $\text{Tr}_R[H_{SR}(s), \rho(0)] = 0$ now implies $\langle B_\alpha(t) \rangle = \text{Tr}_R(B_\alpha(t)\rho^R) = 0$, which is consistent with the assumption made, earlier in this chapter, on the vanishing of single operator averages. This leads to the following form of the Born-Markov equation obtained earlier

$$\frac{d}{dt}\rho^S(t) = \sum_{\omega,\omega'}\sum_{\alpha,\beta} e^{i(\omega'-\omega)t}\Gamma_{\alpha,\beta}(\omega)[A_\beta(\omega)\rho^S(t)A_\alpha^\dagger(\omega') - A_\alpha^\dagger(\omega')A_\beta(\omega)\rho^S(t)] + h.c.$$

(3.86)

Here $h.c.$ denotes Hermitian conjugation and

$$\Gamma_{\alpha,\beta}(\omega) = \int_0^\infty ds e^{i\omega s}\langle B_\alpha^\dagger(t)B_\beta(t-s)\rangle,$$

(3.87)

is the one-sided Fourier transform of reservoir correlation functions

$$\langle B_\alpha^\dagger(t)B_\beta(t-s)\rangle = \text{Tr}_R(B_\alpha^\dagger(t)B_\beta(t-s)\rho^R).$$

(3.88)

If ρ^R is a stationary state of the reservoir, that is, $[H_R, \rho^R] = 0$, the reservoir correlation functions are homogeneous in time

$$\langle B_\alpha^\dagger(t)B_\beta(t-s)\rangle = \langle B_\alpha^\dagger(s)B_\beta(0)\rangle.$$

(3.89)

In the above evolution equation, $|\omega - \omega'|^{-1}$ defines the typical time-scale associated with the intrinsic evolution of the system. If the systematic evolution of the system is very quick, then it goes through many cycles during the relaxation time. Thus the *non-secular terms*, that is, those for which $\omega' \neq \omega$, may be neglected. This is the *rotating wave approximation* . With this, the evolution equation becomes

$$\frac{d}{dt}\rho^S(t) = \sum_\omega\sum_{\alpha,\beta}\Gamma_{\alpha,\beta}(\omega)[A_\beta(\omega)\rho^S(t)A_\alpha^\dagger(\omega) - A_\alpha^\dagger(\omega)A_\beta(\omega)\rho^S(t)] + h.c. \quad (3.90)$$

In the above equation, the term $\Gamma_{\alpha,\beta}$ can be rearranged as

$$\Gamma_{\alpha,\beta}(\omega) = \frac{1}{2}\gamma_{\alpha,\beta}(\omega) + iS_{\alpha,\beta}(\omega),$$

(3.91)

where

$$\gamma_{\alpha,\beta}(\omega) = \Gamma_{\alpha,\beta}(\omega) + \Gamma_{\beta,\alpha}^*(\omega) = \int_{-\infty}^\infty ds e^{i\omega s}\langle B_\alpha^\dagger(s)B_\beta(0)\rangle,$$

(3.92)

and

$$S_{\alpha,\beta}(\omega) = \frac{1}{2i}(\Gamma_{\alpha,\beta}(\omega) - \Gamma_{\beta,\alpha}^*(\omega)).$$

(3.93)

With these, the evolution equation of the system of interest S can be written as

$$\frac{d}{dt}\rho^S(t) = -i[H_{LS}, \rho^S(t)] + \mathcal{D}(\rho^S(t)),$$

(3.94)

where

$$H_{LS} = \sum_\omega \sum_{\alpha,\beta} S_{\alpha,\beta} A_\alpha^\dagger(\omega) A_\beta(\omega), \tag{3.95}$$

is called the Lamb shift as it leads to a Lamb-type renormalization of the unperturbed energy levels due to the $S-R$ coupling and provides a Hamiltonian contribution to the dynamics. The term $\mathcal{D}(\rho^S(t))$ is called the dissipator and takes the form

$$\mathcal{D}(\rho^S(t)) = \sum_\omega \sum_{\alpha,\beta} \gamma_{\alpha,\beta} \left(A_\beta(\omega)\rho^S A_\alpha^\dagger(\omega) - \frac{1}{2}\{A_\alpha^\dagger(\omega)A_\beta(\omega), \rho^S\} \right). \tag{3.96}$$

Here $\{A, B\} = AB + BA$. The term $\gamma_{\alpha,\beta}$, in the dissipator is the Fourier transform of the homogeneous reservoir correlation functions , is positive by Bochner's theorem and hence can be diagonalized. With that the evolution equation, (3.94), takes the form of the standard Lindblad equation

$$\frac{d}{dt}\rho^S(t) = \mathcal{L}\rho^S(t)$$

$$= -i[H, \rho^S(t)] + \sum_{j=1}^{N^2-1} \gamma_j \left(A_j \rho^S A_j^\dagger - \frac{1}{2}\{A_j^\dagger A_j, \rho^S\} \right). \tag{3.97}$$

The operators A_j are called the Lindblad operators. We thus stress that the physical assumptions underlying the LGKS form of the master equation are the *Born* (weak coupling), *Markov* (memoryless) and *Rotating Wave Approximation* (fast system dynamics compared to the relaxation time). These physical assumptions result in implementing the semigroup dynamics, from the initial Liouvillean dynamics, on the system of interest. This is thus a concrete example of the general program of nonequilibrium (quantum) statistical mechanics.

3.7.2   Examples

(a). Dissipative Two-Level System
    We illustrate the LGKS evolution by means of a practical example, that is, the decay of a two-level system interacting with a radiation field (bath) in the weak Born-Markov, rotating wave approximation. Consider a system Hamiltonian $H_S = \frac{1}{2}\hbar\omega_0\sigma_z$, ω_0 is the transition frequency. The system interacts with a bath (reservoir/environment) of harmonic oscillators via the atomic dipole operator (in the interaction picture)

$$\vec{D}(t) = \vec{d}\sigma_- e^{-i\omega t} + \vec{d}^* \sigma_+ e^{i\omega t}, \tag{3.98}$$

where $\vec{d} = \langle g|\vec{D}|e\rangle$ is the transition matrix elements of the dipole operator D and the $S - R$ coupling term is

$$H_{SR} = -\vec{D}.\vec{E}. \tag{3.99}$$

Here $\vec{E}$ is the electric field operator, which in the Schrödinger picture is

$$\vec{E} = i \sum_{\vec{k}} \sum_{\lambda=1,2} \sqrt{\frac{2\pi\hbar\omega_k}{V}} \vec{e}_\lambda(\vec{k}) \left(b_\lambda(\vec{k}) - b_\lambda^\dagger(\vec{k}) \right). \tag{3.100}$$

The field modes are represented by $\vec{k}$ and the two corresponding, transverse unit polarization vectors $\vec{e}_\lambda(\vec{k})$. The Pauli operators σ_-, σ_+ satisfy $[H_S, \sigma_-] = -\omega_0\sigma_-$, $[H_S, \sigma_+] = \omega_0\sigma_+$, that is, they lower/raise the atomic energy by $\mp\omega_0$. Comparing with the structure of the basic Lindbladian equation, (3.94), we see that this process has two Lindblad operators: $\vec{A}(\omega_0) \equiv \vec{A} = \vec{d}\sigma_-$, $\vec{A}(-\omega_0) \equiv \vec{A}^\dagger = \vec{d}^*\sigma_+$.

The LGKS master equation for the reduced density matrix operator in the interaction picture (neglecting the so called Lamb shift terms) becomes

$$\begin{aligned}
\frac{d}{dt}\rho^S(t) &= \gamma_0(N_{th} + 1) \left(\sigma_-\rho^S(t)\sigma_+ - \frac{1}{2}\sigma_+\sigma_-\rho^S(t) - \frac{1}{2}\rho^S(t)\sigma_+\sigma_- \right) \\
&+ \gamma_0 N_{th} \left(\sigma_+\rho^S(t)\sigma_- - \frac{1}{2}\sigma_-\sigma_+\rho^S(t) - \frac{1}{2}\rho^S(t)\sigma_-\sigma_+ \right). \tag{3.101}
\end{aligned}$$

Here γ_0 is spontaneous emission rate

$$\gamma_0 = \frac{4\omega^3|\vec{d}|^2}{3\hbar c^3}, \tag{3.102}$$

and σ_+, σ_- the standard raising and lowering operators, respectively, and are

$$\sigma_+ = |1\rangle\langle 0| = \frac{1}{2}(\sigma_x + i\sigma_y); \quad \sigma_- = |0\rangle\langle 1| = \frac{1}{2}(\sigma_x - i\sigma_y). \tag{3.103}$$

The first term on the *RHS* of Eq. (3.101) containing $\gamma_0(N_{th} + 1)$ is responsible for spontaneous (γ_0) plus thermal ($\gamma_0 N_{th}$) emission while the second term containing $\gamma_0 N_{th}$ is responsible for thermal absorption. The master equation (3.101) leads to the so called optical Bloch equations. We can see from Eq. (3.101) that even at zero temperature ($T = 0$ and hence $N_{th} = 0$), the dynamics is irreversible, that is, not of the unitary von Neumann type and is controlled by the spontaneous emission term. The master equation (3.101) may be expressed in a manifestly LGKS form [81, 82]

$$\frac{d}{dt}\rho^s(t) = \sum_{j=1}^{2} \left(2R_j\rho^s R_j^\dagger - R_j^\dagger R_j\rho^s - \rho^s R_j^\dagger R_j \right), \tag{3.104}$$

where $R_1 = (\gamma_0(N_{th} + 1)/2)^{1/2}\sigma_-$, $R_2 = (\gamma_0 N_{th}/2)^{1/2}\sigma_+$. (If $T = 0$, a single Lindblad operator suffices.) Also, $N_{th} = \dfrac{1}{\left(e^{\frac{\hbar\omega}{k_B T}} - 1\right)}$ is the Planck distribution

giving the number of thermal photons at the frequency ω. We will return to this equation in Chapter 8, where we will see that it has many applications in Quantum Information.

Problem 4: Sketch the steps needed to reach Eq. (3.104) from Eq. (3.101).

(b). Dissipative Harmonic Oscillator

Another example that could be discussed here would be the dissipative harmonic oscillator, $H_S = \omega_0 a^\dagger a$. Now the Lindblad operators would be $\vec{A} = a$, $\vec{A^\dagger} = a^\dagger$. Using this in Eq. (3.97) we get a master equation whose form is similar to Eq. (3.101) and is given, in the Schrödinger picture by,

$$
\begin{aligned}
\frac{d}{dt}\rho^S(t) &= -i\omega_0\left[a^\dagger a, \rho^S(t)\right] \\
&+ \gamma_0(N_{th}+1)\left(a\rho^S(t)a^\dagger - \frac{1}{2}a^\dagger a\rho^S(t) - \frac{1}{2}\rho^S(t)a^\dagger a\right) \\
&+ \gamma_0 N_{th}\left(a^\dagger\rho^S(t)a - \frac{1}{2}aa^\dagger\rho^S(t) - \frac{1}{2}\rho^S(t)aa^\dagger\right).
\end{aligned}
\tag{3.105}
$$

This equation has been used to model the damping of an electromagnetic field mode in a cavity [83], where a, $a^\dagger$ correspond to the annihilation and creation operators of the cavity mode and the damping is mediated by modes outside the cavity at a rate γ_0.

Since the problem of the damped harmonic oscillator is very important and elucidates some of the key techniques of open quantum systems, we will exclusively devote the next chapter to it.

3.7.3   Connection to the Pauli Master Equation

We now briefly show the connection between the LGKS master equation and the Pauli master equation. For this, we use the Lindblad equation for modeling the equation of motion of the populations, the diagonal elements of the system density matrix. We choose a basis $\{|n\rangle\}$ which diagonalizes the Hamiltonian. Since the Hamiltonian's presence in the master equation is in the von Neumann form, it is evident that in this basis the Hamiltonian will not participate in the motion. The typical Lindblad operators, including the ones used in the above example, have at most one non-zero entry in each row or column. Thus they connect each basis state to at most one other basis state. Then the LGKS equation for the diagonal elements $\rho_{nn}^s(t)$ is

$$
\frac{\partial \rho_{nn}(t)}{\partial t} = \sum_j \left[(L_j)_{nm_j}\rho_{m_jm_j}(L_j^\dagger)_{m_jn} - |L_{j,nm_j}|^2\rho_{nn}\right] = \sum_j |L_{j,nm_j}|^2(\rho_{m_jm_j} - \rho_{nn}).
\tag{3.106}
$$

Here we assume that the Lindblad operator L_j couples state n only to state m_j. Thus, the Eq. (3.106) can be recast into the equation of the population $P(n,t) = \rho_{nn}(t)$ as

$$
\frac{\partial P(n,t)}{\partial t} = \sum_j \left[w_{jn}P_j(t) - w_{nj}P_n(t)\right].
\tag{3.107}
$$

This equation has the standard form of the Pauli master equation (3.58) with $w_{jn} = \sum_j |L_{j,nm_j}|^2\delta_{n,m_j}$.

3.8   Quantum Non-Demolition Master Equations

We now consider a very special class of master equations. Let us begin with the generic Hamiltonian

$$H = H_S + H_R + H_{SR}$$

$$= H_S + \sum_k \hbar\omega_k b_k^\dagger b_k + H_S \sum_k g_k(b_k + b_k^\dagger) + H_S^2 \sum_k \frac{g_k^2}{\hbar\omega_k}, \qquad (3.108)$$

where S is the system of interest, R the reservoir (bath), $S - R$ the interaction between them. Here $[H_S, H_{SR}] = 0$ which implies dephasing without dissipation. Use is made of the above Hamiltonian in the context of the influence of dephasing in quantum computation [84, 85]. This form has also been used by [86] in the context of engineered reservoir.

Quantum non-demolition (QND) measurement of observable $\hat{A}$ would be a sequence of precise measurements of $\hat{A}$ such that each measurement is completely predictable from the result of the first measurement, i.e., the system to be measured is independent of the backaction of the measuring apparatus. This implies $[\hat{A}, \hat{H}_{int}] = 0$, where $\hat{H}_{int}$ is the interaction term between the observable and the measuring apparatus. Historically, it was introduced to design gravitational-wave antennas [87].

Further, $[\hat{A}(t_i), \hat{A}(t_j)] = 0$ for all times t_i, t_j. This would protect $\hat{A}$ from contamination by noncommuting (with $\hat{A}$) observables. This is guaranteed if $\hat{A}$ is a constant of the free evolution, i.e., $[\hat{A}, \hat{H}_S] = 0$, where $\hat{H}_S$ is the Hamiltonian responsible for the free evolution of $\hat{A}$ [88]. If $\hat{H}_{int} = \kappa\hat{A}\hat{P}_R$, where κ is a constant and $\hat{P}_R \in \mathcal{H}_R$, $\mathcal{H}_R$ being the Hilbert space of the apparatus or probe, then the evolution of $\hat{A}$ with coupling turned on is identical to its free evolution and is free from contamination. Then $\hat{A}$ is the *pointer observable* and the interaction $\hat{H}_{int}$ corresponds to a measurement of $\hat{A}$. For $\hat{A} = \hat{H}_S$, this would correspond to the measurement of energy.

The system-plus-reservoir composite is closed and hence obeys a unitary evolution given by

$$\rho(t) = e^{-iHt/\hbar}\rho(0)e^{iHt/\hbar}, \qquad (3.109)$$

where

$$\rho(0) = \rho^s(0)\rho_R(0), \qquad (3.110)$$

i.e., we assume separable initial conditions. In order to obtain the reduced dynamics of the system alone, we trace over the reservoir variables. The matrix elements of the reduced density matrix in the system eigenbasis is [89]

$$\rho_{nm}^s(t) = e^{-i(E_n - E_m)t/\hbar}\, e^{-i(E_n^2 - E_m^2)/\hbar \sum_k (g_k^2 t/\hbar\omega_k)}$$

$$\times Tr_R\left[e^{-i\hat{H}_n t/\hbar}\rho_R(0)e^{i\hat{H}_m t/\hbar}\right]\rho_{nm}^s(0). \qquad (3.111)$$

In Eq. (3.111), E_n is the eigenvalue of the system Hamiltonian. Here $\rho_R(0)$ is the initial density matrix of the reservoir which we take to be a squeezed thermal bath given by

$$\rho_R(0) = S(r, \Phi)\rho_{th}S^\dagger(r, \Phi), \tag{3.112}$$

where

$$\rho_{th} = \prod_k \left[1 - e^{-\beta\hbar\omega_k}\right] e^{-\beta\hbar\omega_k b_k^\dagger b_k} \tag{3.113}$$

is the density matrix of the thermal bath at temperature T, with $\beta \equiv 1/(k_B T)$, k_B being the Boltzmann constant, and

$$S(r_k, \Phi_k) = \exp\left[r_k\left(\frac{b_k^2}{2}e^{-2i\Phi_k} - \frac{b_k^{\dagger 2}}{2}e^{2i\Phi_k}\right)\right] \tag{3.114}$$

is the squeezing operator with r_k, Φ_k being the squeezing parameters [90]. In Eq. (3.111),

$$H_n = \sum_k \left[\hbar\omega_k b_k^\dagger b_k + E_n g_k(b_k + b_k^\dagger)\right]. \tag{3.115}$$

We have earlier encountered the density matrix Eq. (3.112), in Chapter 2 (Eq. (54)). Using this, the reduced density matrix of the system is obtained as

$$\rho_{nm}^s(t) = \exp\left[-i(E_n - E_m)t/\hbar\right] \exp\left[-i(E_n^2 - E_m^2)\sum_k (g_k^2 \sin(\omega_k t)/\hbar^2\omega_k^2)\right]$$

$$\times \exp\left[-\frac{1}{2}(E_m - E_n)^2 \sum_k \frac{g_k^2}{\hbar^2\omega_k^2}\coth\left(\frac{\beta\hbar\omega_k}{2}\right)\right.$$

$$\times \left.\left|(e^{i\omega_k t} - 1)\cosh(r_k) + (e^{-i\omega_k t} - 1)\sinh(r_k)e^{2i\Phi_k}\right|^2\right]\rho_{nm}^s(0). \tag{3.116}$$

Differentiating Eq. (3.116) with respect to time we obtain the master equation giving the system evolution under the influence of the environment as

$$\dot{\rho}_{nm}^s(t) = \left[-\frac{i}{\hbar}(E_n - E_m) + i\dot{\eta}(t)(E_n^2 - E_m^2) - (E_n - E_m)^2\dot{\gamma}(t)\right]\rho_{nm}^s(t), \tag{3.117}$$

where

$$\eta(t) = -\sum_k \frac{g_k^2}{\hbar^2\omega_k^2}\sin(\omega_k t), \tag{3.118}$$

and

$$\gamma(t) = \frac{1}{2}\sum_k \frac{g_k^2}{\hbar^2\omega_k^2}\coth\left(\frac{\beta\hbar\omega_k}{2}\right)\left|(e^{i\omega_k t} - 1)\cosh(r_k) + (e^{-i\omega_k t} - 1)\sinh(r_k)e^{2i\Phi_k}\right|^2. \tag{3.119}$$

For the case of zero squeezing, $r = \Phi = 0$, and $\gamma(t)$ given by Eq. (3.119) reduces to the case of a thermal bath. It can be seen that $\eta(t)$ (3.118) is independent of the bath initial conditions and hence remains the same as for the thermal bath. The term responsible for decoherence in the QND case is $\dot{\gamma}(t)$. It is interesting to note that in contrast to the case of quantum Brownian motion, here there is no dissipation. The QND process is thus a purely quantum mechanical effect, having *decoherence (dephasing) without dissipation.* We will encounter this again in Chapter 8.

The nontrivial terms in the master equation (3.117) are encoded in $\eta(t)$, $\gamma(t)$. We will now calculate them explicitly. For this, we assume the bath to have large number of degrees of freedom such that information going out of the system of interest does not return to it. This is effected by taking a *quasi-continuous* bath spectrum with spectral density $I(\omega)$ such that

$$\sum_k \frac{g_k^2}{\hbar^2} f(\omega_k) \longrightarrow \int_0^\infty d\omega I(\omega) f(\omega), \qquad (3.120)$$

which, in the case of an Ohmic bath has the spectral density

$$I(\omega) = \frac{\gamma_0}{\pi} \omega e^{-\omega/\omega_c}, \qquad (3.121)$$

where γ_0 and ω_c two bath parameter. Then, $\eta(t)$ and $\gamma(t)$ turn out to be

$$\eta(t) = -\frac{\gamma_0}{\pi} \tan^{-1}(\omega_c t), \qquad (3.122)$$

and $\gamma(t)$ at $T = 0$

$$\gamma(t) = \frac{\gamma_0}{2\pi} \ln(1 + \omega_c^2 t^2), \qquad (3.123)$$

where $t > 2a$, and for high T

$$\gamma(t) = \frac{\gamma_0 k_B T}{\pi \hbar \omega_c} \left[2\omega_c t \tan^{-1}(\omega_c t) + \ln\left(\frac{1}{1 + \omega_c^2 t^2}\right) \right], \qquad (3.124)$$

where, again, $t > 2a$. Here we have taken, for simplicity, the squeezed bath parameters as

$$\cosh(2r(\omega)) = \cosh(2r), \quad \sinh(2r(\omega)) = \sinh(2r), \quad \Phi(\omega) = a\omega, \qquad (3.125)$$

where a is a constant depending upon the squeezed bath.

3.9   Projection Operator Techniques

We will now focus on some prominent techniques that have been developed to tackle classes of master equations more general than the ones discussed in the last two sections. In general, the reduced dynamics of Open Systems is non-Markovian. Hence the scope of the dynamics generated by the Lindblad master equation, which is Markovian, is restricted. Projection operator techniques provide a systematic way to address the non-Markovian features of the dynamics [59, 60, 61].

The basic idea is to regard the tracing over the reservoir as a formal projection $\rho \longmapsto P\rho$ in the state space of the total system. The superoperator P has the property of a projection operator $P^2 = P$. The density matrix $P\rho$ is the relevant part of ρ. Corresponding to it is the superoperator L (also a projection operator) such that $L\rho$ is the irrelevant part of ρ with $P + L = I$. Thus $\rho \longmapsto P\rho = \text{Tr}_R(\rho) \otimes \rho_R \equiv \rho^S \otimes \rho_R$, where ρ_R is some reference state of the environment. Also $L\rho = \rho - P\rho$ and $PL = LP = 0$.

Let $\rho(t)$ be the density matrix for the total system with the Hamiltonian $H = H_0 + \alpha H_I$, where H_0 stands for the uncoupled system and reservoir Hamiltonians while H_I denotes the interaction between the two and α is a dimensionless expansion parameter.

The equation of motion for the density matrix, in the interaction picture, is

$$\frac{d}{dt}\rho(t) = -i\alpha[H_I, \rho(t)] \equiv \alpha\mathcal{L}(t)\rho(t), \tag{3.126}$$

where $\mathcal{L}(t)$ is the Liouville super-operator and $H_I(t) = e^{iH_0 t}H_I e^{-iH_0 t}$.

3.9.1 Nakajima-Zwanzig Technique

In the Nakajima-Zwanzig technique, the general program of obtaining irreversible dynamics from the Liouvillean is effected by the application of the relevant projection operators. This is achieved by deriving a closed equation for the relevant part of the density matrix $P\rho(t)$. We start with the application of the projection operators P and L to the Liouville-von Neumann equation

$$\begin{aligned}
\frac{d}{dt}P\rho(t) &= \alpha P\mathcal{L}(t)\rho(t), \\
\frac{d}{dt}L\rho(t) &= \alpha L\mathcal{L}(t)\rho(t).
\end{aligned} \tag{3.127}$$

Inserting $P + L = I$ into the above equations, we get

$$\frac{d}{dt}P\rho(t) = \alpha P\mathcal{L}(t)P\rho(t) + \alpha P\mathcal{L}(t)L\rho(t), \tag{3.128}$$

$$\frac{d}{dt}L\rho(t) = \alpha L\mathcal{L}(t)P\rho(t) + \alpha L\mathcal{L}(t)L\rho(t). \tag{3.129}$$

Eq. (3.129) can be solved for $L\rho(t)$ such that

$$L\rho(t) = \mathcal{G}(t, t_0)L\rho(t_0) + \alpha \int_{t_0}^{t} ds\mathcal{G}(t, s)L\mathcal{L}(s)P\rho(s),$$

where

$$\mathcal{G}(t, s) \equiv T_c \exp\left[\alpha \int_{s}^{t} ds' L\mathcal{L}(s')\right],$$

is the propagator and T_c is the chronological time ordering operator, i.e., time arguments increase from right to the left. Substituting $L\rho(t)$ into Eq. (3.128), we get

$$
\begin{aligned}
\frac{d}{dt}P\rho(t) &= \alpha P\mathcal{L}(t)\mathcal{G}(t,t_0)L\rho(t_0) + \alpha P\mathcal{L}(t)P\rho(t) \\
&+ \alpha^2 \int_{t_0}^t ds P\mathcal{L}(t)\mathcal{G}(t,s)L\mathcal{L}(s)P\rho(s).
\end{aligned}
\tag{3.130}
$$

This is the Nakajima-Zwanzig (NZ) equation. The inhomogeneous term as well as the third term in the *RHS* of the NZ equation, involving an integral over the past history of the system, make the evolution non-Markovian.

   We assume that ρ_R is some stationary Gaussian state of the environment with vanishing odd moments of the interaction Hamiltonian such that

$$
\mathrm{Tr}_R\{\rho_R H_I(t_1)\cdots H_I(t_{2k+1})\} = 0.
$$

This implies that the second term in the *RHS* of the NZ equation (3.130) vanishes. For separable (factorizing) initial conditions $\rho(t_0) = \rho_S(t_0) \otimes \rho_R$, $P\rho(t_0) = \rho(t_0)$ and $L\rho(t_0) = 0$. This implies that the first term in the *RHS* of the NZ equation also vanishes. The NZ equation then becomes

$$
\frac{d}{dt}P\rho(t) = \int_{t_0}^t ds \mathcal{K}(t,s)P\rho(s),
\tag{3.131}
$$

where the memory kernel $\mathcal{K}(t,s) = \alpha^2 P\mathcal{L}(t)\mathcal{G}(t,s)L\mathcal{L}(s)P$. To second order in the coupling constant, $\mathcal{K}(t,s) = \alpha^2 P\mathcal{L}(t)L\mathcal{L}(s)P$ and thus to this order, the NZ equation is

$$
\frac{d}{dt}P\rho(t) = \alpha^2 \int_{t_0}^t ds P\mathcal{L}(t)\mathcal{L}(s)P\rho(s),
\tag{3.132}
$$

where $P\mathcal{L}(t)P = 0$ is used. Using the explicit expressions for P and $\mathcal{L}(t)$, Eq. (3.132) can be written as

$$
\frac{d}{dt}\rho^S(t) = -\alpha^2 \int_{t_0}^t ds \mathrm{Tr}_R[H_I(t),[H_I(s),\rho^S(s) \otimes \rho_R]],
\tag{3.133}
$$

which is the equation obtained earlier for evolution in the Born approximation.

3.9.2   Time-Convolutionless Technique

In general, the time convolution in the memory kernel in the NZ equation makes it difficult for applications. To make it more useful for applications, the time convolution in the master equation is removed. This is achieved by the time-convolutionless projection operator technique (TCL) [62, 63, 64].

   The basic idea is to remove the dependence of the future time evolution on the history of the system from the NZ equation and obtain an equation local in time. The starting point is the introduction of the backward propagator such that

$$\rho(s) = G(t, s)(P + L)\rho(t), \tag{3.134}$$

where $G(t, s)$ is the backward propagator of the composite system, the inverse of the unitary time evolution of the total system and is

$$G(t, s) = T_a \exp\left[-\alpha \int_s^t ds' \mathcal{L}(s')\right]. \tag{3.135}$$

Here T_a indicates antichronological time ordering, i.e., time arguments increase from left to right. With the help of this, the irrelevant part of the density matrix, obtained earlier, can be written as

$$
\begin{aligned}
L\rho(t) &= \mathcal{G}(t, t_0)L\rho(t_0) + \alpha \int_{t_0}^t ds \mathcal{G}(t, s)L\mathcal{L}(s)PG(t, s)(P + L)\rho(t) \\
&\equiv \mathcal{G}(t, t_0)L\rho(t_0) + \Sigma(t)(P + L)\rho(t),
\end{aligned}
\tag{3.136}
$$

where $\Sigma(t) = \alpha \int_{t_0}^t ds \mathcal{G}(t, s)L\mathcal{L}(s)PG(t, s)$ is a superoperator containing both the forward $\mathcal{G}$ and backward G propagators. Next, the irrelevant part is rearranged as

$$L\rho(t) = [1 - \Sigma(t)]^{-1}\Sigma(t)P\rho(t) + [1 - \Sigma(t)]^{-1}\mathcal{G}(t, t_0)L\rho(t_0), \tag{3.137}$$

where it is assumed that $1 - \Sigma(t)$ can be inverted. This would be possible for not very large couplings and for small times $t - t_0$. When the above form of $L\rho(t)$ is fed into the equation for $P\rho(t)$, the relevant part, the TCL equation is obtained

$$\frac{d}{dt}P\rho(t) = \mathcal{K}(t)P\rho(t) + \mathcal{I}(t)L\rho(t_0), \tag{3.138}$$

where $\mathcal{K}(t) = \alpha P\mathcal{L}(t)[1 - \Sigma(t)]^{-1}P$ is the time-local TCL generator and $\mathcal{I}(t) = \alpha P\mathcal{L}(t)[1 - \Sigma(t)]^{-1}\mathcal{G}(t, t_0)L$ is the inhomogeneous term, which for factorizing initial conditions does not contribute because then $L\rho(t_0) = 0$. The TCL equation forms the starting point for a systematic approximation method by expanding $\mathcal{K}(t)$ and $\mathcal{I}(t)$ in powers of the coupling strength α.

Assuming that $[1 - \Sigma(t)]$ can be inverted and

$$[1 - \Sigma(t)]^{-1} = \sum_{n=0}^{\infty}[\Sigma(t)]^n. \tag{3.139}$$

This is substituted in the TCL generator to yield

$$\mathcal{K}(t) = \alpha \sum_{n=0}^{\infty} P\mathcal{L}(t)[\Sigma(t)]^n P = \sum_{n=1}^{\infty} \alpha^n \mathcal{K}_n(t). \tag{3.140}$$

To determine the contribution of $\mathcal{K}_n(t)$ of n-th order in the coupling constant α to the TCL generator, the superoperator $\Sigma(t)$ is also expanded in powers of

α

$$\Sigma(t) = \sum_{n=1}^{\infty} \alpha^n \Sigma_n(t). \tag{3.141}$$

This is inserted into the expansion of the TCL generator and a comparison of equal powers of α yields

$$\mathcal{K}_1(t) = P\mathcal{L}(t)P,$$
$$\mathcal{K}_2(t) = P\mathcal{L}(t)\Sigma_1(t)P,$$
$$\mathcal{K}_3(t) = P\mathcal{L}(t)\{[\Sigma_1(t)]^2 + \Sigma_2(t)\}P,$$
$$\mathcal{K}_4(t) = P\mathcal{L}(t)\{[\Sigma_1(t)]^3 + \Sigma_1(t)\Sigma_2(t) + \Sigma_2(t)\Sigma_1(t) + \Sigma_3(t)\}P. \tag{3.142}$$

The forward and backward propagators in the superoperator in these expressions are expanded in powers of the coupling α. Explicitly,

$$\mathcal{K}_1(t) = P\mathcal{L}(t)P = 0,$$

$$\Sigma_1(t) = \int_0^t dt_1 L\mathcal{L}(t_1)P,$$

$$\mathcal{K}_2(t) = \int_0^t dt_1 P\mathcal{L}(t)\mathcal{L}(t_1)P,$$

$$\Sigma_2(t) = \int_0^t dt_1 \int_0^{t_1} dt_2 [L\mathcal{L}(t_1)L\mathcal{L}(t_2)P - L\mathcal{L}(t_2)P\mathcal{L}(t_1)],$$

$$\mathcal{K}_3(t) = P\mathcal{L}(t)\Sigma_2(t)P$$
$$= \int_0^t dt_1 \int_0^{t_1} dt_2 P\mathcal{L}(t)\mathcal{L}(t_1)\mathcal{L}(t_2)P = 0,$$

$$\mathcal{K}_4(t) = P\mathcal{L}(t)[\Sigma_2(t)\Sigma_1(t) + \Sigma_3(t)]P$$
$$= \int_0^t dt_1 \int_0^{t_1} dt_2 \int_0^{t_2} dt_3 \left[P\mathcal{L}(t)\mathcal{L}(t_1)\mathcal{L}(t_2)\mathcal{L}(t_3)P \right. \tag{3.143}$$
$$- P\mathcal{L}(t)\mathcal{L}(t_1)P\mathcal{L}(t_2)\mathcal{L}(t_3)P - P\mathcal{L}(t)\mathcal{L}(t_2)P\mathcal{L}(t_1)\mathcal{L}(t_3)P$$
$$\left. - P\mathcal{L}(t)\mathcal{L}(t_3)P\mathcal{L}(t_1)\mathcal{L}(t_2)P \right]. \tag{3.144}$$

The second order generator $\mathcal{K}_2(t)$ in the TCL master equation gives the following equation for the reduced density matrix

$$\frac{d}{dt}\rho^S(t) = -\alpha^2 \int_{t_0}^t ds \mathrm{Tr}_R[H_I(t), [H_I(s), \rho^S(t) \otimes \rho_R]]. \tag{3.145}$$

Now consider a generic $S - R$ interaction

$$H_I = \sum_k F_k \otimes Q_k, \tag{3.146}$$

where F_k and Q_k act on the system and reservoir's Hilbert space, respectively. If the reservoir is assumed to be in a Gaussian state, all moments of the interaction w.r.t. ρ_R can be expressed in terms of the moments of second order

$$\nu_{ij}(t_1, t_2) = Re\left\{ \text{Tr}_R\{Q_i(t_1)Q_j(t_2)\rho_R\} \right\}, \tag{3.147}$$

$$\eta_{ij}(t_1, t_2) = Im\left\{ \text{Tr}_R\{Q_i(t_1)Q_j(t_2)\rho_R\} \right\}, \tag{3.148}$$

where Re and Im stand for the real and imaginary parts, respectively.

To proceed further, and make contact of the theory with applications, it is convenient to introduce a shorthand notation [2] $\hat{0} = F_{i_0}(t)$, $\hat{1} = F_{i_1}(t_1)$, $\hat{2} = F_{i_2}(t_2)$, $\cdots$, and

$$\nu_{01} = \nu_{i_0 i_1}(t, t_1), \ \nu_{12} = \nu_{i_1 i_2}(t_1, t_2), \cdots \tag{3.149}$$

$$\eta_{01} = \eta_{i_0 i_1}(t, t_1), \ \eta_{12} = \eta_{i_1 i_2}(t_1, t_2), \cdots \tag{3.150}$$

With this notation, the moments can be expressed in a convenient manner. Thus, for example, the second-order moment is

$$\text{Tr}_R\{H_I(t)H_I(t_1)\rho^S \otimes \rho_R\} = \sum_{i_0, i_1} (\nu_{01} + i\eta_{01})\hat{0}\hat{1}\rho^S, \tag{3.151}$$

while the fourth-order moment is

$$\text{Tr}_R\{H_I(t)H_I(t_1)H_I(t_2)H_I(t_3)\rho^S \otimes \rho_R\} = \sum_{i_0, \ldots, i_3} \Big((\nu_{01} + i\eta_{01})(\nu_{23} + i\eta_{23})$$

$$+ (\nu_{02} + i\eta_{02})(\nu_{13} + i\eta_{13})$$

$$+ (\nu_{03} + i\eta_{03})(\nu_{12} + i\eta_{12}) \Big) \hat{0}\hat{1}\hat{2}\hat{3}\rho^S. \tag{3.152}$$

Then the TCL generator to second-order is

$$\mathcal{K}_2(t)\rho^S \otimes \rho^R = -\sum_{i_0, i_1} \int_0^t dt_1 (\nu_{01}[\hat{0}, [\hat{1}, \rho^S]] + i\eta_{01}[\hat{0}, \{\hat{1}, \rho^S\}]) \otimes \rho^R. \tag{3.153}$$

An application: Dissipative Harmonic Oscillator.

We now use the above formalism, of the TCL technique, to the dissipative oscillator [91]. The Hamiltonian is $H = H_S + H_R + H_{SR}$, where

$$H_S = \frac{1}{2}(P^2 + \Omega^2 X^2),$$

$$H_R = \sum_n \frac{1}{2}(\frac{1}{m_n}p_n^2 + m_n \omega_n^2 x_n^2),$$

$$H_{SR} = -X \sum_n c_n q_n = -X \otimes Q. \tag{3.154}$$

Using the notations introduced, the real and imaginary parts of the reservoir correlation functions are seen to be

$$
\nu(t) = Re\left\{ \mathrm{Tr}_R\{Q(t)Q\rho_R\} \right\} = \int_0^\infty \frac{d\omega}{\pi} I(\omega) \coth(\frac{\omega}{2k_B T}) \cos(\omega t),
$$

$$
\eta(t) = Im\left\{ \mathrm{Tr}_R\{Q(t)Q(t)\rho_R\} \right\} = \int_0^\infty \frac{d\omega}{\pi} I(\omega) \sin(\omega t), \tag{3.155}
$$

which are the noise and dissipation kernels, respectively and $I(\omega)$ is the spectral density of the reservoir. The frequency Ω in H_S, Eq. (3.154), is the bare frequency. The observed frequency Ω_r is different, due to the interaction with the reservoir, and is

$$
\Omega^2 = \Omega_r^2 + \Omega_c^2, \tag{3.156}
$$

where

$$
\Omega_c^2 = \sum_n \frac{c_n^2}{m_n \omega_n^2}. \tag{3.157}
$$

The bare frequency Ω, due to the above relation with the frequency Ω_c, depends on the coupling strength and is of the order α. A consistent expansion in terms of the coupling strength would thus be needed to take into account the potential renormalization, explicitly. For the case of factorizing initial conditions, by taking upto the fourth order in the TCL generator $\mathcal{K}(t)$, the TCL master equation yields

$$
\begin{aligned}
\frac{d}{dt}\rho^S(t) =\ & -i[H_S, \rho(t)] - \frac{i}{2}\Delta(t)[X^2, \rho^S(t)] - i\lambda(t)[X, \{P, \rho^S(t)\}] \\
& - D_{PP}(t)[X, [X, \rho^S(t)]] + 2D_{PX}(t)[X, [P, \rho^S(t)]]. \tag{3.158}
\end{aligned}
$$

This has the form of a generalized Fokker-Planck equation. To second order in the coupling strength

$$
\Delta^{(2)}(t) = 2\int_0^t ds\, \eta(s) \cos(\Omega_r s),
$$

$$
\lambda^{(2)}(t) = -\frac{1}{\Omega_r} \int_0^t ds\, \eta(s) \sin(\Omega_r s),
$$

$$
D_{PP}^{(2)}(t) = \int_0^t ds\, \nu(s) \cos(\Omega_r s),
$$

$$
D_{PX}^{(2)}(t) = \frac{1}{2\Omega_r} \int_0^t ds\, \nu(s) \sin(\Omega_r s). \tag{3.159}
$$

In a similar fashion, the terms to the fourth order in the coupling strength can also be obtained. A comparison with the exact results show that the TCL results match with them for sufficietly high temperatures at any coupling strength and for low temperatures for moderate to weak couplings.

3.10   Guide to advanced literature

In this chapter we have introduced, formally, the notion of non-equilibrium (quantum) statistical mechanics. The formal approch is to begin with the Liouville equation and then develop schemes for obtaining irreversibility in the resultant dynamics. There exists specialized literature where this is handled in detail. A very relevant work pertinent to this is [55]. Further details about the Langevin and Fokker-Planck equations can be obtained from [56, 32, 34, 72, 73, 74, 75]. The Boltzmann equation finds prolific use in the study of transport phenomena, see for e.g. [92]. Master equations have become ubiquitous with various studies related to open quantum systems. A very nice treatment of the Lindblad equations can be found in [2, 93]. For further details on the uniquely quantum process, the quantum non-demolition process, the reader is ecouraged to read [89]. Details pertaining to the Nakajima-Zwanzig equation can be obtained from [56, 55] and for more information on the time-convolutionless projection operator technique (TCL), [62, 63, 64] could be read profitably.

Chapter 4

Influence Functional Approach to Open Quantum Systems

4.1 Introduction

The study of the dynamics of Open Quantum Systems , as already introduced before, essentially involves the dynamics of the system of interest taking into account the effect of the ambient environment [94, 1, 2]. Information from the system *leaks* into the environment and is interpreted as dissipation as well as loss of coherence, decoherence. When the dynamics of the system is undeterministic, the most successful approach to study the system dynamics is the functional (path) integral formalism. The dissipative system is considered as interacting with a complex environment and the complete system plus environment composite is assumed to be closed allowing for standard quantization rules to be applied. For equations of motion linear in the environmental coordinates, the environmental coordinates can be easily eliminated and one obtains closed equations for the damped system alone. In the functional integral description, the environment reveals itself through the influence functional (IF).

The functional integral treatment has been used to tackle a vast diversity of problems in the field of open quantum systems. Among the exactly solvable problems are the damped harmonic oscillator and linear quantum Brownian motion [47, 48, 95, 96, 10, 27] wherein the quantum mechanical system is taken as a harmonic oscillator coupled linearly via its displacement x to a fluctuating environment. Quantum Brownian motion (QBM) serves as a paradigm of quantum open systems in that it provides a model wherein the concepts of the system plus reservoir are elucidated. Quantum Brownian motion being a generalization of classical Brownian motion into the quantum regime, gives us a physical realization of dissipation reconciled with quantization. Interest in this has been motivated by observation of macroscopic effects in quantum systems such as dissipation in tunneling and problems of quantum measurement theory (for example, the loss of quantum coherence due to a system's interaction with

© Hindustan Book Agency 2018 and Springer Nature Singapore Pte Ltd. 2018
S. Banerjee, *Open Quantum Systems*, Texts and Readings in Physical Sciences 20,
https://doi.org/10.1007/978-981-13-3182-4_4

its environment). This has also been used to gain useful insight into problems which are not exactly solvable.

As a signature of the diversity of the functional integral treatment of quantum open systems, this has been used in recent years to address issues in quantum gravity [97, 98, 99]. In these investigations detailed analysis was made of quantum Brownian motion to see the interconnection of some basic quantum statistical processes such as decoherence, dissipation, particle creation, noise and fluctuation. The understanding of many quantum statistical processes in the early universe and black holes [98] requires an extension of the existing framework of quantum field theory in the setup of quantum open systems represented by the quantum Brownian motion [1]. These ideas have been applied to the analysis of some basic issues in quantum cosmology [100, 101], effective field theory [102] and the foundation of quantum mechanics, such as the uncertainty principle [103] and decoherence [104, 105, 106] in the quantum to classical transition problem.

The general plan of the influence functional formalism of open quantum systems is to first have an idea about the influence functional, characterizing the environment, also called the reservoir or the bath, influencing the system of interest, such as the harmonic oscillator. Once that is done, one obtains what is known as the propagator. The propagator can then be used as a capsule to generate the final state of the system of interest, given its initial state, hence the name propagator. To extract the physics behind the problem, it is very useful to obtain the master equation, the non-unitary counterpart of the Schrödinger-von-Neumann equation. Here, after a fairly detailed introduction to the influence functional formalism, we solve it explicitly for the case of linear quantum Brownian motion starting from the state where the system was originally decoupled from its ambient environment. We do this by two different methods. We then obtain the corresponding propagator. How to obtain a master equation is then discussed in detail. We also indicate how to obtain the Wigner equation, a very useful tool for probing the quantum-classical connection, from the master equation. We end the chapter with a brief guide to more advanced literature.

4.2   A Primer to the Influence Functional (IF) formalism

In the IF formalism [48, 95, 96, 10] adapted to the open systems, one deals with a system S, in contact with its environment (reservoir) E, with an interaction SE between the two. The object of interest here is the reduced density matrix of the system alone obtained by tracing over the environment variables, i.e.,

$$\rho_r^S(x, x', t) = \int d\mathbf{q}\; \rho(x, \mathbf{q}; x', \mathbf{q}, t). \tag{4.1}$$

Here ρ_r stands for the reduced density matrix of the system obtained by tracing over the reservoir coordinates. x stands for the system coordinate and $\mathbf{q}$ is an N-component vector $\mathbf{q} = (q_1, q_2, ..., q_N)$ denoting the environmental coordinates.

Now action S for the total system can be written as

$$S = S_S + S_E + S_{SE}. \qquad (4.2)$$

Since the $S + E$ composite constitutes a closed system, it is subject to unitary evolution. Using this, the reduced density matrix becomes

$$\rho_r^S(x, x', t) \;=\; \int dx_i dx_i' d\mathbf{q}_i d\mathbf{q}_i' d\mathbf{q}_f \; K(x, \mathbf{q}_f, t; x_i, \mathbf{q}_i, 0)$$
$$\times \rho_0(x_i, \mathbf{q}_i, x_i', \mathbf{q}_i', 0) K^*(x', \mathbf{q}_f, t; x_i', \mathbf{q}_i', 0). \qquad (4.3)$$

This form of the density matrix can be seen to be directly coming from the construction of the path integral approach to the evolution of the density matrix presented earlier in Chapter II, with the addition that now explicit use is made of the reservoir variables as well. Further, due to the operation of taking a trace over the reservoir coordinates, the forward and backward propagators , K and K^*, get coupled to each other. This will be encoded in the functional called the Influence Functional, introduced below, and is responsible for the open system effects . In Eq. (4.3) the sum is over all the paths $x(s), \mathbf{q}(s), x'(s), \mathbf{q}'(s)$ in real time s with $0 \le s \le t$,

$$x(0) = x_i, x(t) = x_f; x'(0) = x_i', x'(t) = x_f';$$
$$\mathbf{q}(0) = \mathbf{q}_i, \mathbf{q}(t) = \mathbf{q}_f; \mathbf{q}'(0) = \mathbf{q}_i', \mathbf{q}'(t) = \mathbf{q}_f'. \qquad (4.4)$$

K stands for the standard expression for the propagator

$$K(x, \mathbf{q}_f, t; x', \mathbf{q}_i', 0) = \int \int D x D \mathbf{q} \, \exp\left\{ \frac{i}{\hbar} S[x, \mathbf{q}] \right\}, \qquad (4.5)$$

as introduced earlier in Chapter II. ρ_0 is the initial density matrix of the $S + E$ composite.

In the conventional Feynman-Vernon theory [47], it was assumed that the system and the environment (reservoir) were initially uncorrelated, a condition called 'separable initial condition' [48] in the literature. In such a situation the initial density matrix factorizes so that

$$\rho(0) = \rho_S(0).\rho_E(0), \qquad (4.6)$$

where $\rho_S(0)$ stands for the initial system density matrix and $\rho_E(0)$ stands for the initial reservoir density matrix. The reduced density matrix becomes

$$\rho_r^S(x, x', t) = \int dx_i dx_i' \; J(x, x', t; x_i x_i', 0) \rho^S(x_i, x_i', 0). \qquad (4.7)$$

Here, for convenience of notation, we have placed S in the superscript, $\rho_r^S(x, x', t)$ is the reduced density matrix of the system of interest, taking into account the effect of the environment and the propagator J is

$$J(x, x', t; x_i, x_i', 0) = \int \int Dx Dx' \exp \left\{ \frac{i}{\hbar} [S_S[x] - S_S[x']] \right\} F[x, y]. \quad (4.8)$$

Here $F[x, y]$ stands for the IF given by

$$
\begin{aligned}
F[x, y] &= \int d\mathbf{q}_i d\mathbf{q}_i' d\mathbf{q}_f \; \rho_R(\mathbf{q}_i, \mathbf{q}_i', 0) \\
&\quad \times \int_{\mathbf{q}_i}^{\mathbf{q}_f} D\mathbf{q} \int_{\mathbf{q}_i'}^{\mathbf{q}_f} D\mathbf{q}' \; \exp \left\{ \frac{i}{\hbar} \left(S_{SE}[x, \mathbf{q}] + S_E[\mathbf{q}] \right. \right. \\
&\quad \left. \left. - S_{SE}[x', \mathbf{q}'] - S_E[\mathbf{q}'] \right) \right\}.
\end{aligned}
\quad (4.9)
$$

The separable initial conditions are not the only kind of initial conditions. The separable initial condition is very different from the equilibrium state of the total system since even far away points are initially quantum mechanically very coherent. Also in many applications, the system and the reservoir are integral parts of the same system and their interaction is not at our disposal. The separable initial conditions assume a sudden switch-on of the interaction between the system and the reservoir at $t = 0$ which leads to unphysical divergences. This can become very severe if one is interested in macroscopic quantum coherence, macroscopic quantum tunneling and related problems since the artificial switch-on of the interaction would very seriously influence the subsequent short-time behavior of the system.

These considerations lead to the introduction of a class of initial conditions, the 'generalized initial conditions' [95, 96, 10, 107]. A very general class of initial conditions are of the form [10, 108]

$$\rho_0 = \sum_j O_j \rho_\beta O_j', \quad (4.10)$$

where

$$\rho_\beta = Z_\beta^{-1} \exp(-\beta H) \quad (4.11)$$

is the canonical density matrix describing the equilibrium of the interacting system in the presence of a time-independent potential V and Z_β^{-1} is the partition function. Here $\beta = (k_B T)^{-1}$, with T being the equilibrium temperature of the interacting system. The operators O_j, O_j' act upon the system coordinate only and leave the environment (reservoir) coordinates unchanged but can be chosen arbitrarily otherwise. This definition of the initial state has the usefulness that it can be used even in situations where ρ_0 is not a proper density matrix.

The simplest initial state of the form Eq. (4.10) is the equilibrium density matrix ρ_β itself. It can be prepared by waiting sufficiently long so that the system reaches equilibrium at time $t = 0$. Then the response of the particle to a time-dependent external force acting at time $t > 0$ can be studied.

A modification of this comes when the system is displaced from equilibrium by applying a constant external force F and ρ_0 describes the modified equilibrium in the presence of this force. When the external force is switched off at time $t = 0_+$, the relaxation towards nonconstrained equilibrium can be studied.

Further, for a system in equilibrium at time $t = 0_-$, a measurement of a dynamical variable of the Brownian particle may be made. This leads to a reduction of the density matrix. The state ρ_0 after the measurement will be of the form Eq. (4.10), where the operators O_j, O_j' describe the effect of the measuring device [108]. For instance, an ideal position measurement with the outcome $q_0 - \delta/2 < q < q_0 + \delta/2$ leads to

$$\rho_0 = P_q \rho_\beta P_q, \tag{4.12}$$

where

$$P_q = N^{-1/2} \int_{q_0-\delta/2}^{q_0+\delta/2} dq \, |q\rangle\langle q| \tag{4.13}$$

projects on the measured interval and N stands for a normalization factor.

Another case where Eq. (4.10) can be used is when a scattering experiment is performed in which the cross-section is related to an equilibrium correlation function of the Brownian particle [109]. An equilibrium correlation function $\langle A(t)B\rangle$ may be viewed as the expectation value of A at time t in the 'initial ensemble' $\rho_0 = B\rho_\beta$. Now $B\rho_\beta$ is not a proper density matrix but is of the form Eq. (4.10).

In the coordinate representation, Eq. (4.10) becomes

$$\rho_0(x, \mathbf{q}, x', \mathbf{q}') = \langle x, \mathbf{q}|\rho_0|x', \mathbf{q}'\rangle$$
$$= \sum_j \int d\bar{x} \int d\bar{x}' \, O_j(x, \bar{x})O_j'(\bar{x}', x')\rho_\beta(\bar{x}, \mathbf{q}, \bar{x}', \mathbf{q}'), \tag{4.14}$$

where

$$O_j(x, \bar{x}) = \langle x|O_j|\bar{x}\rangle \text{ and } O_j'(\bar{x}', x') = \langle \bar{x}'|O_j'|x'\rangle.$$

Thus the initial states are of the form

$$\rho_0(x, \mathbf{q}, x', \mathbf{q}') = \int d\bar{x} \int d\bar{x}' \, \lambda_0(x, \bar{x}, x', \bar{x}')\rho_\beta(\bar{x}, \mathbf{q}, \bar{x}', \mathbf{q}'), \tag{4.15}$$

where

$$\lambda_0(x, \overline{x}, x', \overline{x}') = \sum_j O_j(x, \overline{x}) O_j'(\overline{x}', x') \tag{4.16}$$

is called the 'preparation function' and it gives the deviation from the equilibrium distribution. Now using the unitary evolution of the entire 'system-environment' composite, the density matrix in the coordinate representation is obtained as

$$\rho(x_f, \mathbf{q}_f, x_f', \mathbf{q}_f', t) = \int dx_i dx_i' d\mathbf{q}_i d\mathbf{q}_i' \ K(x_f, \mathbf{q}_f, t; x_i, \mathbf{q}_i, 0)$$
$$\times \rho_0(x_i, \mathbf{q}_i, x_i', \mathbf{q}_i', 0) K^*(x_f', \mathbf{q}_f', t; x_i', \mathbf{q}_i', 0). \tag{4.17}$$

The functional integral representation of ρ_β is

$$\rho_\beta(\overline{x}, \overline{\mathbf{q}}, \overline{x}', \overline{\mathbf{q}}') = Z_\beta^{-1} \int D\overline{x} D\overline{\mathbf{q}} \exp\left[-\frac{1}{\hbar} S^{EQ}[\overline{x}, \overline{\mathbf{q}}]\right], \tag{4.18}$$

where the integral is over all paths $\overline{x}(\tau)$, $\overline{\mathbf{q}}(\tau)$ with $0 \leq \tau \leq \hbar\beta$. Also $\overline{x}(0) = \overline{x}'$, $\overline{\mathbf{q}}(0) = \overline{\mathbf{q}}'$, $\overline{x}(\hbar\beta) = \overline{x}$, $\overline{\mathbf{q}}(\hbar\beta) = \overline{\mathbf{q}}$. Here S^{EQ} stands for the Euclidean action which arises from the Euclidean functional integral representation of the correlation that exists between the system and the environment initially. Thus Eq. (4.17) becomes

$$\rho(x_f, \mathbf{q}_f, x_f', \mathbf{q}_f', t) = \int dx_i dx_i' d\overline{x} d\overline{x}' d\mathbf{q}_i d\mathbf{q}_i' \ \lambda_0(x_i, \overline{x}, x_i', \overline{x}') Z_\beta^{-1}$$
$$\times \int Dx D\mathbf{q} Dx' D\mathbf{q}' D\overline{x} D\overline{\mathbf{q}}$$
$$\times \exp\left\{\frac{i}{\hbar}(S[x, \mathbf{q}] - S[x', \mathbf{q}']) - \frac{1}{\hbar} S^{EQ}[\overline{x}, \overline{\mathbf{q}}]\right\}. \tag{4.19}$$

Here the sum is over all the paths $x(s), \mathbf{q}(s), x'(s), \mathbf{q}'(s)$ in real time s, $0 \leq s \leq t$ (see Eq. (4.4)), and over the paths $\overline{x}(\tau), \overline{\mathbf{q}}(\tau)$ in imaginary time τ, $0 \leq \tau \leq \hbar\beta$, with

$$\overline{x}(0) = \overline{x}', \overline{x}(\hbar\beta) = \overline{x}; \overline{\mathbf{q}}(0) = \mathbf{q}_i', \overline{\mathbf{q}}(\hbar\beta) = \mathbf{q}_i. \tag{4.20}$$

It is evident from the above that the reservoir endpoints $\mathbf{q}_f$ and $\mathbf{q}_f'$ are connected by a continuous path.

To get the reduced density matrix of the system alone, a trace is taken over the reservoir coordinates to obtain

$$\rho_r^S(x_f, x_f', t) = \int d\mathbf{q}_f \ \rho(x_f, \mathbf{q}_f, x_f', \mathbf{q}_f, t). \tag{4.21}$$

Thus the reduced density matrix involves integrations over only the closed paths of the environment. This can then be written as

$$\rho_r^S(x_f, x_f', t) = \int dx_i dx_i' d\bar{x} d\bar{x}' \; \lambda_0(x_i, \bar{x}, x_i', \bar{x}') Z^{-1} \int D x D x' D \bar{x}$$

$$\times \exp\left\{\frac{i}{\hbar}(S_S[x] - S_S[x']) - \frac{1}{\hbar}S_S^{EQ}[\bar{x}]\right\} \tilde{F}[x, x', \bar{x}], \qquad (4.22)$$

where $\tilde{F}$ is the 'generalized IF' and is given by

$$\tilde{F}[x, x', \bar{x}] = \int d\mathbf{q}_f d\mathbf{q}_i d\mathbf{q}_i' Z_R^{-1} \int D\mathbf{q} D\mathbf{q}' D\bar{\mathbf{q}}$$

$$\times \exp\left\{\frac{i}{\hbar}(S_E[\mathbf{q}] + S_{SE}[x, \mathbf{q}] - S_E[\mathbf{q}'] - S_{SE}[x', \mathbf{q}'])\right.$$

$$\left. -\frac{1}{\hbar}(S_E^{EQ}[\bar{\mathbf{q}}] + S_{SE}^{EQ}[\bar{x}, \bar{\mathbf{q}}])\right\}. \qquad (4.23)$$

Here $Z = Z_\beta/Z_R$ and Z_R is a normalization constant such that the IF is equal to one for vanishing interactions.

Below, we will work out, in detail, the IF for linear QBM starting from separable initial conditions and provide a guidance to the literature dealing with more general IFs, including those dealing with correlated initial conditions.

4.3 Influence Functionals: An Explicit Evaluation

From the above discussion, it is evident that the Influence Functional (IF) is a central object in the considerations of a number of problems, such as, quantum Brownian motion (QBM). Here we will evaluate the IF in two different ways; the first one is a conventional derivation, while the next is a somewhat unconventional one, involving ideas from a number of areas of physics. We will concentrate on the case of separable initial conditions.

4.3.1 Conventional Derivation of IF

The IF is given by, see Eq. (4.9),

$$F[x, y] = \int d\mathbf{q}_i d\mathbf{q}_i' d\mathbf{q}_f \; \rho_R(\mathbf{q}_i, \mathbf{q}_i', 0)$$

$$\times \tilde{F}[x, \mathbf{q}_f, \mathbf{q}_i] \tilde{F}^*[x', \mathbf{q}_f, \mathbf{q}_i']. \qquad (4.24)$$

Concentrating on the QBM model, a harmonic oscillator interacting with a
bath of harmonic oscillators, we have the following Hamiltonian

$$H = H_S + H_E + H_{SE}, \tag{4.25}$$

where

$$H_S = \frac{1}{2} M \left[\dot{x}^2 + \omega_0^2 x^2 \right] \tag{4.26}$$

is the system Hamiltonian,

$$H_E = \sum_{n=1}^{N} \frac{1}{2} m_n \left[\dot{q}_n^2 + \omega_n^2 q_n^2 \right] \tag{4.27}$$

is the environment (reservoir) Hamiltonian, and

$$H_{SE} = - \sum_{n=1}^{N} \left[c_n x q_n \right] + x^2 \sum_{n=1}^{N} \frac{c_n^2}{2 m_n} \omega_n^2 \tag{4.28}$$

is the system-environment interaction Hamiltonian. In H_{SE}, the last term on
the *RHS* is a coupling induced normalization of the potential. It can be absorbed in the potential term in H_S. Here c_n is the system-reservoir coupling
constant. In Eq. (4.24),

$$\tilde{F}[x, \mathbf{q}_f, \mathbf{q}_i] = \int_{\mathbf{q}_i}^{\mathbf{q}_f} \exp \left\{ \frac{i}{\hbar} \left(S_{SE}[x, \mathbf{q}] + S_E[\mathbf{q}] \right) \right\}. \tag{4.29}$$

Here

$$S_E[\mathbf{q}] = \exp \left\{ \frac{i}{\hbar} \sum_{n=1}^{N} \int_0^t ds \left(\frac{1}{2} m_n \left[\dot{q}_n^2 - \omega_n^2 q_n^2 \right] \right) \right\}, \tag{4.30}$$

and

$$S_{SE}[x, \mathbf{q}] = \exp \left\{ \frac{i}{\hbar} \sum_{n=1}^{N} \int_0^t ds \left(c_n x q_n - x^2 \frac{c_n^2}{2 m_n \omega_n^2} \right) \right\}. \tag{4.31}$$

For a reservoir of harmonic oscillators, Eq. (5.88), the functional integral in Eq.
(4.29) factorizes into individual contributions from each reservoir oscillator, i.e.,
$\tilde{F}[x, \mathbf{q}_f, \mathbf{q}_i] = \prod_{k=1}^{N} \tilde{F}_k[x, \mathbf{q}_{k_f}, \mathbf{q}_{k_i}]$, where

$$\tilde{F}_k[x, \mathbf{q}_{k_f}, \mathbf{q}_{k_i}] = \int D\mathbf{q} \exp \left\{ \frac{i}{\hbar} \int_0^t ds \left(\frac{1}{2} m_n \left[\dot{q}_n^2 - \omega_n^2 q_n^2 \right] + c_n x q_n - x^2 \frac{c_n^2}{2 m_n \omega_n^2} \right) \right\}. \tag{4.32}$$

We now need to evaluate the functional integral in Eq. (4.32) where we have to
sum over all paths $q_n(s)$ of the nth environmental oscillator with $q_n(0) = q_{n_i}$
and $q_n(t) = q_{n_f}$. Since the functional integral is Gaussian , its dependence on

the boundary values q_{n_i}, q_{n_f}, can be obtained by expanding about the path $\tilde{q}_n(s)$ minimizing the action in the exponent of Eq. (4.32), the classical path. Expressing

$$q_n(s) = \tilde{q}_n(s) + \alpha_n(s), \tag{4.33}$$

where $\tilde{q}_n(s)$ is the solution of the stationary value of the action and $\alpha_n(s)$ is the quantum fluctuation around this path, $\tilde{q}_n(0) = q_{n_i}$ and $\tilde{q}_n(t) = q_{n_f}$. Eq. (4.32) can be written as

$$\tilde{F}_k[x, \mathbf{q}_{k_f}, \mathbf{q}_{k_i}] = \exp\left\{ \frac{i}{\hbar} \int_0^t ds \left(\frac{1}{2} m_n \left[\dot{\tilde{q}}_n^2 - \omega_n^2 \tilde{q}_n^2 \right] + c_n x \tilde{q}_n - x^2 \frac{c_n^2}{2 m_n \omega_n^2} \right) \right\}$$

$$\times \int D\alpha_n \exp\left\{ \frac{i}{\hbar} \int_0^t ds \left(\frac{1}{2} m_n \dot{\alpha}_n^2 - \omega_n^2 \alpha_n^2 \right) \right\}. \tag{4.34}$$

The functional integral over $\alpha_n(s)$ sums over all paths $\alpha_n(s)$ with $\alpha_n(0) = \alpha_n(t) = 0$ so that the dependence on q_{n_i} and q_{n_f} is completely included in the first exponential. To calculate the functional integral in Eq. (4.34) we made use of the fact that $\tilde{q}_n(s)$ is a trajectory that minimizes the action. Consider the first term in Eq. (4.34). The Euler-Lagrangian equation of motion reads

$$m_n \ddot{\tilde{q}}_n(s) + m_n \omega_n^2 \tilde{q}_n(s) = c_n x(s). \tag{4.35}$$

Here the term on the *RHS* can be interpreted as a time-dependent force acting on the environmental oscillator due to its coupling to the Brownian particle. The solution of Eq. (4.35) satisfying the boundary conditions is

$$\tilde{q}_n(s) = \frac{\beta_n(s)}{\beta_n(t)} q_{n_f} + \frac{\beta_n(t-s)}{\beta_n(t)} q_{n_i} - c_n \left(\frac{\beta_n(s)}{\beta_n(t)} \int_0^t du \beta_n(t-u) x(u) - \int_0^s du \beta_n(s-u) x(u) \right),$$

$$\tag{4.36}$$

where $\beta_n(s) = \frac{1}{m_n \omega_n} \sin(\omega_n s)$. To compute the action of the trajectory $\tilde{q}_n(s)$ we perform an integration by parts and use the equation of motion (4.35). This yields

$$\int_0^t ds \left(\frac{1}{2} m_n \left[\dot{\tilde{q}}_n^2 - \omega_n^2 \tilde{q}_n^2 \right] + c_n x \tilde{q}_n - x^2 \frac{c_n^2}{2 m_n \omega_n^2} \right) = \frac{1}{2} m_n (\tilde{q}_n(t) \dot{\tilde{q}}_n(t)$$

$$- \tilde{q}_n(0) \dot{\tilde{q}}_n(0))$$

$$- \frac{1}{2} c_n \int_0^t ds\, x(s) \tilde{q}_n(s)$$

$$- \frac{c_n^2}{2 m_n \omega_n^2} \int_0^t ds\, x^2(s). \tag{4.37}$$

The *RHS* of Eq. (4.36) is then inserted in Eq. (4.37). We now want to evaluate the contribution to the functional integral, Eq. (4.32), from the quantum fluctuations $\alpha_n(s)$, see Eq. (4.33). This can be obtained by expanding the α_n dependent part of the functional integral, Eq. (4.34), into a Fourier series, using $\alpha_n(0) = \alpha_n(t) = 0$

$$\alpha_n(s) = \sum_{\theta=1}^{\infty} \alpha_n^{\theta} \sin(\nu_\theta s); \quad \nu_\theta = \pi\theta/t. \tag{4.38}$$

Due to the orthogonality of the sine functions, the integrand in the second term in the *RHS* of Eq. (4.34) becomes

$$\exp\left\{\frac{i}{\hbar} \int_0^t ds \left(\frac{1}{2} m_n \dot{\alpha}_n^{\;2} - \omega_n^2 \alpha_n^2\right)\right\} = \exp\left\{i\frac{m_n t}{4\hbar} \sum_{\theta=1}^{\infty} \left((\alpha_n^{\theta})^2 (\nu_\theta^2 - \omega_n^2)\right)\right\}, \tag{4.39}$$

while the integration measure becomes

$$\int D\alpha_n \cdots = \prod_{\theta=1}^{\infty} \left(N^{-1} \int_{-\infty}^{\infty} d\alpha_n^{\theta} \cdots\right). \tag{4.40}$$

Here N is a constant independent of ω_n which arises from the Jacobian of the transformation in Eq. (4.38). From the Eqs. (4.39) and (4.40), it can be seen that the functional integral factorizes into regular Gaussian integrals over the Fourier components α_n^{θ} which can be done separately. We then have

$$f_n(t) = \int D\alpha_n \exp\left\{\frac{i}{\hbar} \int_0^t ds \left(\frac{1}{2} m_n \dot{\alpha}_n^{\;2} - \omega_n^2 \alpha_n^2\right)\right\} = C \prod_{\theta=1}^{\infty} \left(1 - \frac{\omega_n^2}{\nu_\theta^2}\right)^{-1/2}. \tag{4.41}$$

The constant C has all the factors independent of ω_n. Using the mathematical identity

$$\prod_{\theta=1}^{\infty} \left(1 - \frac{\omega_n^2}{\nu_\theta^2}\right) = \frac{\sin(\omega_n t)}{\omega_n t}, \tag{4.42}$$

we get

$$f_n(t) = C \left(\frac{\omega_n t}{\sin(\omega_n t)}\right)^{1/2}. \tag{4.43}$$

The constant C can be determined by evaluating the Jacobian of the transformation, Eq. (4.38). Another way of doing this is by comparing Eq. (4.41) with the corresponding result for the case of a free particle, see Chapter 2.

$$f_n(t, \omega = 0) = C = \left(\frac{m_n}{2\pi i \hbar t}\right)^{1/2}. \tag{4.44}$$

Using Eqs. (4.44), (4.43) and the result obtained by substituting the RHS of Eq. (4.36) in the RHS of Eq. (4.37), we get

$$\tilde{F}_n[x, \mathbf{q}_{n_f}, \mathbf{q}_{n_i}] = \left(\frac{m_n \omega_n}{2\pi i \hbar \sin(\omega_n t)} \right)^{1/2} \exp \left\{ \frac{i}{\hbar} \Phi_n(x, \mathbf{q}_{n_f}, \mathbf{q}_{n_i}) \right\}, \qquad (4.45)$$

where

$$
\begin{aligned}
\Phi_n(x, \mathbf{q}_{n_f}, \mathbf{q}_{n_i}) \;=\; & \frac{m_n \omega_n}{2\sin(\omega_n t)} \left\{ (q_{n_i}^2 + q_{n_f}^2) \cos(\omega_n t) - 2 q_{n_i} q_{n_f} \right\} \\
& + \frac{q_{n_i} c_n}{\sin(\omega_n t)} \int_0^t ds \sin(\omega_n(t-s)) x(s) + \frac{q_{n_f} c_n}{\sin(\omega_n t)} \int_0^t ds \sin(\omega_n s) x(s) \\
& - \frac{c_n^2}{m_n \omega_n \sin(\omega_n t)} \int_0^t ds \int_0^s du \sin(\omega_n(t-s)) \sin(\omega_n u) x(s) x(u) \\
& - \frac{c_n^2}{2 m_n \omega_n^2} \int_0^t ds\, x^2(s). \qquad\qquad (4.46)
\end{aligned}
$$

Using Eq. (4.45) and its conjugate in Eq. (4.24) along with the initial state of the reservoir

$$\rho_R(\mathbf{q}_i, \mathbf{q}_i', 0) = \left[1 - \exp\left(\frac{-\hbar\omega}{k_B T} \right) \right] \sum_n \exp\left(\frac{-n\hbar\omega}{k_B T} \right) |n\rangle\langle n|, \qquad (4.47)$$

i.e., a thermal density matrix at temperature T, given here in the representation of the Fock basis, and doing the Gaussian integrals over the intermediate coordinates q_{n_i}, q_{n_i}' and q_{n_f}, the IF becomes

$$F[x, y] = e^{-\frac{i}{\hbar} \tilde{\Phi}(x,y)}, \qquad (4.48)$$

where

$$
\begin{aligned}
\tilde{\Phi}(x, y) \;=\; & \int_0^t ds \int_0^s du (x(s) - y(s)) \left\{ K(s-u) x(u) - K^*(s-u) y(u) \right\} \\
& + i \int_0^t ds \frac{\mu}{2} (x^2(s) - y^2(s)), \qquad\qquad (4.49)
\end{aligned}
$$

$$K(s) = \sum_{n=1}^N \frac{c_n^2}{2 m_n \omega_n} \frac{\cosh\left[\omega_n \frac{\hbar\beta}{2} - is \right]}{\sinh(\frac{1}{2}\omega_n \hbar\beta)}, \qquad (4.50)$$

$$\mu = \sum_{n=1}^{N} \frac{c_n^2}{2m_n \omega_n}. \tag{4.51}$$

It should be noted that the pre-exponential factors in Eq. (4.45), its complex conjugate and the partition function of the unperturbed reservoir combine with the coupling-independent factors arising from the Gaussian integrals over q_{n_i}, q'_{n_i} and q_{n_f} to give 1. This is consistent with the fact that the influence functional should be 1 for vanishing coupling. In general, the influence functional is a highly nonlocal object. Not only does it depend on the time history, it also irreducibly mixes the two sets of histories in the path integral representing the propagator, see Eq. (4.133). Note that the histories x and x' could be interpreted as moving forward and backward in time, respectively.

4.3.2 Basis Independent Derivation of IF

We discussed in detail the derivation of the QBM influence functional in the representation or basis dependent form, Eq. (4.9). Now, we present a novel basis independent derivation of the influence functional. This would provide a broader view to the subject and at the same time illustrate how a number of concepts originally developed in a different setup are all brought together to address the problem in hand. This method is also powerful enough to treat the problem of parametric QBM, i.e., with time dependent couplings and has been applied to problems dealing with quantum statistical mechanics, quantum optics and cosmology [99], thereby highlighting the ubiquity of the technique as well as the subject under study.

We take the following model Hamiltonian

$$H = H_S + H_E + H_{SE}, \tag{4.52}$$

where

$$H_S = \frac{1}{2} M \left[\dot{x}^2 + \Omega^2 x^2 \right] \tag{4.53}$$

is the system Hamiltonian,

$$H_E = \sum_{n=1}^{N} \frac{1}{2} m_n \left[\dot{q_n}^2 + \omega_n^2 q_n^2 \right] \tag{4.54}$$

is the environment (reservoir) Hamiltonian, and

$$H_{SE} = \sum_{n=1}^{N} [c_n x q_n] \tag{4.55}$$

is the system-environment interaction Hamiltonian. This is a simplified form of a linear QBM Hamiltonian [48]. Here, as in the rest of this book, we will stick to the factorizable initial condition

$$\rho(0) = \rho_S(0)\rho_E(0). \tag{4.56}$$

where

$$\rho_E(0) = \hat{S}(r, \Phi)\hat{\rho}_{th}\hat{S}^\dagger(r, \Phi), \tag{4.57}$$

i.e., we have a squeezed thermal initial state. The squeezed thermal state generalizes the usual thermal state in that it reduces to the thermal state when the squeezing parameters r and Φ are set to zero. Squeezing, a concept introduced in quantum optics [29], is a form of a quantum correlation and has the potential to enhance the performance of quantum operations in the presence of external noise [110]. Here

$$\hat{\rho}_{th} = \prod_k \left[1 - e^{-\beta\hbar\omega_k}\right] e^{-\beta\hbar\omega_k \hat{b}_k^\dagger \hat{b}_k} \tag{4.58}$$

is the density matrix of the thermal bath, and

$$\hat{S}(r_k, \Phi_k) = \exp\left[r_k\left(\frac{\hat{b}_k^2}{2}e^{-i2\Phi_k} - \frac{\hat{b}_k^{\dagger 2}}{2}e^{i2\Phi_k}\right)\right] \tag{4.59}$$

is the squeezing operator with r_k, Φ_k being the squeezing parameters [90].

The IF for the linear QBM, discussed here, is

$$F[x, x'] = \prod_n F_n[x, x'], \tag{4.60}$$

i.e., it can be written as a product over the contributions from the n environmental harmonic oscillators. From now we will take the case of a single mode. The IF for a *single* mode is

$$
\begin{aligned}
F_n[x, x'] &= \int_{-\infty}^{\infty} dq_f \int_{-\infty}^{\infty} dq_i \int_{-\infty}^{\infty} dq_i' \int_{q_i}^{q_f} Dq \int_{q_i}^{q_f} Dq' \\
&\quad \times \exp\left[\frac{i}{\hbar}\{S_E[q] + S_{SE}[x, q] - S_E[q'] - S_{SE}[x', q']\}\right]\rho_E(q_i, q_i', 0) \\
&= \exp\left[\frac{i}{\hbar}\delta A[x, x']\right]
\end{aligned} \tag{4.61}
$$

where $\delta A[x, x']$ is the influence action. Here we have suppressed the index n from the subscript of the environmental oscillators q.

The actions S_E, S_{SE} come from the corresponding Hamiltonians in Eqs. (4.54), (4.55), respectively. The IF can be written in the representation independent form as

$$F_n[x, x'] = \text{Tr}[\hat{U}(t, 0)\hat{\rho}_E(0)\hat{U}'^\dagger(t, 0)], \tag{4.62}$$

where $\hat{U}(t)$ and $\hat{U}'(t)$ are quantum propagators for the actions $S_E[q] + S_{SE}[x, q]$ and $S_E[q] + S_{SE}[x', q]$, respectively. This is the key observation behind the present method. It can be easily deduced from the conventional form of the IF, Eq. (4.9), where the IF can be seen to involve the environment and system-environment interaction actions, the initial environment density matrix and a trace is made over the environment (reservoir) coordinates. We will use the above form to evaluate the IF.

$$S_E[q] + S_{SE}[x, q] = \int_0^t ds \left[\sum_n \left\{ \frac{1}{2} m[\dot{q}^2 - \omega^2 q^2] \right\} + \sum_n (-)cx(s)q(s) \right].$$

(4.63)

It should be reiterated that here we have suppressed the subscript denoting the oscillator index, since we are working with a single mode. The Lagrangian L for the above action is

$$L = \frac{1}{2} m[\dot{q}^2 - \omega^2 q^2] - cx(t)q(t).$$

(4.64)

The canonical momentum is

$$p_c = \frac{\partial L}{\partial \dot{q}} = m\dot{q}.$$

(4.65)

Thus the Hamiltonian is

$$H(t) = p_c \dot{q} - L(t) = \frac{p_c^2}{2m} + \frac{1}{2} m\omega^2 q^2 + cx(t)q(t).$$

(4.66)

This equation is quantized by promoting q, p_c to operators obeying the commutation relation

$$[\hat{q}, \hat{p}_c] = i\hbar,$$

(4.67)

where

$$\hat{q} = \sqrt{\frac{\hbar}{2\kappa}}(\hat{a} + \hat{a}^\dagger), \quad \hat{p}_c = i\sqrt{\frac{\hbar\kappa}{2}}(\hat{a}^\dagger - \hat{a}).$$

(4.68)

The κ appearing in the above equation is given by $m\omega$. Using Eq. (4.68) in (4.66) we have

$$\hat{H}(t) = f(t)\hat{A} + f^*(t)\hat{A}^\dagger + h(t)\hat{B} + d(t)\hat{a} + d^*(t)\hat{a}^\dagger + g(t),$$

(4.69)

where

$$f(t) = f^*(t) = \frac{\hbar}{2} \left\{ \frac{m\omega^2}{\kappa} - \frac{\kappa}{m} \right\},$$

(4.70)

$$h(t) = \frac{\hbar}{2} \left\{ \frac{m\omega^2}{\kappa} + \frac{\kappa}{m} \right\},$$

(4.71)

$$d(t) = \sqrt{\frac{\hbar}{2\kappa}} cx(t),$$

(4.72)

$$g(t) = 0, \tag{4.73}$$

$$\hat{A} = \frac{\hat{a}^2}{2}, \quad \hat{A}^\dagger = \frac{\hat{a}^{\dagger 2}}{2}, \quad \hat{B} = \hat{a}^\dagger \hat{a} + \frac{1}{2}. \tag{4.74}$$

As shown in [111] the ansatz

$$\hat{U}(t,0) = e^{x(t)\hat{B}} \; e^{y(t)\hat{A}} \; e^{z(t)\hat{A}^\dagger} \; e^{q(t)\hat{a}} \; e^{p(t)\hat{a}^\dagger} \; e^{r(t)} \tag{4.75}$$

for the time-evolution operator generated by Eq. (4.69) is global. Now using Eqs. (4.69) and (4.75) in

$$\hat{H}(t)\hat{U}(t,0) = i\hbar \frac{\partial}{\partial t}\hat{U}(t), \tag{4.76}$$

we get

$$f = i\hbar[\dot{y}e^{-2x} + \dot{z}y^2 e^{-2x}], \tag{4.77}$$

$$f^* = i\hbar[\dot{z}e^{2x}], \tag{4.78}$$

$$h = i\hbar[\dot{x} + \dot{z}y], \tag{4.79}$$

$$d = i\hbar[\dot{q}(1 - yz)e^{-x} + \dot{p}ye^{-x}], \tag{4.80}$$

$$d^* = i\hbar[\dot{p}e^x - \dot{q}ze^x], \tag{4.81}$$

$$g = i\hbar[\dot{p}q + \dot{r}]. \tag{4.82}$$

Using

$$x = \ln\alpha, \quad y = -\beta x, \quad z = \frac{\beta^*}{\alpha}, \quad \alpha = e^{-i\theta}\cosh(r), \quad \beta = -e^{-i2\varphi}\sinh(r), \tag{4.83}$$

Eq. (4.75) can be written as

$$\hat{U}(t,0) = \hat{S}(r,\Phi) \; \hat{R}(\theta) \; e^{q\hat{a}} \; e^{p\hat{a}^\dagger} e^r, \tag{4.84}$$

where

$$2\Phi = 2\varphi - \theta, \quad \hat{R}(\theta) = e^{-i\theta\hat{B}}, \quad \hat{S}(r,\Phi) = \exp\left[r\left(\hat{A}e^{-i2\Phi} - \hat{A}^\dagger e^{i2\Phi}\right)\right]. \tag{4.85}$$

Here $\hat{S}$ and $\hat{R}$ are the squeeze and rotation operators, respectively. These operators play an important role in many aspects of quantum optics [29]. They also generate canonical transformations [112], i.e., transformations preserving the canonical commutation relations.

Now substituting Eq. (4.83) in Eqs. (4.77) to (4.82) we have

$$\hbar\dot{\alpha} = -if^*\beta - i\hbar\alpha, \tag{4.86}$$

$$\hbar\dot{\beta} = -i\hbar\beta + if\alpha, \tag{4.87}$$

$$\hbar\dot{p} = -i(d\beta^* + d^*\alpha^*), \tag{4.88}$$

$$\dot{q} = -\dot{p}^*, \tag{4.89}$$

$$\hbar\dot{r} = -ig - \hbar\dot{p}q. \tag{4.90}$$

Using the Campbell-Baker-Hausdorff identity [29] we have

$$\hat{U}(t,0) = \hat{S}(r,\Phi)\hat{R}(\theta)\hat{D}(p)e^{-pp^*/2}e^r, \tag{4.91}$$

where

$$\hat{D}(p) = \exp[p\hat{a}^\dagger - p^*\hat{a}] \tag{4.92}$$

is the displacement operator,

$$p(t,0) = -\frac{i}{\hbar}\int_0^t ds\left[d(s)\beta^*(s) + d^*(s)\alpha^*(s)\right], \tag{4.93}$$

$$r(t,0) = \int_0^t ds\, \dot{p}(s)p^*(s). \tag{4.94}$$

From this we find that the evolution operator can be written as

$$\hat{U}(t,0) = \hat{S}(r,\Phi)\hat{R}(\theta)\hat{D}(p)\exp\left[-\frac{pp^*}{2} + \int_0^t ds\int_0^s ds'\, \dot{p}(s)\, \dot{p}^*(s')\right]. \tag{4.95}$$

Substituting Eq. (4.95) into Eq. (4.62) we have the IF for a single mode as

$$F_n[x, x'] = Tr\left[\hat{S}\ \hat{R}\ \hat{D}(p)\exp\left\{-\frac{pp^*}{2} + \int\limits_0^t ds \int\limits_0^s ds'\dot{p}(s)\dot{p}^*(s')\right\}\right.$$

$$\times \hat{\rho}_E(0)\hat{D}^\dagger(p')\hat{R}'^\dagger\hat{S}'^\dagger\exp\left\{-\frac{p'p'^*}{2} + \int\limits_0^t ds \int\limits_0^s ds'\ p'(s)\dot{p}'^*(s')\right\}\right]$$

$$= Tr\left[\hat{S}\ \hat{R}\ \hat{D}(p)\ \hat{\rho}_E(0)\hat{D}^\dagger(p')\hat{R}^\dagger\ \hat{S}^\dagger\right.$$

$$\times \exp\left\{-\frac{pp^*}{2} - \frac{p'p'^*}{2}\right\}\exp\left\{\int\limits_0^t ds \int\limits_0^s ds'\left[\dot{p}(s)\dot{p}^*(s') + \dot{p}'(s)\dot{p}'^*(s')\right]\right\}$$

$$\tag{4.96}$$

where $\hat{S}'^\dagger = \hat{S}^\dagger$ and $\hat{R}'^\dagger = \hat{R}^\dagger$, since $\hat{S}$ and $\hat{R}$ are independent of the coordinate.

By a straightforward application of the Campbell-Baker-Hausdorff identity, we see that

$$\hat{D}(p)\hat{D}(p') = \hat{D}(p + p')\exp\left[\frac{1}{2}(pp'^* - p^*p')\right].\tag{4.97}$$

Problem 1: Prove the above identity.

From this,

$$F_n[x, x'] = Tr[\hat{\rho}_E(0)\hat{D}(p - p')]$$

$$\times \exp\left[\frac{1}{2}\int\limits_0^t ds \int\limits_0^s ds'\left\{[\dot{p}(s) - \dot{p}'(s)][\dot{p}'^*(s')\right.\right.$$

$$\left.\left. + \dot{p}^*(s')] + [\dot{p}(s') + \dot{p}'(s')][\dot{p}'^*(s) - \dot{p}^*(s)]\right\}\right].\tag{4.98}$$

Here we have made use of the following functional integral identity

$$\int_m^n dt\ \alpha(t) \int_m^n dt\ \beta(t) = \int_m^n \int_m^t dt'dt\ [\alpha(t)\beta(t') + \beta(t')\alpha(t)].\tag{4.99}$$

Problem 2: Prove the above identity.

For the squeezed thermal initial state of the reservoir, we have

$$\hat{\rho}_E(0) = \hat{S}(r, \Phi)\hat{\rho}_{th}\hat{S}^\dagger(r, \Phi), \tag{4.100}$$

where

$$\hat{\rho}_{th} = \left[1 - \exp\left(-\frac{\hbar\omega}{2k_BT}\right)\right]\sum_n \exp\left(-\frac{n\hbar\omega}{k_BT}\right)|n\rangle\langle n|, \tag{4.101}$$

and

$$\hat{S}(r, \Phi) = \exp\left[r(\hat{A}e^{-i2\Phi} - \hat{A}^\dagger e^{i2\Phi})\right]. \tag{4.102}$$

The terms are as explained before. Now

$$\begin{aligned} Tr[\hat{\rho}_E(0)\hat{D}(p - p')] &= Tr[\hat{S}(r, \Phi)\hat{\rho}_{th}\hat{S}^\dagger(r, \Phi)\hat{D}(p - p')] \\ &= Tr[\hat{\rho}_{th}\hat{S}^\dagger(r, \Phi)\hat{D}(p - p')\hat{S}(r, \Phi)]. \end{aligned} \tag{4.103}$$

Using Eq. (4.85) and the Campbell-Baker-Hausdorff identity , we have

$$\hat{S}^\dagger(r, \Phi)\hat{D}(p)\hat{S}(r, \Phi) = \hat{D}\left(p\cosh(r) + p^*\sinh(r)e^{i2\Phi}\right). \tag{4.104}$$

Problem 3: Prove the above identity.

Using a very useful identity regarding trace with respect to thermal reservoirs, which we have encountered earlier in Chapter 2,

$$Tr\left[\hat{\rho}_{th}\exp(x\hat{a}^\dagger + y\hat{a})\right] = \exp\left[\frac{xy}{2}\coth\left(\frac{\hbar\omega}{2k_BT}\right)\right], \tag{4.105}$$

we get

$$Tr\left[\hat{\rho}_E(0)\hat{D}(p-p')\right] = \exp\left\{-\frac{1}{2}\coth\left(\frac{\hbar\omega}{2k_BT}\right)\right. \tag{4.106}$$

$$\left. \times \left|(p-p')\cosh(r)+(p-p')^*\sinh(r)e^{i2\Phi}\right|^2\right\}$$

$$= \exp\left\{-\frac{1}{2\hbar^2}\coth\left(\frac{\hbar\omega}{2k_BT}\right)\int_0^t ds \int_0^s ds'\right. \tag{4.107}$$

$$\times \left\{\cosh(2r)\left[U(s)U^*(s')\right.\right.$$

$$\left.+ U^*(s)U(s')\right] - \sinh(2r)e^{-i2\Phi}U(s)U(s')$$

$$\left.\left. - \sinh(2r)e^{i2\Phi}U^*(s)U^*(s')\right\}\Delta(s)\Delta(s')\right\}$$

$$= \exp\left\{-\frac{1}{\hbar}\int_0^t ds \int_0^s ds'\,\Delta(s)\nu(s,s')\Delta(s')\right\}. \tag{4.108}$$

Problem 4: Derive the above equation.

Here

$$\Delta(s) = [x(s) - x'(s)], \tag{4.109}$$

and we have used

$$p(s) = -\frac{i}{\hbar}\int_0^s du\,[U(u)x(u)]. \tag{4.110}$$

Using Eq. (4.72) in Eq. (4.93) and comparing with Eq. (4.110) we have

$$U(s) = [u\beta^*(s) + u^*\alpha^*(s)], \tag{4.111}$$

with

$$u = \sqrt{\frac{\hbar}{2\kappa}}c. \tag{4.112}$$

The argument in the exponential on the *RHS* of Eq. (4.98) is

$$-\frac{2i}{\hbar} \int_0^t ds \int_0^s ds' \left(\frac{i}{2\hbar}\right) \left[U(s)U^*(s') - U(s')U^*(s)\right] \Delta(s)\Sigma(s')$$

$$= -\frac{2i}{\hbar} \int_0^t ds \int_0^s ds' \, \Delta(s)\mu(s,s')\Sigma(s'),$$

$$\tag{4.113}$$

where

$$\Sigma(s) = \frac{1}{2}[x(s) + x'(s)]. \tag{4.114}$$

Using Eq. (4.113) and Eq. (4.108) in Eq. (4.98) we have the IF for a single mode n, reverting back to the mode subscripts in the notation,

$$F_n[x,x'] = \exp\left[-\frac{2i}{\hbar} \int_0^t ds \int_0^s ds' \Delta(s)\mu_n(s,s')\Sigma(s') \right.$$

$$\left. -\frac{1}{\hbar} \int_0^t ds \int_0^s ds' \Delta(s)\nu_n(s,s')\Delta(s') \right]. \tag{4.115}$$

The full IF is then obtained by substituting Eq. (4.115) in Eq. (4.60). We get

$$F[x,x'] = \exp\left[-\frac{2i}{\hbar} \int_0^t ds \int_0^s ds' \Delta(s)\mu(s,s')\Sigma(s') \right.$$

$$\left. -\frac{1}{\hbar} \int_0^t ds \int_0^s ds' \Delta(s)\nu(s,s')\Delta(s') \right], \tag{4.116}$$

where

$$\mu(s,s') = -\int_0^\infty d\omega \, I(\omega) \times \sin(\omega(s-s')), \tag{4.117}$$

is the dissipation kernel,

$$\nu(s,s') = \int_0^\infty d\omega \, I(\omega) \coth(\frac{\hbar\omega}{2k_B T})\{\cosh(2r(\omega)) \cos[\omega(s-s')]$$

$$- \sinh(2r(\omega)) \cos[\omega(s+s') - 2\Phi(\omega)] \tag{4.118}$$

is the noise kernel, and

$$I(\omega) = \sum_n \frac{c_n^2}{2\kappa_n} \delta(\omega - \omega_n) \tag{4.119}$$

is the spectral density of the reservoir. Note that the idea behind the Eq. (4.119) is based on the fact that the environment (reservoir) is composed of a large number of oscillators such that within the time scales of interest, the information that escapes the system into the environment does not return back to it; *hence the irreversibility*. Thus in the dissipation and noise kernels, which involve summation of the reservoir oscillator frequencies and square of the coupling constants over all the reservoirs, the summations can be converted into integrals, in the mathematical limit of an infinite number of reservoir oscillators. The spectral density (4.119) characterizes this transition. In practical calculations, the reservoir is handled by assuming phenomenological forms of the spectral density, a very well known form being the Ohmic density which has a linear frequency dependence. We will make use of the spectral density in a number of model calculations in this book.

Note that the second term on the *RHS* of the noise kernel (4.118) has a non-stationary contribution. This is a result of the squeezing inherent in the bath initial conditions and can lead to interesting physical consequences such as preservation of quantum coherences for a longer time [113]. In the long time limit these non-stationary contributions are washed out leading towards a thermal distribution. From the above expressions the corresponding expressions for simpler cases can be obtained. For example, the situation where we have a thermal reservoir instead of a squeezed thermal reservoir is obtained by setting r and Φ to zero. In Eq. (4.116),

$$\Delta(s) = [x(s) - x'(s)], \quad \Sigma(s) = \frac{1}{2} [x(s) + x(s')], \tag{4.120}$$

the well known center of mass and relative co-ordinates.

4.3.3   Semiclassical Interpretation of the Influence Functional

It is fruitful to consider the semi-classical behaviour of the open quantum system represented by the IF, Eq. (4.116). This would be useful for elucidating the meaning of the dissipation and noise kernels in (4.116). Consider an action that generates the same IF as in Eq. (4.116)

$$S = \int_0^t ds \, [\mathcal{L}(x, \dot{x}, s) + x\zeta(s)], \tag{4.121}$$

where $\zeta(s)$ is a Gaussian stochastic force with a non-zero mean. This generates the IF

$$F[\Sigma, \Delta] = \left\langle \exp\left[\frac{i}{\hbar} \int_0^t ds \, \zeta(s)\Delta(s)\right]\right\rangle. \tag{4.122}$$

Here Σ, Δ are as in Eq. (4.120) and the average is a functional integral over $\zeta(s)$ which is distributed according to a normalized probability density functional $P[\zeta(s), \Sigma(s)]$. The averaging can be performed to yield

$$F[\Sigma, \Delta] = \exp\left[\frac{i}{\hbar} \int_0^t ds \, \Delta(s)\langle\zeta(s)\rangle - \frac{1}{\hbar^2} \int_0^t ds \int_0^s ds' \Delta(s)\Delta(s')C_2(s, s')\right]. \tag{4.123}$$

Here $C_2(s, s')$ is the second cumulant of the stochastic force ζ. The equation of motion generated by Eq. (4.121) is

$$\frac{\partial \mathcal{L}}{\partial x} - \frac{d}{dt} \frac{\partial \mathcal{L}}{\partial \dot{x}} + \langle \zeta(t) \rangle = -\overline{\zeta}(t). \tag{4.124}$$

Here $\overline{\zeta}(t)$ is a Gaussian stochastic force with zero mean and $\langle \overline{\zeta}(t)\overline{\zeta}(t') \rangle = C_2(t, t')$. By comparing Eq. (4.123) with Eq. (4.116), we find that

$$\langle \zeta(t) \rangle \equiv -2 \int_0^s ds' \, \mu(s, s') \Sigma(s'), \quad C_2(s, s') \equiv \hbar \nu(s, s'). \tag{4.125}$$

On substituting Eq. (4.125) in Eq. (4.124), we find the equation of motion, in the semi-classical limit, to be

$$\frac{\partial \mathcal{L}}{\partial x} - \frac{d}{dt} \frac{\partial \mathcal{L}}{\partial \dot{x}} - 2 \int_0^t ds \, \mu(t, s) x(s) = -\overline{\zeta}(t), \tag{4.126}$$

with $\langle \overline{\zeta}(t)\overline{\zeta}(t') \rangle = \hbar \nu(t, t')$. In some special cases μ becomes the derivative of a delta function. Then the above equation of motion becomes a typical dissipative evolution equation, generating local dissipation. In general, we have non-local dissipation along with noise ν.

Problem 5: Do the Gaussian integration to derive Eq. (4.123) from the Eq. (4.122).

4.4 Propagator for linear Quantum Brownian Motion

Having discussed the influence functional for QBM in some detail, we use it to get to the next step, i.e., the propagator. The bath is assumed to have started in a special initial state called the squeezed thermal state which can have tangible physical consequences. We take the problem of linear QBM , depicted by the Hamiltonians in Eqs. (4.52) to (4.55) with the initial states being defined by Eqs. (4.56) and (4.57).

We are interested in the reduced dynamics of the 'open' system of interest S, which is obtained by tracing over the bath degrees of freedom. For this we need the propagator which can be written as:

$$\begin{aligned}
J_r(x_f, x_f', t; x_i, x_i', 0) &= \int_{x_i}^{x_f} \int_{x_i'}^{x_f'} Dx \, Dx' \exp\left[\frac{i}{\hbar}\{S_S[x] - S_S[x']\}\right] F[x, x'] \\
&= \int_{x_i}^{x_f} \int_{x_i'}^{x_f'} Dx \, Dx' \exp\left[\frac{i}{\hbar} A[x, x']\right]
\end{aligned} \tag{4.127}$$

where $F[x, x']$ is the influence functional (IF), and $A[x, x']$ is the effective action of the open system. As explained above, the IF can be written in the representation independent form as

$$F[x, x'] = Tr[\hat{U}(t, 0)\hat{\rho}_E(0)\hat{U}'^{\dagger}(t, 0)], \tag{4.128}$$

where $\hat{U}(t)$ and $\hat{U}'(t)$ are quantum propagators for the actions $S_E[q] + S_{SE}[x, q]$ and $S_E[q] + S_{SE}[x', q]$ respectively. We will use the above form to evaluate the IF. In the above equation:

$$S_E[q] + S_{SE}[x, q] = \int_0^t ds \left[\sum_n \left\{ \frac{1}{2}m[\dot{q}^2 - \omega^2 q^2] \right\} \right.$$
$$\left. + \sum_n (-)cx(s)q(s) \right]. \tag{4.129}$$

Here we have suppressed the subscript denoting the oscillator index. For the Hamiltonians in Eqs. (4.52) to (4.55), the IF (4.128) can be evaluated, for the initial conditions given by Eqs. (4.56, 4.57), see Eq. (4.116). Using the IF we obtain the propagator as

$$J_r(\Sigma_f, \Delta_f, t; \Sigma_i, \Delta_i, 0) = \int_{\Sigma_i}^{\Sigma_f} D\Sigma \int_{\Delta_i}^{\Delta_f} D\Delta \, \exp\left\{ \frac{i}{\hbar} A[\Sigma(s), \Delta(s)] \right\}, \tag{4.130}$$

where the action

$$A[\Sigma(s), \Delta(s)] = \int_0^t ds[M\dot{\Sigma}(s)\dot{\Delta}(s) - M\Omega^2\Sigma(s)\Delta(s)$$
$$-2\int_0^s ds' \, \Delta(s)\mu(s, s')\Sigma(s') + i\int_0^s ds' \, \Delta(s)\nu(s, s')\Delta(s')]$$
$$= \int_0^t ds \, L. \tag{4.131}$$

Here L stands for the Lagrangian. By tracing over the bath we obtain the reduced density matrix of the system which is encapsulated in the propagator. The procedure to solve the functional integral in Eq. (4.130) is similar to the ones adapted earlier to workout the functional integral, Eq. (4.32), see also Eqs. (4.33) and (4.34). The functional integrals being Gaussian, we can expand the paths $\Sigma(s)$ and $\Delta(s)$ about the stationary paths $\Sigma_{cl}(s)$ and $\Delta_{cl}(s)$, see Eqs. (5.98) and (5.99) below,

$$\Sigma(s) = \Sigma_{cl}(s) + \alpha_+(s)$$
$$\Delta(s) = \Delta_{cl}(s) + \alpha_-(s), \tag{4.132}$$

where $\alpha_{\pm}(s)$ are the quantum corrections. Using Eq. (4.132) in Eq. (4.130), we get

$$J_r(\Sigma_f, \Delta_f, t; \Sigma_i, \Delta_i, 0) = N(t, t_i) \exp\left\{ \frac{i}{\hbar} A[\Sigma_{cl}(s), \Delta_{cl}(s)] \right\}. \tag{4.133}$$

$N(t, t_i)$ is part of the functional containing quantum fluctuations around the classical paths and is

$$N(t, t_i) = \int_{\alpha_+=0;t_i}^{\alpha_+=0;t} D\alpha_+ \int_{\alpha_-=0;t_i}^{\alpha_-=0;t} D\alpha_- \exp \left\{ \frac{i}{\hbar} A[\alpha_+(s), \alpha_-(s)] \right.$$
$$\left. - \frac{1}{\hbar} \int_{t_i}^{t} ds \int_{t_i}^{t} ds' [\alpha_-(s)\Delta_{cl}(s)\nu(s, s')] \right\}. \tag{4.134}$$

Using Eqs. (4.133) and (4.134), we get, for $t_i = 0$

$$J_r(\Sigma_f, \Delta_f, t; \Sigma_i, \Delta_i, 0) = N(t, 0) \exp \left[\frac{i}{\hbar} \{ b_1 \Sigma_f \Delta_f - b_2 \Sigma_f \Delta_i + b_3 \Sigma_i \Delta_f - b_4 \Sigma_i \Delta_i \} \right]$$
$$\times \exp \left[\frac{-1}{\hbar} \{ a_{11}\Delta_i^2 + a_{12}\Delta_i\Delta_f + a_{22}\Delta_f^2 \} \right], \tag{4.135}$$

where

$$b_1(t) = M\dot{u}_2(t), b_2(t) = M\dot{u}_2(0), \tag{4.136}$$

$$b_3(t) = M\dot{u}_1(t), b_4(t) = M\dot{u}_1(0), \tag{4.137}$$

$$a_{mn}(t) = \frac{1}{1 + \delta_{mn}} \int_0^t ds \int_0^t ds' \, v_m(s)\nu(s, s')v_n(s'). \tag{4.138}$$

The factor $N(t, 0)$, in Eq. (4.134), can be obtained by the method adapted earlier, see for example, the solution of the integrand in the second term in the RHS of Eq. (4.34). Another, straightforward way, to do this would be to obtain this factor from the normalization of the reduced density matrix, see Eq. (4.146) below.

$$N(t, 0) = \frac{b_2(t)}{2\pi\hbar}. \tag{4.139}$$

Problem 6: Do this.

In Eqs. (4.136) to (4.138), u_1, u_2, v_1, v_2 come from the solutions of the equations

$$\ddot{\Sigma}_{cl}(s) + \Omega^2\Sigma_{cl}(s) + \frac{2}{M} \int_0^s ds' \, \mu(s, s')\Sigma_{cl}(s') = 0, \tag{4.140}$$

and

$$\ddot{\Delta}_{cl}(s) + \Omega^2\Delta_{cl}(s) + \frac{2}{M} \int_s^t ds' \, \mu(s', s)\Delta_{cl}(s') = 0, \tag{4.141}$$

with $I(\omega)$ being the bath spectral density. Now the solutions of the equations (4.140), (4.141) can be parametrized in terms of u and v as

$$\Sigma_{cl}(s) = \Sigma_i u_1(s) + \Sigma_f u_2(s), \tag{4.142}$$

$$\Delta_{cl}(s) = \Delta_i v_1(s) + \Delta_f v_2(s), \tag{4.143}$$

where in order that the classical solutions satisfy proper boundary conditions
we have

$$u_1(0) = 1 = u_2(t), u_1(t) = 0 = u_2(0), \tag{4.144}$$

$$v_1(0) = 1 = v_2(t), v_1(t) = 0 = v_2(0). \tag{4.145}$$

Problem 7: Consider the evolution of the reduced density matrix

$$\rho_r^S(x, x', t) = \int dx_i dx'_i \ J(x, x', t; x_i x'_i, 0) \rho^S(x_i, x'_i, 0),$$

where the propagator J is given by Eq. (4.135). Assuming that the system
harmonic oscillator starts from the Gaussian state

$$\rho^S(x_i, x'_i, 0) = \tilde{C} e^{-\xi x_i^2 + \chi x_i x'_i - \xi^* x_i'^2},$$

find out the final system state $\rho_r^S(x, x', t)$ at time t. Here ξ, χ are arbitrary
complex numbers and use may be made of Eq. (4.120).

4.5   Master Equation for Quantum Brownian Motion

The next step in the sequel is to obtain the master equation for a damped
harmonic oscillator , a paradigm model of QBM, starting from a separable
initial condition. This will unravel the physics behind QBM and also shed light
on the subject of quantum dissipation, in general.

The state of the system at any time t is given by

$$\rho_r^S(x_f, x'_f, t) = \int dx_i dx'_i \ J_r(x_f, x'_f, t; x_i, x'_i, 0) \rho^S(x_i, x'_i, 0). \tag{4.146}$$

The following, introduced in [114], illustrates a neat way to get the master
equation, without having to solve Eq. (4.146). The basic idea is to differentiate
both sides of (4.146), making use of Eq. (4.135) to get

$$
\begin{aligned}
\dot{\rho}_r^S(\Sigma_f, \Delta_f, t) &= \left[\frac{\dot{Z}}{Z} + \frac{i}{\hbar} \dot{b}_1 \Sigma_f \Delta_f - \frac{\dot{a}_{22}}{\hbar} \Delta_f^2 \right] \rho_r^S(\Sigma_f, \Delta_f, t) \\
&\quad + \frac{i}{\hbar} \dot{b}_3 \Delta_f \int d\Delta_i d\Sigma_i \ \Sigma_i J_r \rho_r^S(\Sigma_i, \Delta_i, 0) \\
&\quad - \frac{1}{\hbar} (i \dot{b}_2 \Sigma_f + \dot{a}_{12} \Delta_f) \int d\Delta_i d\Sigma_i \ \Delta_i J_r \rho_r^S(\Sigma_i, \Delta_i, 0) \\
&\quad - \frac{i}{\hbar} \dot{b}_4 \int d\Delta_i d\Sigma_i \ \Sigma_i \Delta_i J_r \rho_r^S(\Sigma_i, \Delta_i, 0) \\
&\quad - \frac{\dot{a}_{11}}{\hbar} \int d\Delta_i d\Sigma_i \ \Delta_i^2 J_r \rho_r^S(\Sigma_i, \Delta_i, 0). \tag{4.147}
\end{aligned}
$$

Now we express, in the above equation, terms inside the integrals and depending upon the initial conditions, as differentials of the final conditions which would enable them to be taken out of the integrand. Thus we have

$$\Delta_i J_r = \frac{i\hbar}{b_2} \frac{\partial J_r}{\partial \Sigma_f} + \frac{b_1}{b_2} \Delta_f J_r, \tag{4.148}$$

$$\Sigma_i J_r = -\frac{i}{b_3} \left[\hbar \frac{\partial J_r}{\partial \Delta_f} + (\Delta_i a_{12} + 2\Delta_f a_{22}) J_r \right] - \frac{b_1}{b_3} \Sigma_f J_r, \tag{4.149}$$

$$\Sigma_i \Delta_i J_r = -\left(\frac{i\hbar}{b_2} \frac{\partial}{\partial \Sigma_f} + \frac{b_1}{b_2} \Delta_f \right) \left(\frac{i\hbar}{b_3} \frac{\partial}{\partial \Delta_f} + \frac{i}{b_3} \left[\Delta_i a_{12} + 2\Delta_f a_{22} \right] + \frac{b_1}{b_3} \Sigma_f \right) J_r, \tag{4.150}$$

$$\Delta_i^2 J_r = -\frac{\hbar^2}{b_2^2} \frac{\partial^2 J_r}{\partial \Sigma_f^2} + 2i\hbar \frac{b_1}{b_2^2} \Delta_f \frac{\partial J_r}{\partial \Sigma_f} + \frac{b_1^2}{b_2^2} \Delta_f^2 J_r. \tag{4.151}$$

Using these we get the master equation as

$$
\begin{aligned}
i\hbar \frac{\partial}{\partial t} \rho_r^S(x, x', t) = & \left\{ \frac{-\hbar^2}{2M} \left(\frac{\partial^2}{\partial x^2} - \frac{\partial^2}{\partial x'^2} \right) + \frac{M}{2} \Omega_{ren}^2(t)(x^2 - x'^2) \right\} \rho_r^S(x, x', t) \\
& - i\hbar \Gamma(t)(x - x') \left(\frac{\partial}{\partial x} - \frac{\partial}{\partial x'} \right) \rho_r^S(x, x', t) \\
& + i D_{pp}(t)(x - x')^2 \rho_r^S(x, x', t) \\
& - \hbar (D_{xp}(t) + D_{px}(t))(x - x') \left(\frac{\partial}{\partial x} + \frac{\partial}{\partial x'} \right) \rho_r^S(x, x', t) \\
& - i\hbar^2 D_{xx}(t) \left(\frac{\partial}{\partial x} + \frac{\partial}{\partial x'} \right)^2 \rho_r^S(x, x', t).
\end{aligned}
\tag{4.152}
$$

Here we have reverted back to the original coordinates and the coefficients are

$$\Omega_{ren}^2(t) = \frac{b_1 \dot{b}_3}{M b_3} - \frac{\dot{b}_1}{M}, \tag{4.153}$$

$$\Gamma(t) = \frac{-1}{2}\left(\frac{\dot{b}_3}{b_3} - \frac{\dot{b}_2}{b_2}\right), \tag{4.154}$$

$$D_{pp}(t) = \frac{b_1^2}{b_2}\left(\frac{a_{12}}{M} - \frac{\dot{a}_{11}}{b_2}\right) + \frac{2b_1}{M}a_{22} - \dot{a}_{22} + \frac{2\dot{b}_3}{b_3}a_{22} + a_{12}\frac{b_1 \dot{b}_3}{b_2 b_3} - \dot{a}_{12}\frac{b_1}{b_2}, \tag{4.155}$$

$$D_{xp}(t) = D_{px}(t) = \frac{-1}{2}\left[\frac{\dot{a}_{12}}{b_2} - \frac{2a_{22}}{M} - \frac{\dot{b}_3}{b_3 b_2}a_{12} - \frac{2b_1}{b_2}\left(\frac{a_{12}}{M} - \frac{\dot{a}_{11}}{b_2}\right)\right], \tag{4.156}$$

$$D_{xx}(t) = \frac{1}{b_2}\left(\frac{a_{12}}{M} - \frac{\dot{a}_{11}}{b_2}\right). \tag{4.157}$$

Here $\Gamma(t)$ is the dissipation term, $D_{pp}(t)$ is the term responsible for de-coherence in x, $D_{xp}(t)$, $D_{px}(t)$ are the so called anomalous diffusion terms (important in low temperature regimes) and $D_{xx}(t)$ causes decoherence in p. This term ($D_{xx}(t)$) is entirely due to the non-stationary effects introduced by the squeezing in the bath and would be absent in a master equation describing an evolution due to a thermal bath, such as in Halliwell and Yu [115]. The coefficients of the master equation (4.152) are in general time-dependent in-dicative of the non-Markovian nature of the problem. This equation therefore describes the physics of linear QBM in a compact fashion. The master equation (4.152) is a generalization of the well-known Caldeira Leggett master equation for quantum dissipation [48], in that it includes the anomalous diffusion terms as well as the $D_{xx}(t)$ term.

Another useful equation that can be obtained from the master equation is the Wigner equation [19]. The Wigner equation may be employed for calculation of various correlation functions in a quasiclassical manner. The Wigner equation serves as a starting point of a number of investigations in transport theory [92]. Also, it has been used to establish the quantum-classical connection, see for e.g. [116, 117, 118, 119]. The Wigner equation is obtained from the master equation (4.152) by the prescription

$$\frac{\partial}{\partial t}W(p, x, t) = \frac{1}{2\pi\hbar}\int_{-\infty}^{\infty} dy\ e^{\frac{i}{\hbar}py}\left\langle x - \frac{1}{2}y\left|\frac{\partial}{\partial t}\rho_r^S\right|x + \frac{1}{2}y\right\rangle. \tag{4.158}$$

Using this and Eq. (4.152) we get

$$
\begin{aligned}
\frac{\partial W}{\partial t} = & -\frac{1}{M}\frac{\partial}{\partial x}pW + M\Omega^2_{ren}(t)\frac{\partial}{\partial p}xW + 2\Gamma(t)\frac{\partial}{\partial p}pW \\
& -\hbar D_{pp}(t)\frac{\partial^2}{\partial p^2}W - \hbar\left(D_{xp}(t) + D_{px}(t)\right)\frac{\partial^2}{\partial x\partial p}W \\
& -\hbar D_{xx}(t)\frac{\partial^2}{\partial x^2}W.
\end{aligned}
\tag{4.159}
$$

Problem 8: Derive the Eq. (4.159) from Eq. (4.152).

The Eq. (4.152) is the generalization of the equation obtained by [115] using a thermal bath and reduces to it by setting the squeezing parameters to zero.

4.6   Guide to advanced literature

In this Chapter, after introducing the general framework of IFs for both separable as well as correlated initial conditions, we have worked out in detail the problem of linear QBM starting from separable initial conditions. However, there has been a number of advances over these simple cases. Thus, there have been studies where the problem of IF for the QBM starting from non-separable initial conditions have been considered [1, 10, 95, 117, 116, 118, 119]. These involve, along with the real time path integrals studied here, imaginary time path integrals. Also, the problem of non-linear QBM, with non-linear system-reservoir couplings, have also been considered [98, 118]. Getting a firm grip of the techniques discussed in this chapter would enable the reader to tackle these, more advanced, applications of the IF technique easily. Bounds on the low energy spectral behavior of the system-bath coupling comes from the van Hove bound, see for e.g., [120].

Chapter 5

Dissipative Harmonic Oscillator

5.1 Introduction

This chapter is devoted to the dissipative harmonic oscillator, a paradigm model of open quantum systems. We present two well known approaches to this problem, *viz.*, the semigroup or the Lindbladian approach and the quantum Brownian approach. The physical conditions under which each regime is operational are spelled out. The models are developed in detail. A brief comparison is made with other master equations studied in the literature. The problem of quantum Brownian motion is tackled both in the perturbative as well as in the non-perturbative regimes. The well known Caldeira-Leggett model is retrieved. An explicit calculation is made of the diagonalization of the dissipative harmonic oscillator, in the non-perturbative quantum Brownian regime. The fluctuation-dissipation theorem is spelled out. The last part of the chapter is devoted to the application of the dissipative harmonic oscillator to foundational issues, such as quantum phase distribution and complementarity between number and phase. In this context, an operator solution of the Lindbladian master equation is presented.

5.2 Lindbladian Approach to the Damped Oscillator

We have already discussed the LGKS master equation in Chapter 3. The standard form of the Lindblad equation, derived there, can be also expressed as

$$\frac{d\rho^S(t)}{dt} = \mathcal{L}\rho^S(t), \tag{5.1}$$

where

$$\mathcal{L}\rho^S(t) = -\frac{i}{\hbar}[H, \rho_S(t)] + \frac{1}{2\hbar}\sum_{j=1}\left([B_j\rho^S(t), B_j^\dagger] + [B_j, \rho^S(t)B_j^\dagger]\right). \tag{5.2}$$

To make contact with the earlier notation, the operators B_j here are related to the operators A_j, there, by $B_j = \gamma_j A_j$. Also, in our earlier discussion of this

© Hindustan Book Agency 2018 and Springer Nature Singapore Pte Ltd. 2018
S. Banerjee, *Open Quantum Systems*, Texts and Readings in Physical Sciences 20,
https://doi.org/10.1007/978-981-13-3182-4_5

equation we had set $\hbar$ to one. Note that the first term on the *RHS* of Eq. (5.2) is the von Neumann term responsible for unitary evolution while the second term depicts dissipation and decoherence or dephasing, both irreversible processes. Markovian master equations have, modulo some rearrangements, the form as in Eq. (5.2), even for unbounded generators, as would be the case for harmonic oscillators. In this context, a general construction of the B_j operators can be affected by assuming that they are functions of x and p, position and momentum operators of the system, such that $[x, p] = i\hbar$. Thus, we can assume B_j and H to be at most first and second degree polynomials in the noncommuting operators x and p, respectively [57, 77]. As a matter of fact, a linear space spanned by such polynomials is invariant under the action of the completely dissipative mapping $\mathcal{L}$, Eq. (5.2). Hence, in general we have

$$B_i = \alpha_i p + \beta_i x, \quad i = 1, 2 \tag{5.3}$$

where α_i, β_i are complex numbers. Also,

$$H = \frac{1}{2m}p^2 + \frac{m\omega^2}{2}x^2 + \frac{\chi}{2}(px + xp). \tag{5.4}$$

With this parametrization, $\mathcal{L}\rho^S(t)$ (5.2) can be written as

$$\mathcal{L}\rho^S(t) = -\frac{i}{\hbar}[H_0, \rho^S(t)] - \frac{i}{2\hbar}(\lambda + \chi)[x, (\rho^S(t)p + p\rho^S(t))]$$
$$+ \frac{i}{2\hbar}(\lambda - \chi)[p, (\rho^S(t)x + x\rho^S(t))] - \frac{D_{pp}}{\hbar^2}[x, [x, \rho^S(t)]]$$
$$- \frac{D_{xx}}{\hbar^2}[p, [p, \rho^S(t)]] + \frac{D_{px}}{\hbar^2}[x, [p, \rho^S(t)]] + \frac{D_{xp}}{\hbar^2}[p, [x, \rho^S(t)]]. \tag{5.5}$$

Here

$$H_0 = \frac{1}{2m}p^2 + \frac{m\omega^2}{2}x^2, \quad D_{xx} = \frac{\hbar}{2}\sum_{i=1}^{2}|\alpha_i|^2, \quad D_{pp} = \frac{\hbar}{2}\sum_{i=1}^{2}|\beta_i|^2,$$

$$D_{px} = D_{xp} = -\frac{\hbar}{2}\text{Re}\sum_{i=1}^{2}\alpha_i^*\beta_i, \quad \lambda = \text{Im}\sum_{i=1}^{2}\alpha_i^*\beta_i. \tag{5.6}$$

The master equation (5.1) with *RHS* given by Eq. (5.5) includes within its ambit, the dissipative interaction of the harmonic oscillator with a squeezed thermal bath, to which we will get familiar with in Chapter 8. Using the x, p commutation relations, the Eq. (5.5) can be rearranged and expressed as

$$\mathcal{L}\rho^S(t) = -\frac{i}{\hbar}[H_0, \rho^S(t)] - \frac{i}{2\hbar}(\lambda - \chi)[\rho^S(t), (px + xp)] - \frac{i}{\hbar}\lambda[x, (p\rho^S(t)$$
$$+ \rho^S(t)p)] - \frac{D_{pp}}{\hbar^2}[x, [x, \rho^S(t)]] - \frac{D_{xx}}{\hbar^2}[p, [p, \rho^S(t)]] \tag{5.7}$$

$$+ \frac{(D_{px} + D_{xp})}{\hbar^2}[p, [x, \rho^S(t)]]. \tag{5.8}$$

Problem 1: Derive Eqs. (5.5) and (5.8).

The diffusion coefficients D_{xx}, D_{pp} and D_{xp} satisfy certain constraints, that is,

$$(a). \quad D_{xx} \quad > 0, \tag{5.9}$$
$$(b). \quad D_{pp} \quad > 0, \tag{5.10}$$
$$(c). \quad D_{xx} \quad D_{pp} - D_{px}^2 \geq \lambda^2 \hbar^2 / 4. \tag{5.11}$$

The Eqs. (5.9) and (5.10) follow in a straightforward manner from Eqs. (5.6). Eq. (5.11) follows from Eq. (5.6) and the Schwartz inequality

$$\left(\text{Re} \sum_{i=1}^{2} \alpha_i^* \beta_i \right)^2 + \left(\text{Im} \sum_{i=1}^{2} \alpha_i^* \beta_i \right)^2 \leq \sum_{i=1}^{2} |\alpha_i|^2 \sum_{i=1}^{2} |\beta_i|^2. \tag{5.12}$$

Problem 2: Derive Eq. (5.11).

Eqs. (5.5) or (5.8) have been used in the literature in various guises [121]. The master equation for the damped harmonic oscillator, discussed in [19], is a particular case of this equation. There λ was equal to χ. Variants of Eq. (5.5) have been used to study collective modes in inelastic collisions of heavy ions, see for example, [122, 123, 124]. The evolution of a general electromagnetic field mode coupled to a squeezed thermal bath, of great relevance to studies in quantum optics and information, and to which we shall return to later in Chapter 8, has been derived and analyzed in [125, 126]. This equation is a simplified version of Eq. (5.5). The damping of a harmonic oscillator, in the context of quantum Brownian motion [127, 128] as well as the damping of quantum coherence [129] have been addressed using master equations that are subsets of Eqs. (5.5) or (5.8), with $\lambda = \chi = \gamma$, $D_{xx} = D_{xp} = D_{px} = 0$ and $D_{pp} = 2\gamma(N_{th} + 1/2)m\omega\hbar$. Here γ is the damping constant, its meaning will become clearer shortly, and N_{th} is the Planck distribution giving the number of thermal photons at a particular frequency, see for example, below Eq. (96) in Chapter 3. Further, in [130], a variant was applied to lasers. Among the models discussed here, there are some which satisfy the constraints laid down in Eqs. (5.9) to (5.11) and others which do not. Those which do satisfy belong to the category of Lindblad evolution while the others do not strictly belong to the Lindblad category.

Now we will use Eq. (5.5) to study the behaviour of the mean and variances of x and p. The quantities of interest are

$$\sigma_x(t) = \text{Tr}\left(x\rho^S(t)\right),$$

$$\sigma_{xx}(t) = \text{Tr}\left(x^2\rho^S(t)\right) - (\sigma_x(t))^2,$$

$$\sigma_{xp}(t) = \text{Tr}\left(\left(\frac{xp + px}{2}\right)\rho^S(t)\right) - \sigma_x(t)\sigma_p(t). \tag{5.13}$$

The other quantities such as $\sigma_p(t)$, $\sigma_{pp}(t)$ can be defined analogously. The equations of motion of the above quantities can be easily derived. Thus, for example,

$$\frac{d\sigma_x(t)}{dt} = \text{Tr}\left(\frac{d\rho^S(t)}{dt}x\right) = \text{Tr}\left((\mathcal{L}\rho^S(t))x\right), \tag{5.14}$$

$$\frac{d\sigma_{xx}(t)}{dt} = \text{Tr}\left((\mathcal{L}\rho^S(t))x^2\right) - 2\sigma_x(t)\frac{d\sigma_x(t)}{dt}. \tag{5.15}$$

The equations for the other quantities can be set up in a similar fashion. Using Eq. (5.5) in Eq. (5.14) we get for the means

$$\frac{dR_1(t)}{dt} = M_1 R_1(t), \tag{5.16}$$

where

$$R_1(t) = \begin{pmatrix} \sigma_x(t) \\ \sigma_p(t) \end{pmatrix}, \quad M_1 = \begin{pmatrix} -(\lambda - \chi) & \frac{1}{m} \\ -m\omega^2 & -(\lambda + \chi) \end{pmatrix}. \tag{5.17}$$

The Eq. (5.16) is a linear homogeneous matrix differential equation and has the solution

$$R_1(t) = V^{-1}e^{M_d t}V R_1(0), \tag{5.18}$$

where $M_1 = V^{-1}M_d V$. Here M_d is the diagonal matrix.

Problem 3: Derive Eq. (5.18).

There are two cases:
(a). *Overdamped Case:*
Here $\chi > \omega$ and

$$M_d = \begin{pmatrix} -(\lambda + \nu) & 0 \\ 0 & -(\lambda - \nu) \end{pmatrix}, \tag{5.19}$$

and

$$V = \begin{pmatrix} m\omega^2 & (\chi + \nu) \\ m\omega^2 & (\chi - \nu) \end{pmatrix}, \tag{5.20}$$

with $\nu^2 = \chi^2 - \omega^2$. With these, we get from Eq. (5.18)

$$\sigma_x(t) = e^{-\lambda t}\left[\left(\cosh(\nu t) + \frac{\chi}{\nu}\sinh(\nu t)\right)\sigma_x(0) + \frac{1}{m\omega}\sinh(\nu t)\sigma_p(0)\right], \qquad (5.21)$$

$$\sigma_p(t) = e^{-\lambda t}\left[-\frac{m\omega^2}{\nu}\sinh(\nu t)\sigma_x(0) + \left(\cosh(\nu t) - \frac{\chi}{\nu}\sinh(\nu t)\right)\sigma_p(0)\right]. \quad (5.22)$$

Problem 4: Derive Eqs. (5.21) and (5.22).

From these solutions it is easy to see that for $\lambda > \nu$, the asymptotic values tend to zero, that is, $\sigma_x(\infty) = \sigma_p(\infty) = 0$, while for $\lambda < \nu$, $\sigma_x(\infty) = \sigma_p(\infty) = \infty$.

(b). *Underdamped Case:*
Here $\chi < \omega$ and

$$M_d = \begin{pmatrix} -(\lambda + i\Omega) & 0 \\ 0 & -(\lambda - i\Omega) \end{pmatrix}, \qquad (5.23)$$

and

$$V = \begin{pmatrix} m\omega^2 & (\chi + i\Omega) \\ m\omega^2 & (\chi - i\Omega) \end{pmatrix}. \qquad (5.24)$$

In these equations, $\Omega^2 = \omega^2 - \chi^2$. With Eqs. (5.23) and (5.24), Eq. (5.18) yields

$$\sigma_x(t) = e^{-\lambda t}\left[\left(\cos(\nu t) + \frac{\chi}{\Omega}\sin(\Omega t)\right)\sigma_x(0) + \frac{1}{m\Omega}\sin(\Omega t)\sigma_p(0)\right], \qquad (5.25)$$

$$\sigma_p(t) = e^{-\lambda t}\left[-\frac{m\omega^2}{\Omega}\sin(\Omega t)\sigma_x(0) + \left(\cos(\Omega t) - \frac{\chi}{\Omega}\sin(\Omega t)\right)\sigma_p(0)\right]. \quad (5.26)$$

Note that $\sigma_x(\infty) = \sigma_p(\infty) = 0$.

Problem 5: Derive Eqs. (5.25) and (5.26).

Following a similar strategy, we use Eq. (5.5) in Eq. (5.15) and get for the variances

$$\frac{d\Theta(t)}{dt} = M_2\Theta(t) + \Gamma, \qquad (5.27)$$

where

$$\Theta(t) = \begin{pmatrix} m\omega\sigma_{xx}(t) \\ \frac{1}{m\omega}\sigma_{pp}(t) \\ \sigma_{px}(t) \end{pmatrix}, \qquad (5.28)$$

$$M_2 = \begin{pmatrix} -2(\lambda - \chi) & 0 & 2\omega \\ 0 & -2(\lambda + \chi) & -2\omega \\ -\omega & \omega & -2\lambda \end{pmatrix}, \qquad (5.29)$$

and

$$\Gamma = \begin{pmatrix} 2m\omega D_{xx} \\ \frac{2}{m\omega} D_{pp} \\ 2D_{px} \end{pmatrix}.$$ (5.30)

Eq. (5.27) is an inhomogeneous matrix linear differential equation and has the solution

$$\Theta(t) = \left(V_2 e^{M_{2d}t} V_2^{-1} \right) \Theta(0) + V_2 \left(e^{M_{2d}t} - 1 \right) M_{2d}^{-1} V_2^{-1} \Gamma,$$ (5.31)

where M_{2d} is the diagonal matrix $M_2 = V_2 M_{2d} V_2^{-1}$.

Problem 6: Derive Eq. (5.31).

Now we turn to the overdamped and underdamped cases.
(a). *Overdamped:*
Here $\chi > \omega$. Also,

$$V_2 = \frac{1}{2\nu} \begin{pmatrix} \chi + \nu & \chi - \nu & 2\omega \\ \chi - \nu & \chi + \nu & 2\omega \\ -\omega & -\omega & -2\chi \end{pmatrix},$$ (5.32)

$$M_{2d} = \begin{pmatrix} -2(\lambda - \chi) & 0 & 0 \\ 0 & -2(\lambda + \chi) & 0 \\ 0 & 0 & -2\lambda \end{pmatrix}.$$ (5.33)

Here $\nu^2 = \chi^2 - \omega^2$.
(b). *Underdamped:*
Now $\chi < \omega$ and

$$V_2 = \frac{1}{2i\Omega} \begin{pmatrix} \chi + i\Omega & \chi - i\Omega & 2\omega \\ \chi - i\Omega & \chi + i\Omega & 2\omega \\ -\omega & -\omega & -2\chi \end{pmatrix},$$ (5.34)

$$M_{2d} = \begin{pmatrix} -2(\lambda - i\Omega) & 0 & 0 \\ 0 & -2(\lambda + i\Omega) & 0 \\ 0 & 0 & -2\lambda \end{pmatrix}.$$ (5.35)

Here $\Omega^2 = \omega^2 - \chi^2$. From Eq. (5.31) it can be seen that

$$\Theta(\infty) = -M_2^{-1} \Gamma.$$ (5.36)

This brings out the asymptotic connection between the variances and various diffusion constants and holds for both the overdamped and underdamped regimes.

5.3   Quantum Brownian Motion

We now come to the important example of quantum Brownian motion of the harmonic oscillator. We have encountered this before, see Eqs. (17) to (20), Chapter 3. The total Hamiltonian is

$$
\begin{aligned}
H &= H_S + H_R + H_{SR} \\
&= \frac{p^2}{2m} + V_{ren}(x) + \sum_i \left[\frac{p_i^2}{2m_i} + \frac{1}{2} m_i \omega_i^2 q_i^2 \right] - \sum_i c_i x q_i. \quad (5.37)
\end{aligned}
$$

Here $V_{ren}(x) = V(x) + \sum_i \frac{c_i^2}{2m_i \omega_i^2} x^2$ is the renormalization of the potential $V(x) = \frac{1}{2} m \omega^2 x^2$ due to interaction with the bath. The term added to the harmonic potential is sometimes called the *counter-term*. The study of quantum Brownian motion can proceed depending upon the regime for which it is to be applied. Based on this, we could have (a). weak coupling, high temperature T master equation and (b). strong coupling, low T master equation.

5.3.1   Weak coupling, high T regime

Here we follow along the lines of the derivation of the LGKS master equation, Section 6.1 in Chapter 4 and begin with Eq. (73) there, representing the Born-Markov (weak coupling, memoryless) regime. Re-written in the Schrödinger picture, we have

$$
\frac{d}{dt} \rho^S(t) = \frac{-i}{\hbar} \left[H_S + H_{counter}, \rho^S(t) \right] + \mathcal{L} \rho^S(t), \quad (5.38)
$$

where $H_{counter} = \sum_i \frac{c_i^2}{2m_i \omega_i^2} x^2$ is the above defined counter-term. Also, H_S is as in Eq. (5.37) and

$$
\mathcal{L} \rho^S(t) = -\frac{1}{\hbar^2} \int_0^\infty ds \, \mathrm{Tr}_R [H_{SR}, [H_{SR}(-s), \rho^S(t) \otimes \rho^R]]. \quad (5.39)
$$

The terms appearing here are as defined in Eq. (5.37). Also, $H_{SR}(-s)$ represents interaction picture operator *w.r.t.* $H_S + H_R$. We will assume that initially the system, here the quantum Brownian oscillator, is separated from its reservoir H_R, of harmonic oscillators. This is often known as the *separable initial condition*. Further, the reservoir is assumed to be in thermal equilibrium at a temperature T, with the reservoir density matrix ρ_R being in accordance with the canonical distribution

$$
\rho_R = \left(\mathcal{Z} \right)^{-1} e^{-\beta H_R}, \quad (5.40)
$$

where $\beta = 1/(k_B T)$ and $\mathcal{Z} = \mathrm{Tr}_R e^{-\beta H_R}$. To proceed further, it is convenient to define the following correlation operators

$$
\begin{aligned}
C_1(s) &= i \langle [R(s), R(0)] \rangle = i \langle [R(0), R(-s)] \rangle, \quad (5.41) \\
C_2(s) &= \langle \{R(s), R(0)\} \rangle = \langle \{R(0), R(-s)\} \rangle. \quad (5.42)
\end{aligned}
$$

Here $R = \sum_i c_i q_i$, Eq. (5.37), is the reservoir operator. The angular brackets in the above equation denote average *w.r.t.* the reservoir density matrix ρ_R. Also, the curly brackets in Eq. (5.42) denote the anticommutation operation. Defining the reservoir spectral density $I(\omega)$ as

$$I(\omega) = \sum_i \frac{c_i^2}{2m_i\omega_i} \delta(\omega - \omega_i). \tag{5.43}$$

With this, the reservoir correlation functions $C_1(s)$ and $C_2(s)$ can be expressed as

$$C_1(s) = 2\hbar \int_0^\infty d\omega I(\omega) \sin(\omega s), \tag{5.44}$$

$$C_2(s) = 2\hbar \int_0^\infty d\omega I(\omega) \coth(\beta\hbar\omega/2) \cos(\omega s). \tag{5.45}$$

Let us sketch the derivation of the above equations. To begin, we note that the reservoir operator $R(s)$ has the form, in the interaction picture,

$$R(s) = \sum_i c_i \sqrt{\frac{\hbar}{2m_i\omega_i}} \left(e^{-i\omega_i t} a_i + e^{i\omega_i t} a_i^\dagger\right), \tag{5.46}$$

where we have used the standard expansion of the position operator q in terms of the annihilation a and creation $a^\dagger$ operators, that is,

$$q_i = \sqrt{\frac{\hbar}{2m_i\omega_i}} \left(a_i + a_i^\dagger\right). \tag{5.47}$$

Now consider $C_1(s)$ (5.41) which becomes, using Eq. (5.46),

$$C_1(s) = i \sum_{i,j} c_i c_j \frac{\hbar}{2\sqrt{m_i\omega_i m_j\omega_j}} \left[a_i + a_i^\dagger, \ e^{i\omega_j s} a_j + e^{-i\omega_j t} a_j^\dagger\right]. \tag{5.48}$$

By the linearity of the commutation operation, this yields four terms, two of which are zero. Using $[a_i, a_j^\dagger] = \delta_{ij}$, we see that $C_1(s)$ reduces to

$$C_1(s) = 2\hbar \sum_i \frac{c_i^2}{2m_i\omega_i} \sin(\omega_i s) = 2\hbar \int_0^\infty d\omega I(\omega) \sin(\omega s). \tag{5.49}$$

In the last step we have made use of Eq. (5.43). Thus, Eq. (5.44) is recovered. Eq. (5.45) can be obtained in a similar fashion. Here the reservoir averages would need to be done. These can be easily done using the identity presented in Eq. (46), Chapter 2.

Problem 7: Derive Eq. (5.45).

Using Eqs. (5.41) and (5.42), Eq. (5.39) can be re-written as

$$\mathcal{L}\rho^S(t) = \frac{1}{2\hbar^2} \int_0^\infty ds \Big(iC_1(s)\big[x, \{x(-s), \rho^S(t)\}\big] - C_2(s)\big[x, [x(-s), \rho^S(t)]\big] \Big).$$
(5.50)

In order to model dissipative dynamics, an irreversible process, the reservoir spectral density is phenomelogically modeled by a continuous distribution of oscillator modes. Thus, the summation over discrete frequencies is replaced by an integral over a continuous ω. A common form of the spectral density $I(\omega)$ is the Ohmic spectrum

$$I(\omega) = \frac{2m\gamma}{\pi}\omega,$$
(5.51)

where γ is a damping constant. Ohmic spectral density simulates what is commonly known as *white noise*. In order to control the high frequency effect of the Ohmic spectrum, it is usually found feasible to modify it with an appropriate upper cut-off frequency Ω_c. Thus, for example, Eq. (5.51) could be modified to

$$I(\omega) = \frac{2m\gamma}{\pi}\omega\frac{\Omega_c^2}{\Omega_c^2 + \omega^2}.$$
(5.52)

Now we make a crucial assumption that defines the regime of quantum Brownian motion and separates it from that of the LGKS master equation. Here we assume that the system has very little time to evolve before the effect of the environment becomes active, that is, $\tau_S \gg \tau_R$, where τ_S and τ_R are the typical system and reservoir time scales, respectively. Hence, the system evolves little, before the environmental influence takes effect. Thus, the evolution of a system operator O is

$$O(t - s) \approx O(t) - s\dot{O}(t),$$
(5.53)

where $\dot{O}(t) = -\frac{i}{\hbar}\big[O(t), H_S(t)\big]$. We make use of this condition in Eq. (5.50), with $x(-s) \approx x - \frac{p}{m}s$, to get

$$\mathcal{L}\rho^S(t) = \frac{i}{2\hbar^2} \int_0^\infty ds\, C_1(s)\big[x, \{x, \rho^S\}\big] - \frac{i}{2\hbar^2 m} \int_0^\infty ds\, s\, C_1(s)\big[x, \{p, \rho^S\}\big]$$
$$- \frac{1}{2\hbar^2} \int_0^\infty ds\, C_2(s)\big[x, [x, \rho^S]\big] + \frac{1}{2\hbar^2 m} \int_0^\infty ds\, s\, C_2(s)\big[x, [p, \rho^S]\big].$$
(5.54)

Using Eq. (5.43) we see that

$$\int_0^\infty ds\, C_1(s) = 2\hbar \sum_i \frac{c_i^2}{2m_i\omega_i^2}.$$
(5.55)

This enables us to express the first term on the *RHS* of Eq. (5.54) as $\frac{i}{\hbar}\big[H_c, \rho^S\big]$. This compensates the corresponding counter term on the *RHS* of Eq. (5.38), resulting in the cancellation of the reservoir induced renormalization of the system Hamiltonian. To tackle the second term on the *RHS* of Eq. (5.54), we observe that

$$\int_0^\infty ds \; s \; \sin(\omega s) = -\pi \delta'(\omega). \tag{5.56}$$

From this, using an Ohmic spectral density (5.51), it is straightforward to see that

$$\int_0^\infty ds \; s \; C_1(s) = 2m\gamma\hbar. \tag{5.57}$$

Hence, the second term on the *RHS* of Eq. (5.54) is seen to be $-i\frac{\gamma}{\hbar}[x, \{p, \rho^S\}]$. Further,

$$\int_0^\infty ds \; C_2(s) = 4m\gamma k_B T. \tag{5.58}$$

The third term on the *RHS* of Eq. (5.54) is thus seen to be $-\frac{2m\gamma k_B T}{\hbar^2}[x, [x, \rho^S]]$. Similarly, the fourth term on the *RHS* of Eq. (5.54) can be shown to be $\frac{2\gamma k_B T}{\hbar^2 \Omega}[x, [p, \rho^S]]$, where Ω is an upper cut-off frequency of the reservoir oscillators. This turns out to be very small compared to the third term and hence can be neglected. Collecting all these terms, and using Eq. (5.54), the quantum Brownian motion master equation is seen to be

$$\frac{d}{dt}\rho^S(t) = \frac{-i}{\hbar}[H_S, \rho^S(t)] - i\frac{\gamma}{\hbar}[x, \{p, \rho^S(t)\}] - \frac{2m\gamma k_B T}{\hbar^2}[x, [x, \rho^S(t)]]. \tag{5.59}$$

This is sometimes called the Caldeira Leggett master equation [131]. Note that since this master equation is derived by invoking the approximation $\tau_S \gg \tau_R$, which is opposite to the rotating wave approximation regime, see Chapter 3.6, it is not surprising that Eq. (5.59) is *not* of the Lindblad type.

   Mean and second moments:

   Using Eq. (5.59), the equations of motion of the mean and variance are

$$\frac{d\langle x \rangle}{dt} = \frac{1}{m}\langle p \rangle, \tag{5.60}$$

$$\frac{d\langle p \rangle}{dt} = -\langle \frac{dV(x)}{dx} \rangle - 2\gamma \langle p \rangle, \tag{5.61}$$

$$\frac{d\langle x^2 \rangle}{dt} = \frac{1}{m}\langle px + xp \rangle, \tag{5.62}$$

$$\frac{d\langle px + xp \rangle}{dt} = \frac{2}{m}\langle p^2 \rangle - 2\langle x\frac{dV(x)}{dx} \rangle - 2\gamma \langle px + xp \rangle, \tag{5.63}$$

$$\frac{d\langle p^2 \rangle}{dt} = -\langle p\frac{dV(x)}{dx} + \frac{dV(x)}{dx}p \rangle - 4\gamma \langle p^2 \rangle + 4m\gamma k_B T. \tag{5.64}$$

For a harmonic potential, $\frac{dV(x)}{dx} = m\omega^2 x$. For the case of a free particle undergoing Brownian motion, $V(x) = 0$. Then Eqs. (5.60) and (5.61) yield, in a trivial manner,

$$\langle x(t) \rangle = \langle x(0) \rangle + \frac{1}{2m\gamma}(1 - e^{-2\gamma t})\langle p(0) \rangle, \tag{5.65}$$

$$\langle p(t) \rangle = e^{-2\gamma t}\langle p(0) \rangle. \tag{5.66}$$

Similarly the second moments can be seen to be

$$
\sigma_x^2(t) = \sigma_x^2(0) + \left(\frac{1-e^{-2\gamma t}}{2\gamma}\right)^2 \frac{\sigma_p^2(0)}{m^2} + \frac{1-e^{-2\gamma t}}{2m\gamma}\sigma_{px}(0)
$$
$$
+\frac{k_B T}{m\gamma^2}\left[\gamma t - \left(1 - e^{-2\gamma t}\right) + \frac{1}{4}\left(1 - e^{-4\gamma t}\right)\right], \tag{5.67}
$$
$$
\sigma_p^2(t) = e^{-4\gamma t}\sigma_p^2(0) + mk_B T\left(1 - e^{-4\gamma t}\right), \tag{5.68}
$$
$$
\sigma_{px}^2(t) = e^{-2\gamma t}\sigma_{px}^2(0) + \frac{1}{m\gamma}\left(1 - e^{-2\gamma t}\right)e^{-2\gamma t}\sigma_p^2(0)
$$
$$
+\frac{k_B T}{\gamma}\left(1 - e^{-2\gamma t}\right)^2. \tag{5.69}
$$

Note that the symbols of the mean and the second moments are as defined in Eq. (5.13). From these expressions it is easily seen that, in the long time limit, $\sigma_x^2(t) \to \frac{k_B T}{m\gamma}t$, that is, it shows *diffusive* behaviour, while the corresponding variance in momentum approaches the stationary value of $mk_B T$.

Problem 8: Derive Eqs. (5.65) to (5.69). Also, find out the mean position and momentum for the case of a Brownian particle moving in a harmonic potential $V(x) = \frac{1}{2}m\omega^2 x^2$.

Equations of Motion Approach:
   The equations of motion approach is based on the study of evolution of the operators, in the Heisenberg picture and compliments the master equation approach, which is basically in the Schrödinger picture, discussed above. Given the Hamiltonian, Eq. (5.37), the equations of motion of the system S and the reservoir operators are

$$
\dot{x}(t) = \frac{i}{\hbar}[H, x(t)] = \frac{1}{m}p, \tag{5.70}
$$
$$
\dot{q}_i(t) = \frac{i}{\hbar}[H, q_i(t)] = \frac{1}{m_i}p_i, \tag{5.71}
$$
$$
\dot{p}(t) = \frac{i}{\hbar}[H, p(t)] = -\frac{dV_{ren}(x(t))}{dx} + \sum_i c_i q_i(t), \tag{5.72}
$$
$$
\dot{p}_i(t) = \frac{i}{\hbar}[H, p_i(t)] = -m_i\omega_i^2 q_i(t) + c_i x(t). \tag{5.73}
$$

From these equations, it is easy to see that the equations of motion of the system and reservoir coordinates are

$$
m\ddot{x}(t) + \frac{dV_{ren}(x(t))}{dx} - \sum_i c_i q_i(t) = 0, \tag{5.74}
$$
$$
m\ddot{q}_i(t) + m_i\omega_i^2 q_i(t) - c_i x(t) = 0. \tag{5.75}
$$

In order to obtain the solution of the system coordinate, the strategy would be to solve Eq. (5.75) and substitute it into Eq. (5.74). The solution of Eq. (5.75) is

$$q_i(t) = \sqrt{\frac{\hbar}{2m_i\omega_i}}\left(e^{-i\omega_i t}a_i + e^{i\omega_i t}a_i^\dagger\right) + \frac{c_i}{m_i\omega_i}\int_0^t du\sin\left(\omega_i(t-u)\right)x(u). \quad (5.76)$$

Here use is made of Eq. (5.47). Eq. (5.76) is now substituted into Eq. (5.74) such that we obtain the equation of motion of the Brownian system coordinate

$$\ddot{x}(t) + \frac{1}{m}\frac{dV_{ren}(x(t))}{dx} - \frac{1}{\hbar m}\int_0^t du C_1(t-u)x(u) = \frac{1}{m}F(t). \quad (5.77)$$

Here $C_1(t)$ is as defined in Eq. (5.44) and $F(t)$ is the interaction picture operator

$$F(t) = \sum_i c_i\sqrt{\frac{\hbar}{2m_i\omega_i}}\left(e^{-i\omega_i t}a_i + e^{i\omega_i t}a_i^\dagger\right). \quad (5.78)$$

Its Schrödinger picture counterpart is $\sum_i c_i q_i(0)$. In the present context, it is convenient to express the dissipation kernel $C_1(u)$ in terms of the damping kernel γ, such that

$$\frac{d}{dt}\gamma(t-u) = -\frac{1}{\hbar m}C_1(t-u). \quad (5.79)$$

This has the advantage that the integral of the dissipation kernel, in Eq. (5.77), is split into two parts

$$-\frac{1}{\hbar m}\int_0^t du C_1(t-u)x(u) = \frac{d}{dt}\int_0^t du\gamma(t-u)x(u) - \gamma(0)x(t). \quad (5.80)$$

The last term in the RHS of the above equation takes care of the counterterm in the potential $V_{ren}(x)$ in Eq. (5.77) such that the Heisenberg equation of motion can be reexpressed as

$$\ddot{x}(t) + \frac{1}{m}\frac{dV(x(t))}{dx} + \frac{d}{dt}\int_0^t du\gamma(t-u)x(u) = \frac{1}{m}F(t). \quad (5.81)$$

This has the form of a stochastic differential equation with damping $\gamma(t-u)$ and fluctuation determined by $F(t)$, which in turn depend upon the initial distribution of the reservoir oscillators, Eq. (5.78). For the case of a harmonic potential $V(x) = \frac{1}{2}m\omega^2 x^2$, Eq. (5.81) becomes

$$\ddot{x}(t) + \omega^2 x(t) + \frac{d}{dt}\int_0^t du\gamma(t-u)x(u) = \frac{1}{m}F(t). \quad (5.82)$$

This equation can be easily solved to yield

$$x(t) = G_1(t)x(0) + G_2(t)\dot{x}(0) + \frac{1}{m}\int_0^t du G_2(t-u)F(u). \quad (5.83)$$

Here G_1 and G_2 are Green's functions providing the solutions of the homogeneous part of Eq. (5.82) [10].

Problem 9: Prove Eq. (5.83).

5.3.2 Strong coupling regime

For the problem of quantum Brownian motion of a harmonic oscillator system, the problem of dissipation can be easily studied in the strong coupling regime as well. For this we resort to the technique of *Path Integration*, in particular, the influence functionals studied in the previous chapter. This enables a *non-perturbative* solution to the problem. For convenience, we recapitulate some of the material required for our present purpose. The Hamiltonian is

$$H = H_S + H_R + H_{SR}, \tag{5.84}$$

where

$$H_S = \frac{1}{2} M \left[\dot{x}^2 + \Omega^2 x^2 \right] \tag{5.85}$$

is the system Hamiltonian,

$$H_R = \sum_{n=1}^{N} \frac{1}{2} m_n \left[\dot{q_n}^2 + \omega_n^2 q_n^2 \right] \tag{5.86}$$

is the environment (reservoir) Hamiltonian, and

$$H_{SR} = - \sum_{n=1}^{N} [c_n x q_n] + x^2 \sum_{n=1}^{N} \frac{c_n^2}{2 m_n \omega_n^2} \tag{5.87}$$

is the system-environment interaction Hamiltonian. We use separable initial conditions, i.e.,

$$\rho(0) = \rho_0^S \rho_0^R, \tag{5.88}$$

where

$$\hat{\rho}^R(0) = \hat{S}(r, \Phi) \hat{\rho}_{th} \hat{S}^\dagger(r, \Phi). \tag{5.89}$$

Here we have a squeezed thermal initial state, see for example, Chapter 4, Section 3.2. Further,

$$\hat{\rho}_{th} = \left[1 - \exp\left(\frac{-\hbar \omega}{k_B T} \right) \right] \sum_n \exp\left(\frac{-n \hbar \omega}{k_B T} \right) |n\rangle \langle n|, \tag{5.90}$$

i.e., a thermal density matrix at temperature T and

$$\hat{S}(r, \Phi) = \exp\left[r(\hat{B} e^{-i2\Phi} - \hat{B}^\dagger e^{i2\Phi}) \right], \tag{5.91}$$

i.e., a squeeze operator [132]. Also,

$$\hat{B} = \frac{\hat{b}^2}{2}, \hat{B}^\dagger = \frac{\hat{b}^{\dagger 2}}{2}, \tag{5.92}$$

and r, Φ are the squeeze parameters. By tracing over the bath we obtain the reduced density matrix of the system which is encapsulated in the propagator

$$J_r(\Sigma_f, \Delta_f, t; \Sigma_i, \Delta_i, 0) = Z(t,0) \exp\left[\frac{i}{\hbar}\{b_1\Sigma_f\Delta_f - b_2\Sigma_f\Delta_i + b_3\Sigma_i\Delta_f - b_4\Sigma_i\Delta_i\}\right]$$

$$\times \exp\left[\frac{-1}{\hbar}\{a_{11}\Delta_i^2 + a_{12}\Delta_i\Delta_f + a_{22}\Delta_f^2\}\right], \tag{5.93}$$

where

$$b_1(t) = M\dot{u}_2(t), \quad b_2(t) = M\dot{u}_2(0), \tag{5.94}$$

$$b_3(t) = M\dot{u}_1(t), \quad b_4(t) = M\dot{u}_1(0), \tag{5.95}$$

$$a_{mn}(t) = \frac{1}{1+\delta_{mn}} \int_0^t ds \int_0^t ds' \, v_m(s)\nu(s,s')v_n(s'), \tag{5.96}$$

and

$$Z(t,0) = \frac{b_2(t)}{2\pi\hbar}. \tag{5.97}$$

Here u_1, u_2, v_1, v_2 come from the solutions of the equations

$$\ddot{\Sigma}_{cl}(s) + \tilde{\Omega}^2 \Sigma_{cl}(s) + \frac{2}{M}\int_0^s ds' \, \mu(s,s')\Sigma_{cl}(s') = 0, \tag{5.98}$$

and

$$\ddot{\Delta}_{cl}(s) + \tilde{\Omega}^2 \Delta_{cl}(s) + \frac{2}{M}\int_s^t ds' \, \mu(s',s)\Delta_{cl}(s') = 0, \tag{5.99}$$

where

$$\tilde{\Omega}^2 = \Omega^2 + \frac{2}{M}\int_0^\infty d\omega \frac{I(\omega)}{\omega}, \tag{5.100}$$

with $I(\omega)$ being the bath spectral density. The second term in the above equation arises as a result of the interaction of the system with the bath. In the above equations

$$\mu(s,s') = -\int_0^\infty d\omega \, I(\omega)\sin\omega(s-s') \tag{5.101}$$

is the dissipation kernel and,

$$\nu(s,s') = \frac{1}{2}\int_0^\infty d\omega \, I(\omega)\coth(\frac{\hbar\omega}{2k_BT})\{\cosh(2r(\omega))2\cos\omega(s-s')$$

$$- \sinh(2r(\omega))e^{-i2\Phi(\omega)}e^{i\omega(s+s')}$$

$$- \sinh(2r(\omega))e^{i2\Phi(\omega)}e^{-i\omega(s+s')}\} \tag{5.102}$$

is the noise kernel. The solutions of the equations (5.98), (5.99) can be parametrized in terms of u and v as

$$\Sigma_{cl}(s) = \Sigma_i u_1(s) + \Sigma_f u_2(s), \tag{5.103}$$

$$\Delta_{cl}(s) = \Delta_i v_1(s) + \Delta_f v_2(s), \tag{5.104}$$

where in order that the classical solutions satisfy proper boundary conditions we have

$$u_1(0) = 1 = u_2(t), u_1(t) = 0 = u_2(0), \tag{5.105}$$

$$v_1(0) = 1 = v_2(t), v_1(t) = 0 = v_2(0). \tag{5.106}$$

In the above equations, $v_1(s) = u_2(t-s)$ and $v_2(s) = u_1(t-s)$. The state of the system at any time t is given by

$$\rho_r^S(x_f, x_f', t) = \int dx_i dx_i' \ J_r(x_f, x_f', t; x_i, x_i', 0)\rho^S(x_i, x_i'). \tag{5.107}$$

Starting with a Gaussian initial state

$$\rho^S(x_i, x_i', 0) = \tilde{C}e^{-\xi x_i^2 + \chi x_i x_i' - \xi^* x_i'^2}, \tag{5.108}$$

we obtain

$$\rho_r^S(x_f, x_f', t) = 2\sqrt{\frac{C}{\pi}}e^{-A\Delta_f^2 - 2iB\Delta_f \Sigma_f - 4C\Sigma_f^2}, \tag{5.109}$$

where

$$A = \frac{a_{22}}{\hbar} + \frac{1}{D}\left(\frac{b_3^2}{\hbar^2}\left[\frac{(2\xi_r + \chi)}{4} + \frac{a_{11}}{\hbar}\right] + (2\xi_i + \frac{b_4}{\hbar})a_{12}b_3 - \frac{a_{12}^2}{\hbar^2}(2\xi_r - \chi)\right), \tag{5.110}$$

$$B = \frac{-b_1}{2\hbar} + \frac{1}{\hbar^2 D}\left[(\xi_i + \frac{b_4}{2\hbar})b_2 b_3 - (2\xi_r - \chi)a_{12}b_2\right], \tag{5.111}$$

$$C = \frac{1}{4\hbar^2 D}[2\xi_r - \chi]b_2^2, \tag{5.112}$$

$$D = 4|\xi|^2 - \chi^2 + \frac{4}{\hbar}(2\xi_r - \chi)a_{11} + \frac{4}{\hbar}\xi_i b_4 + \frac{b_4^2}{\hbar^2}. \tag{5.113}$$

Here $\Delta_f = x_f - x_f'$, $\Sigma_f = \frac{1}{2}(x_f + x_f')$ and $\xi = \xi_r + i\xi_i$. Now we diagonalize Eq. (5.109). We have the following eigenvalue equation

$$\int_{-\infty}^{\infty} dx_f' \ \rho_r^S(x_f, x_f', t)\varphi_n(x_f') = p_n\varphi_n(x_f). \tag{5.114}$$

We use the ansatz

$$\varphi_n(x) = N H_n\left[2(AC)^{\frac{1}{4}}x\right]e^{-[2(AC)^{\frac{1}{2}} + iB]x^2}, \tag{5.115}$$

where H_n is the Hermite polynomial. Substituting Eq. (5.115) in Eq. (5.114) and using

$$\int_{-\infty}^{\infty} dt\, e^{-(t-z)^2} H_n(\alpha t) = \pi^{\frac{1}{2}}(1-\alpha^2)^{\frac{n}{2}} H_n\left(\frac{\alpha}{(1-\alpha^2)^{\frac{1}{2}}} z\right), \qquad (5.116)$$

we see that the eigenvalue equation is satisfied by

$$p_n = \frac{2\sqrt{C}}{(\sqrt{A}+\sqrt{C})}\left(\frac{\sqrt{A}-\sqrt{C}}{\sqrt{A}+\sqrt{C}}\right)^n. \qquad (5.117)$$

Using this, we can calculate the von Neumann and linear entropies as

$$\begin{aligned}
S_{von} &= -tr\rho^S(t)\ln(\rho^S(t)) = -\sum_n p_n \ln p_n \\
&= -\ln p_0 - \frac{q}{p_0}\ln q, \qquad (5.118)
\end{aligned}$$

where $p_n = p_0 q^n$ with $p_0 = \frac{2\sqrt{C}}{(\sqrt{A}+\sqrt{C})}$ and $q = \left(\frac{\sqrt{A}-\sqrt{C}}{\sqrt{A}+\sqrt{C}}\right)$. Also,

$$\begin{aligned}
S_{lin} &= 1 - tr\rho^{S^2}(t) = 1 - \sum_n p_n^2 \\
&= 1 - \sqrt{\frac{C}{A}}. \qquad (5.119)
\end{aligned}$$

This could serve as a starting point for studying the approach to equilibrium of a dissipative harmonic oscillator.

5.3.3   Fluctuation-Dissipation Theorem

Consider the standard Brownian motion scenario, a harmonic oscillator acted upon by a force $F(t)$. Then the response of the coordinate x to the force is

$$\langle x \rangle_t = \int_0^t ds\, \chi(t-s) F(s). \qquad (5.120)$$

Here, as usual, the angular brackets signify average over an appropriate distribution. Along similar lines the equilibrium autocorrelation function is

$$C(t) = \langle x(t)x \rangle. \qquad (5.121)$$

The structure of $C(t)$ is such that it can be written as

$$C(t) = S(t) + iA(t), \qquad (5.122)$$

that is, split into a real $S(t) = 1/2\langle x(t)x + xx(t) \rangle$ and imaginary $A(t) = 1/(2i)\langle x(t)x - xx(t) \rangle$ part, the symmetric and anti-symmetric parts of the autocorrelation function, respectively [10]. The anti-symmetric part is directly related to the response function as $A(t) = -\hbar/2\chi(t)$. The symmetric and anti-symmetric parts are themselves not independent. Their dependence is the formal statement of fluctuation-dissipation theorem and can be expressed as

$$S(\omega) = \hbar \coth(\frac{\beta\hbar\omega}{2})\tilde{\chi}(\omega). \tag{5.123}$$

Here $S(\omega)$ is the Fourier transform of $S(t)$, that is, $S(\omega) = \int_{-\infty}^{\infty} dt S(t)e^{i\omega t}$ and $\tilde{\chi}(\omega) = \frac{i}{2}[\hat{\chi}(i\omega) - \hat{\chi}(-i\omega)]$, where $\hat{\chi}(\omega) = \int_0^{\infty} e^{-\omega t}\chi(t)$ is the Laplace transform of $\chi(t)$. The statement of the fluctuation-dissipation theorem presented in Eq. (5.123) is valid for arbitrary strong coupling of the linear harmonic oscillator with its reservoir.

5.4 Foundational Issues

The problem of dissipative harmonic oscillator, as should be evident by now, is ubiquitous to physics in general and open quantum systems in particular. Now we will use the theory developed so far to some pertinent problems that fall in the realms of *foundational issues*. The first will be the problem of phases in the context of quantum mechanics. This will be followed by the closely related problem of number, phase complementarity. These studies will be made in the backdrop of the Lindbladian theory of dissipative harmonic oscillators.

5.4.1 Quantum Phase Distribution

One of the earliest investigations in this direction was made by Dirac [133]. In a similar spirit, Pegg and Barnett [134], carried out a polar decomposition of the annihilation operator and defined a Hermitian phase operator in a finite-dimensional Hilbert space. However, there are some technical issues in trying to define a Hermitian phase operator in an infinite-dimensional Hilbert space [135, 136]. To circumvent this problem, the concept of phase distribution for the quantum phase was introduced [135, 137]. Here, a phase distribution is associated to a given state such that the average of a function of the phase operator in the state, computed with the phase distribution, reproduces the results of Pegg and Barnett.

We start by taking the harmonic oscillator to be the system of our interest S with the Hamiltonian

$$H_S = \hbar\omega\left(a^\dagger a + \frac{1}{2}\right). \tag{5.124}$$

The number states $\{|n\rangle\}$ serve as an appropriate basis for the system Hamiltonian and the system energy eigenvalue (5.124) in this basis is

$$E_n = \hbar\omega(n + \frac{1}{2}), \tag{5.125}$$

i.e., $H_S|n\rangle = E_n|n\rangle$. A phase distribution $\mathcal{P}(\theta)$ for a given density operator $\hat{\rho}$ associated with a state $|\theta\rangle$ was defined as [137]

$$\mathcal{P}(\theta) = \frac{1}{2\pi}\langle\theta|\rho|\theta\rangle, \; 0 \le \theta \le 2\pi,$$

$$= \frac{1}{2\pi}\sum_{m,n=0}^{\infty}\rho_{m,n}e^{i(n-m)\theta}, \tag{5.126}$$

where the states $|\theta\rangle$ are the analogues of the Susskind-Glogower [138] phase operator and are defined in terms of the number states $|n\rangle$ as

$$|\theta\rangle = \sum_{n=0}^{\infty}e^{in\theta}|n\rangle. \tag{5.127}$$

The sum in Eq. (5.126) is assumed to converge and the phase distribution normalized to unity.

Now we will obtain the quantum phase distribution of a harmonic-oscillator, $H_s = \hbar\omega(a^\dagger a + \frac{1}{2})$, in a dissipative interaction with a squeezed thermal bath. The reduced density matrix operator of the system S, in the interaction picture, is given by [29, 2]

$$\frac{d}{dt}\rho^s(t) = \gamma_0(N+1)\left(a\rho^s(t)a^\dagger - \frac{1}{2}a^\dagger a\rho^s(t) - \frac{1}{2}\rho^s(t)a^\dagger a\right)$$

$$+ \; \gamma_0 N\left(a^\dagger\rho^s(t)a - \frac{1}{2}aa^\dagger\rho^s(t) - \frac{1}{2}\rho^s(t)aa^\dagger\right)$$

$$+ \; \gamma_0 M\left(a^\dagger\rho^s(t)a^\dagger - \frac{1}{2}(a^\dagger)^2\rho^s(t) - \frac{1}{2}\rho^s(t)(a^\dagger)^2\right)$$

$$+ \; \gamma_0 M^*\left(a\rho^s(t)a - \frac{1}{2}(a)^2\rho^s(t) - \frac{1}{2}\rho^s(t)(a)^2\right). \tag{5.128}$$

This is an extension of the master equation in Eq. (97), Chapter 3 to include effects of reservoir squeezing and belongs to the family of master equations represented by Eq. (5.1) with the *RHS* given by Eq. (5.5). In the above equation, N, M are bath parameters, given below and γ_0 is a parameter which depends upon the system-bath coupling strength. The Eq. (5.128) can be solved using a variety of methods (cf. [2], [29]). However, the solutions obtained thus are not amenable to treatment of the quantum phase distribution by use of Eq. (5.126). For this purpose we again briefly detail the solution of Eq. (5.128) in an operator form [139, 140]. The following transformations are introduced [141]:

$$\rho'^s(t) = S^\dagger(\zeta)\rho^s(t)S(\zeta), \; a' = S^\dagger(\zeta)aS(\zeta), \tag{5.129}$$

where

$$S(\zeta) = e^{\frac{1}{2}(\zeta^* a^2 - \zeta a^{\dagger 2})}. \tag{5.130}$$

Using Eqs. (5.129) we get

$$a^{'} = \cosh(|\zeta|)a - \frac{\zeta}{|\zeta|}\sinh(|\zeta|)a^{\dagger}. \tag{5.131}$$

Using Eqs. (5.129) and (5.131), Eq. (5.128) gets transformed to

$$\frac{d}{dt}\rho^{'s}(t) = \left[\alpha K_{+} + \beta K_{-} + (\alpha + \beta)K_{0} + \frac{\gamma_0}{2}\right]\rho^{'s}(t), \tag{5.132}$$

where

$$\alpha = \gamma_0 N \cosh(2|\zeta|) + \gamma_0 \cosh^2(|\zeta|) - \frac{\gamma_0}{2|\zeta|}\sinh(2|\zeta|)(M\zeta^* + M^*\zeta),$$

$$\beta = \gamma_0 N \cosh(2|\zeta|) + \gamma_0 \sinh^2(|\zeta|) - \frac{\gamma_0}{2|\zeta|}\sinh(2|\zeta|)(M\zeta^* + M^*\zeta). \tag{5.133}$$

The parameters involved in the above equation need to satisfy the following consistency condition:

$$\frac{|\zeta|}{\zeta}M\coth(|\zeta|) + \frac{\zeta}{|\zeta|}M^*\tanh(|\zeta|) = 2N + 1. \tag{5.134}$$

It can be seen that

$$
\begin{aligned}
M &= \frac{1}{2}\sinh(2r)(2N_{\text{th}} + 1)e^{i\Phi}, \\
N &= N_{\text{th}}(\cosh^2(r) + \sinh^2(r)) + \sinh^2(r), \\
N_{\text{th}} &= \frac{1}{e^{\frac{\hbar\omega}{k_B T}} - 1}, \quad \zeta = re^{i\Phi},
\end{aligned} \tag{5.135}
$$

satisfy Eq. (5.134). In Eq. (5.132), K_+, K_- and K_0 are *superoperators*, i.e., operators in the space of operators, satisfying the following rules:

$$K_+\rho^{'s} = a\rho^{'s}a^{\dagger}, \; K_-\rho^{'s} = a^{\dagger}\rho^{'s}a, \; K_0\rho^{'s} = -\frac{1}{2}(a^{\dagger}a\rho^{'s} + \rho^{'s}a^{\dagger}a + \rho^{'s}). \tag{5.136}$$

These superoperators can be seen to satisfy:

$$[K_-, K_+]\rho^{'s} = 2K_0\rho^{'s}, \; [K_0, K_\pm]\rho^{'s} = \pm K_\pm\rho^{'s}, \tag{5.137}$$

which coincides with the commutation relations of the $su(1,1)$ Lie algebra. This brings out the intimate connection between the solutions of the master equation (5.128) and the generators of the $su(1,1)$ Lie algebra. Using the disentangling theorems of the $su(1,1)$ Lie algebra, basically an application of the *BCH* identities discussed in Chapter 2, Eq. (5.132) can be solved to yield:

$$\rho^{'s}(t) = e^{\frac{\gamma_0 t}{2}}e^{y_-(t)K_-}e^{\ln(y_0(t))K_0}e^{y_+(t)K_+}\rho^{'s}(0), \tag{5.138}$$

where

$$y_0(t) = \left(\frac{\alpha e^{\frac{\gamma_0 t}{2}} - \beta e^{-\frac{\gamma_0 t}{2}}}{\gamma_0} \right)^2,$$

$$y_+(t) = \frac{\alpha(e^{-\gamma_0 t} - 1)}{(\beta e^{-\gamma_0 t} - \alpha)},$$

$$y_-(t) = \frac{\beta(e^{-\gamma_0 t} - 1)}{(\beta e^{-\gamma_0 t} - \alpha)}. \tag{5.139}$$

Using Eqs. (5.138), (5.129), the solution of Eq. (5.128) can be written as

$$\rho^s(t) = S(\zeta) \left\{ e^{\frac{\gamma_0 t}{2}} e^{y_-(t) K_-} e^{\ln(y_0(t)) K_0} e^{y_+(t) K_+} S^\dagger(\zeta) \rho^s(0) S(\zeta) \right\} S^\dagger(\zeta). \tag{5.140}$$

This is the form of solution of the master equation appropriate for investigation of the quantum phase distribution. We will use a special initial state of the system, the squeezed coherent state,

$$\rho^s(0) = |\zeta, \eta\rangle\langle\eta, \zeta|, \tag{5.141}$$

where

$$|\zeta, \eta\rangle = S(\zeta)D(\eta)|0\rangle. \tag{5.142}$$

Here $|0\rangle$ is the vacuum state and $D(\eta)$ is the standard displacement operator. Substituting Eq. (5.141) in Eq. (5.140), the solution of the Eq. (5.128) starting from the initial state (5.141) is obtained as [140]

$$\rho^s(t) = \frac{1}{(1 + \tilde{\beta}(t))} e^{-\tilde{\beta}(t)|\tilde{\eta}(t)|^2} \sum_{k=0}^{\infty} \left(\frac{\tilde{\beta}(t)}{(1 + \tilde{\beta}(t))} \right)^k \frac{1}{k!}$$

$$\times \sum_{l,p=0}^{k} \binom{k}{l} \binom{k}{p} \sqrt{l!p!} (\tilde{\eta}^*(t))^{k-l} (\tilde{\eta}(t))^{k-p} |\zeta, \tilde{\eta}(t), l\rangle \langle p, \tilde{\eta}(t), \zeta|, \tag{5.143}$$

where

$$|\zeta, \tilde{\eta}(t), l\rangle = S(\zeta)|\tilde{\eta}(t), l\rangle = S(\zeta)D(\tilde{\eta}(t))|l\rangle, \tag{5.144}$$

and

$$\tilde{\beta}(t) = \frac{\beta}{\gamma_0}(1 - e^{-\gamma_0 t}), \quad \tilde{\eta}(t) = \eta \frac{e^{-\frac{\gamma_0 t}{2}}}{(1 + \tilde{\beta}(t))}. \tag{5.145}$$

Here β is given by Eq. (5.133). In Eq. (5.144), $D(\tilde{\eta}(t)) = e^{\tilde{\eta}(t)a^\dagger - \tilde{\eta}^*(t)a}$ and $D(\tilde{\eta}(t))|l\rangle$ is known as the generalized coherent state (GCS) [142, 143] and thus the state $|\zeta, \tilde{\eta}(t), l\rangle$ would be the generalized squeezed coherent state (GSCS) [143]. It can be seen from Eqs. (5.143) and (5.141) that under the action of the master equation (5.128), a harmonic oscillator starting in a squeezed coherent state ends in a mixture that can be expressed as a sum over GSCS. Thus the above case can be thought of as a concrete physical realization of GSCS.

Making use of the Fock-space representation of GCS [142]

$$|n, \alpha(t)\rangle = e^{-\frac{|\alpha(t)|^2}{2}} \sum_{l=0}^{\infty} \left(\frac{n!}{l!}\right)^{\frac{1}{2}} L_n^{l-n}(|\alpha(t)|^2)[\alpha(t)]^{l-n}|l\rangle, \qquad (5.146)$$

where $L_n^{l-n}(x)$ is the generalized Laguerre polynomial, and substituting Eq. (5.143) in Eq. (5.126), reverting back to the Schrödinger picture, we obtain the quantum phase distribution of a dissipative harmonic oscillator starting in a squeezed coherent state (5.141) as

$$\mathcal{P}(\theta) = \frac{1}{2\pi} e^{-|\tilde{\eta}(t)|^2} \frac{e^{-\tilde{\beta}(t)|\tilde{\eta}(t)|^2}}{(1 + \tilde{\beta}(t))} \sum_{m,n} e^{-i\omega(m-n)t} e^{i(n-m)\theta}$$

$$\times \sum_{u,v,k} G_{u,m}^*(\zeta) G_{v,n}(\zeta) \left(\frac{\tilde{\beta}(t)}{(1 + \tilde{\beta}(t))}\right)^k \frac{1}{k!} \sum_{l,p=0}^{k} \binom{k}{l}\binom{k}{p}$$

$$\times \frac{l!p!}{\sqrt{(u!v!)}} (\tilde{\eta}^*(t))^{v-p+k-l} (\tilde{\eta}(t))^{u-l+k-p} L_l^{u-l}(|\tilde{\eta}(t)|^2) L_p^{*v-p}(|\tilde{\eta}(t)|^2).$$

$$(5.147)$$

In the above equation, $G_{m,n}(\zeta) = \langle m|S(\zeta)|n\rangle$ and is explicitly given, with $\zeta = r_1 e^{i\phi}$, as [143]

$$G_{2m,2p} = \frac{(-1)^p}{(p)!(m)!} \left(\frac{(2p)!(2m)!}{\cosh(r_1)}\right)^{\frac{1}{2}} \exp\left(i(m-p)\phi\right)$$

$$\times \left(\frac{\tanh(r_1)}{2}\right)^{(m+p)} F_1^2\left[-p, -m; \frac{1}{2}; -\frac{1}{(\sinh(r_1))^2}\right]. \qquad (5.148)$$

Similarly $G_{2m+1,2p+1}(\zeta)$ is given by

$$G_{2m+1,2p+1} = \frac{(-1)^p}{(p)!(m)!} \left(\frac{(2p+1)!(2m+1)!}{\cosh^3(r_1)}\right)^{\frac{1}{2}} \exp\left(i(m-p)\phi\right)$$

$$\times \left(\frac{\tanh(r_1)}{2}\right)^{(m+p)} F_1^2\left[-p, -m; \frac{3}{2}; -\frac{1}{(\sinh(r_1))^2}\right]. \qquad (5.149)$$

$G_{m,n}$ is nonzero only for either m, n both even or both odd. For convenience it is sometimes assumed that ϕ is zero and $z = r_1$ is real. Here $r_1 = r$, due to the initial condition (5.141) and F_1^2 is the Gauss hypergeometric function [144].

In Fig. (5.1), we depict the quantum phase distributions $\mathcal{P}(\theta)$ for a harmonic oscillator system starting in a squeezed coherent state (5.141), for dissipative system-bath interaction (Eq. (5.147)). The phase distribution is normalized.

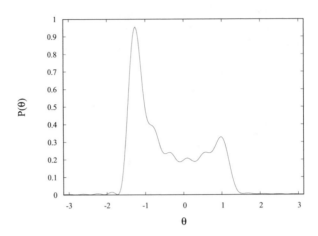

Figure 5.1: Quantum phase distributions $\mathcal{P}(\theta)$, for a harmonic oscillator system starting in a squeezed coherent state, for dissipative system-bath interaction (Eq. (5.147)). Temperature (in units where $\hbar \equiv k_B \equiv 1$) $T = 0$, the squeezing parameters $r = r_1 = 1$, bath exposure time $t = 0.1$, $\gamma_0 = 0.025$, $\omega = 1$ and $\Phi = 0$.

5.4.2 Number-Phase Complementarity

Two observables A and B of a d-level system are called complementary if knowledge of the measured value of A implies maximal uncertainty of the measured value of B, and vice versa [145]. Complementarity is an aspect of the Heisenberg uncertainty principle, which says that for any state ψ, the probability distributions obtained by measuring A and B cannot both be arbitrarily peaked if A and B are sufficiently non-commuting. Expressed in terms of measurement entropy the Heisenberg uncertainty principle takes the form:

$$H(A) + H(B) \geq \log d, \qquad (5.150)$$

where $H(A)$ and $H(B)$ are the Shannon entropy of the measurement outcomes of a d-level quantum system [38, 146, 147].

An extension of Eq. (5.150) to the case where A or B is not discrete is considered in [148]. The problem that Shannon entropy of a continuous random variable may be negative is circumvented by using relative entropy (also called Kullbäck-Leibler divergence, which is always positive) [149] with respect to a uniform distribution. This quantity is a measure of knowledge [148]. This generalization of the entropic uncertainty principle to cover discrete-continuous systems still suffers from the restriction that the system must be finite dimensional, since in the case of an infinite-dimensional system, such as an oscillator, entropic knowledge of the number distribution can diverge, making it unsuitable for infinite-dimensional systems. Therefore to set up an entropic version of the uncertainty principle, that unifies and is applicable to all systems, including infinite dimensional and/or continuous-variable systems, it may be advantageous to use a combination of entropy and knowledge, in particular, the difference between entropy of the discrete, infinite observable and between phase knowledge.

We have already discussed the quantum phase distribution, Eq. (5.126). Its complementary number distribution is

$$p(m) = \langle m|\rho|m \rangle, \tag{5.151}$$

where $|m\rangle$ is the number (Fock) state.

We define entropic knowledge $R[f]$ of random variable f as its relative entropy with respect to the uniform distribution $\frac{1}{d}$, i.e.,

$$R[f] \equiv S\left(f(j)||\frac{1}{d}\right) = \sum_j f(j)\log(df(j)). \tag{5.152}$$

This enables the recasting of the Heisenberg uncertainty principle in terms of entropy H and knowledge R, as

$$X(A, B) \equiv H(A) - R(B) \geq 0. \tag{5.153}$$

Here A and B are two Hermitian observables *w.r.t.* which we wish to study the complementary behaviour. $X(A, B)$ could be called the *entropy excess*. For an infinite dimensional system such as a harmonic oscillator, the problem of interest here would be to study the entropy excess of the harmonic oscillator undergoing a dissipative interaction. To this effect we define entropic knowledge by the functional [140]

$$R[\mathcal{P}(\theta)] = \int_0^{2\pi} d\theta \, \mathcal{P}(\theta)\log[2\pi\mathcal{P}(\theta)], \tag{5.154}$$

where the $\log(\cdot)$ refers to the binary base.

It is at first not obvious that Eq. (5.153) holds for infinite dimensional systems. Based on a result due to [150] for an oscillator system, which in turn uses the concept of the (p, q)-norm of the Fourier transformation [151] for all values of p, for an oscillator system, it can be shown that it is indeed the case. In particular,

$$-\int_{-\pi}^{\pi} d\theta P(\theta)\log(P(\theta)) - \sum_{m=0}^{\infty} p_m\log(p_m) \geq \log(2\pi). \tag{5.155}$$

Setting the 'number variable' m in Eq. (5.155) as A, and the phase variable θ as B, and noting that the first term in the *L.H.S* of Eq. (5.155), using Eq. (5.154), is just $\log(2\pi) - R[P(\theta)]$, we obtain

$$X[m, \theta] \equiv H[m] - R[\theta] \geq 0. \tag{5.156}$$

Hence, $X \geq 0$ is a description of the Heisenberg uncertainty relation applied to an infinite-dimensional system.

Principle of entropy excess applied to a dissipative harmonic oscillator:

The initial state of the system is a superposition of coherent states which are 180° out of phase with respect to each other [152].

$$|\psi\rangle = A^{1/2}(|\alpha\rangle + e^{i\phi}|-\alpha\rangle), \tag{5.157}$$

where $\alpha = |\alpha|e^{i\phi_0}$ and

$$A = \frac{1}{2}[1 + \cos(\phi)e^{-2|\alpha|^2}]^{-1}. \tag{5.158}$$

The state $|\psi\rangle$ for $\phi = 0$ would be an even coherent state and for $\phi = \pi$ would be an odd coherent state. The reduced density matrix can be shown to have the following form [153]

$$\rho(t) = \sum_{n,m=0}^{\infty} \rho_{n,m}(t)|n\rangle\langle m|, \tag{5.159}$$

where

$$\rho_{n,m}(t) = \frac{A}{N(t)+1} \left(\frac{e^{-\gamma_0 t/2}}{N(t)+1}\right)^{m+n} Q_n Q_m e^{i(n-m)\phi_0}$$

$$\times \sum_{l=0}^{\infty} \left(1 - \frac{e^{-\gamma_0 t/2}}{N(t)+1}\right)^l \frac{|\alpha|^{2l}}{l!} \left(1 + (-1)^{n+m} + (-1)^l[(-1)^n e^{i\phi} + (-1)^m e^{-i\phi}]\right)$$

$$\times F_1^2 \left[-m, -n; l+1; 4N(t)(N(t)+1)(\sinh(\gamma_0 t/2))^2\right]. \tag{5.160}$$

Here F_1^2 is the Gauss hypergeometric function [144], γ_0 is a parameter which depends upon the system-reservoir coupling strength,

$$Q_n = \frac{|\alpha|^n}{\sqrt{n!}} e^{-\frac{|\alpha|^2}{2}}, \tag{5.161}$$

and,

$$N(t) = N_{th}(1 - e^{-\gamma_0 t}), \quad N_{th} = \left(e^{\frac{\hbar\omega}{k_B T}} - 1\right)^{-1}. \tag{5.162}$$

The phase distribution is given by

$$\mathcal{P}(\theta) = \frac{1}{2\pi} \sum_{m,n=0}^{\infty} \rho_{m,n} e^{i(n-m)\theta}, \tag{5.163}$$

where $\rho_{m,n}$ can be obtained from Eq. (5.160).

The corresponding complementary number distribution is obtained, using Eq. (5.151), as

$$p(m) = \rho_{m,m}(t), \tag{5.164}$$

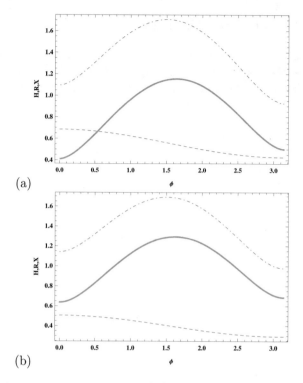

Figure 5.2: Plot of number entropy $H[m]$ (dot-dashed line), phase knowledge $R[\theta]$ (dashed line) and entropy excess $X[m,\theta]$ (bold line) for a harmonic oscillator, initially in a coherent state superposition (5.157), as a function of the state parameter ϕ (5.157). Figure (a) pertains to the pure state case. Figure (b) represents the system subjected to a dissipative interaction with the environment for an evolution time $t = 0.1$ and temperature $T = 2$. The parameters used are $\omega = 1.0$, $\gamma_0 = 0.026$, $|\alpha|^2 = 2.2$, ϕ_0 (5.157) $= 0$.

where $\rho_{m,m}$ is as in Eq. (5.160).

Using $\mathcal{P}(\theta)$ (5.163) in Eq. (5.154) to get the phase knowledge, $p(m)$ (5.164) to get the number entropy and using these in Eq. (5.156) we get the entropy excess which are plotted in Figs. (5.2). Figure (5.2) (a), pertains to unitary evolution, and depicts the variation of number entropy, phase knowledge and entropy excess as a function of the parameter ϕ ranging from the even cat (coherent) state ($\phi = 0$), to the odd cat (coherent) state ($\phi = \pi$). The Fig. (5.2) (b) shows that the effect of the dissipative environment causes phase to become randomized, leading to an increased entropy excess at all ϕ (5.157). The principle of entropy excess, Eq. (5.156), is clearly seen to be satisfied, for both unitary as well as dissipative evolution [140].

5.5   Guide to advanced literature

In this chapter we have studied the problem of dissipative harmonic oscillator. This is one of the *workhorses* of open quantum systems. As already pointed out, this problem is of relevance to a large number of investigations spanning many domains in physics, from nuclear to atomic physics and quantum optics. It is also very pertinent to the study of mesoscopic phenomena. The interested reader can, after digesting the material in this chapter, go on to any specific subject of his/her liking. Most books on quantum optics give a detailed exposure to Lindbladian master equations and various techniques for handling them, see for e.g., [29]. Two very good references in this regard would be [2, 1]. The later book [1] is particularly suited for someone interested in pursuing the relevance of the problem studied in this chapter to the mesoscopic regime.

Chapter 6

Dissipative Two-State System

6.1 Introduction

A number of phenomena in nature can be explained by means of a two-state model wherein a particle tunnels between two different localized states. Such a system is usually strongly affected by the ambient environment. The existence of two-level systems in glasses was proposed in [154, 155] in order to help understand low temperature anomalies of specific heat in them. The problem discussed in this chapter has relevance to the tunneling of light particles in solids. It is also relevant to studies related to the Kondo effect [156] which points out the anomalous temperature dependence of muon diffusion in host metals at low temperatures. Electron transfer reactions [157] are abundant in chemical and biological systems. An electron localized at a donor site tunnels to the acceptor site. This is influenced by the environment and can be described by a variant of the spin-Boson model, the subject of this chapter. Further examples where the discussions in this chapter would be of relevance are the inversion resonance of the NH_3 molecule, strangeness oscillations of neutral K meson, a topic to which we will return to in a later chapter, rf SQUID ring threaded by an external flux near half a flux quantum. The last example has been used to discuss macroscopic quantum coherence [68].

In this chapter, we take up the problem of the dissipative two-level system. This is sometimes known as the spin-Boson problem. After introducing the spin-Boson model we take up examples of two-state systems, qubits, based on Josephson tunnel junctions. We then discuss the thermodynamics and subsequently the dynamics of the spin-Boson model. Here we principally use influence functional techniques, introduced earlier in Chapter IV. We also discuss, briefly, an approximation of the exact dynamics, the noninteracting-blip approximation. A quantum mechanical model for electron transfer using a two-level system coupled to an intermediate harmonic oscillator, representing the reaction coordinate, which in turn is coupled to a bath of harmonic oscillators is then discussed with a specific example worked out. This is then used to obtain the

© Hindustan Book Agency 2018 and Springer Nature Singapore Pte Ltd. 2018
S. Banerjee, *Open Quantum Systems*, Texts and Readings in Physical Sciences 20,
https://doi.org/10.1007/978-981-13-3182-4_6

analytical expression of the asymptotic value of the operator corresponding to the population.

6.2 Spin-Boson Model

Two-level models find many applications in the natural sciences and are extensively studied in the literature [1]. The system is described by a generalized co-ordinate, for example, a spin, in an effective potential energy with two separate minima, say in a double-well system, hence the nomenclature *two-state system* (TSS). For thermal energy lesser than the spacing between the states, only the ground states of the two wells are involved, resulting in the dynamics taking place in a two-dimensional (2-*d*) Hilbert space. The TSS is frequently used to exhibit quantum interference effects and the dissipative TSS model has been used to understand the concept of macroscopic quantum coherence [68].

In the nomenclature of quantum information [38], a two-level system is called a *qubit* and could be envisaged as a particle of spin-1/2. A realistic scenario would involve its interaction with the ambient environment, the subject we are dealing with in this book. In a number of cases, the environment can be thought of as a bosonic bath of harmonic oscillators. Then, the model describing the dynamics of the TSS with a bosonic bath (environment/reservoir) is called the *spin-Boson model* [158].

6.2.1 Hamiltonian

The problem under consideration could be thought of as truncation of a spatially extended double-well system to a TSS, see Fig. (6.1). In the figure, ζ and Δ_0 denote the detuning energy, characterizing the asymmetric (biased) double-well, and tunnel splitting, of the symmetric double-well, respectively. In the semiclassical limit, the tunnel splitting is determined by the action coming from the instanton path. The notion of instanton is introduced in the next chapter where it is seen to play a crucial role in quantum tunneling. Further details of this method can be obtained from [159, 160]. For a standard double-well potential of the quartic form $V(x) = \frac{M\omega_0^2}{2x_0^2}\left(x^2 - x_0^2/4\right)^2/2x_0^2$, the tunnel splitting term Δ_0 can be shown to depend on the barrier height $V_b = M\omega_0^2 x_0^2/32$. For

$$V_b \gg \hbar\omega_0 \gg \hbar\Delta_0, \hbar|\zeta|, k_B T, \qquad (6.1)$$

the system gets effectively restricted to a $2 - d$ Hilbert space spanned by, for example, the basis $\{|\pm\rangle\} \equiv \{|R\rangle, |L\rangle\}$, i.e., states localized in the right and left wells, respectively. Since the Pauli matrices, along with identity operator, form a natural basis for a $2 - d$ Hilbert space, it is natural to express the corresponding Hamiltonian of the TSS in terms of the Pauli matrices, in the computational basis [38], as

$$H_{TSS} = -\frac{\hbar}{2}\Delta_0\sigma_x - \frac{\hbar}{2}\zeta\sigma_z = \frac{\hbar}{2}\begin{pmatrix} -\zeta & -\Delta_0 \\ -\Delta_0 & \zeta \end{pmatrix}. \qquad (6.2)$$

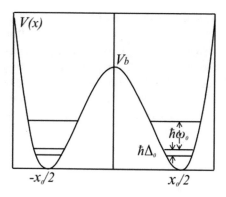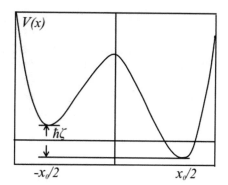

Figure 6.1: Double Well representing the TSS; (a). Symmetric, (b). Asymmetric Well. Figure adapted from [1].

The *generalized* position operator would be $\hat{x} = \frac{\hbar}{2}x_0\sigma_z$ with the eigenvalues $\pm\frac{1}{2}x_0$ corresponding to the positions of the localized states. The localized states $|R\rangle$ and $|L\rangle$ are related to the eigenstates $|e\rangle$ (excited) and $|g\rangle$ (ground) states of H_{TSS} (6.2) as

$$\begin{aligned} |R\rangle &= \cos\left(\frac{\theta}{2}\right)|g\rangle + \sin\left(\frac{\theta}{2}\right)|e\rangle, \\ |L\rangle &= \sin\left(\frac{\theta}{2}\right)|g\rangle - \cos\left(\frac{\theta}{2}\right)|e\rangle. \end{aligned} \tag{6.3}$$

Here

$$\sin(\theta) = \frac{\Delta_0}{\Delta_b}, \quad \cos(\theta) = \frac{\zeta}{\Delta_b}, \quad \tan(\theta) = \frac{\Delta_0}{\zeta}. \tag{6.4}$$

The tunnel splitting energy of the biased TSS is

$$E_e - E_g = \hbar\Delta_b = \hbar\sqrt{\Delta_0^2 + \zeta^2}. \tag{6.5}$$

The interaction of the above defined TSS with a bath can be modelled as

$$H_{SR} = -\sigma_z\mathcal{P}(t), \quad \mathcal{P}(t) = \frac{x_0}{2}\sum_{i=1}^{N} c_i q_i(t). \tag{6.6}$$

Let us try to motivate the form of the above interaction. Let $|\psi_\pm\rangle$ be the eigenstates (up to order Δ_0/ω_0) in respectively the right and the left wells. As σ_x and σ_y have only non-diagonal nonzero elements, they change $|\psi_\pm\rangle$ to $|\psi_\mp\rangle$. Thus, any interaction proportional to σ_x and σ_y will also be proportional to the overlap of $|\psi_+\rangle$ and $|\psi_-\rangle$. This will be of order $\hbar\Delta_0$ and hence relatively small. Only the interaction containing σ_z is directly proportional to $|\psi_+\rangle$ and $|\psi_-\rangle$.

Here the interaction could be envisaged as a coupling of the heat bath to the position of the TSS, represented by a collective bath mode $\mathcal{P}(t)$. The interaction with the bath results in a fluctuating force $\sum_{i=1}^{N} c_i q_i(t)$ which causes a fluctuation in the double-well bias and could be called the *polarization energy* $\mathcal{P}(t)$. If the bath stochastics follow a Gaussian distribution then it can be modelled by a Bosonic bath of harmonic oscillators, discussed in detail in the earlier chapters. The Hamiltonian of the TSS coupled to its bath via the interaction term in Eq. (6.6) is

$$
\begin{aligned}
H &= -\frac{\hbar}{2}\Delta_0\sigma_x - \frac{\hbar}{2}\zeta\sigma_z - \sigma_z\mathcal{P}(t) + H_R \\
&= -\frac{\hbar}{2}\Delta_0\sigma_x - \frac{\hbar}{2}\zeta\sigma_z + \frac{1}{2}\sum_i \left(\frac{p_i^2}{m_i} + m_i\omega_i^2 q_i^2 - x_0\sigma_z c_i q_i \right). \quad (6.7)
\end{aligned}
$$

This Hamiltonian is known as the *spin-Boson* Hamiltonian. Here H_R is the usual bath of harmmonic oscillators. In the eigenbasis of the TSS Hamiltonian (6.2), this can be expressed as

$$
H = -\frac{\hbar}{2}\Delta_b\sigma_z + \left(\cos(\theta)\sigma_z - \sin(\theta)\sigma_x \right)\mathcal{P}(t) + H_R. \quad (6.8)
$$

In this form it is clear that the system-bath coupling has a transverse and a longitudnal part, i.e., the terms proportional to $\sin(\theta)$ and $\cos(\theta)$, respectively. Due to the nature of the Pauli operators, only the transverse part can induce spin flips. Resorting to second quantization of the harmonic oscillator, i.e., expressing the harmonic oscillator coordinates in terms of the creation and annihilation operators, see Chapter II (Section 5), the spin-Boson Hamiltonian can be expressed as

$$
H = H_{TSS} - \frac{1}{2}\sigma_z \sum_{i=1}^{N} \hbar\lambda_i(a_i + a_i^\dagger) + \sum_{i=1}^{N} \hbar\omega_i a_i^\dagger a_i. \quad (6.9)
$$

The effect of the bath is quantified by the spectral density of the coupling I_{SB}

$$
I_{SB}(\omega) = \sum_{i=1}^{N} \lambda_i^2 \delta(\omega - \omega_i) = \frac{x_0^2}{2\hbar} \sum_{i=1}^{N} \frac{c_i^2}{m_i\omega_i} \delta(\omega - \omega_i) = \frac{x_0^2}{\pi\hbar} I(\omega). \quad (6.10)
$$

Here $I(\omega)$ is the usual spectral density of a continuous model of the environment.

6.2.2   Shifted Oscillators

In the adiabatic limit, the modes of heat bath instantaneously adapt to the position of the particle. Thus, the oscillator part of the spin-Boson Hamiltonian,

Eq. (6.7), for $\sigma_z = \pm 1$ becomes

$$
\begin{aligned}
H_{osc\pm} &= \frac{1}{2} \sum_i \left(\frac{p_i^2}{m_i} + m_i \omega_i^2 q_i^2 \mp x_0 c_i q_i \right) \\
&= \sum_i \left(\hbar \omega_i a_i^\dagger a_i \mp \frac{1}{2} \hbar \lambda_i (a_i + a_i^\dagger) \right).
\end{aligned}
\tag{6.11}
$$

The relation between λ_i and a_i can be easily seen from Eq. (6.10). Introducing the shifted operators

$$
a_{\pm,i} = a_i \mp \frac{\lambda_i}{2\omega_i},
\tag{6.12}
$$

the terms linear in the creation and annihilation operators, in Eq. (6.11), cancel and $H_{osc\pm}$ can be written in a normal ordered form as

$$
H_{osc\pm} = \sum_i H_{\pm,i} \quad \text{with} \quad H_{\pm,i} = \hbar \left(\omega_i a_{\pm,i}^\dagger a_{\pm,i} - \frac{\lambda_i^2}{4\omega_i} \right).
\tag{6.13}
$$

As the shifted operators obey the same commutation relations as the original ones, these are examples of canonical transformations [161, 162]. The vacuum of $H_{\pm,i}$ can be transformed into each other via

$$
|0_{\pm,i}\rangle = e^{i\Omega_{\mp,i}} |0_{\mp,i}\rangle \quad \text{with} \quad \Omega_{\mp,i} = \pm i \frac{\lambda_i}{\omega_i} \left(a_{\mp,i}^\dagger - a_{\mp,i} \right).
\tag{6.14}
$$

By making use of the commutation relation between $a_{\pm,i}$ and $a_{\pm,i}^\dagger$, the vacuum of one shifted oscillator can be seen to be related to the coherent state [29] of the other shifted oscillator

$$
|0_{\pm,i}\rangle = \exp\left(-\frac{\lambda_i^2}{2\omega_i^2} \right) \sum_{n=0}^{\infty} \frac{(\pm 1)^n}{\sqrt{n!}} \left(\frac{\lambda_i}{\omega_i} \right)^n |n_{\mp,i}\rangle.
\tag{6.15}
$$

Problem 1: Derive Eq. (6.15).

This relation is useful as it enables the computation of useful quantities. Thus, for e.g., the probability to excite n Bosons with energy $\hbar \omega_i$ in a sudden transition from the ground state at $\sigma_z = -1$ to the state $\sigma_z = 1$ is given by

$$
p_{n,i} = |\langle 0_{-,i} | n_{+,i} \rangle|^2 = \exp\left(-\frac{\lambda_i^2}{\omega_i^2} \right) \frac{1}{n!} \left(\frac{\lambda_i}{\omega_i} \right)^{2n}.
\tag{6.16}
$$

This is a Poissonian distribution with mean particle number $\sum_i n p_{n,i} = \lambda_i^2 / \omega_i^2$.

6.2.3 Polaron Transformation

In a number of applications , it is useful to transform the spin-Boson Hamiltonian to a basis of dressed states. This is effected by a unitary transformation

$$U = \exp\left(-\frac{1}{2}i\sigma_z\Omega\right), \quad \text{where} \quad \Omega = i\sum_i \frac{\lambda_i}{\omega_i}\left(a_i^\dagger - a_i\right). \tag{6.17}$$

The *polaron* transformation $\tilde{H} = U^\dagger H U$ diagonalizes the last three terms in the Hamiltonian Eq. (6.9) resulting in

$$\tilde{H} = -\frac{\hbar\Delta}{2}\left(|R\rangle\langle L|e^{i\Omega} + |L\rangle\langle R|e^{-i\Omega}\right) - \frac{\hbar\zeta}{2}\sigma_z + \sum_i \hbar\omega_i a_i^\dagger a_i. \tag{6.18}$$

Here $|R\rangle$ and $|L\rangle$ are as in Eq. (6.3). Instead of transforming the Hamiltonian one may transform instead the tunneling operator as $\tilde{\sigma}_x = U\sigma_x U^\dagger$. Physically, this transformation entails the transformation of the particle from one localized state to the other and the reservoir oscillators get shifted by $x_0 c_i/m_i\omega_i^2$. In the dressed basis, the particle could be imagined to drag behind it a polaronic cloud.

6.3 Examples of two-state systems based on Josephson tunnel junctions

Superconducting circuits [163] based on the Josephson tunnel junctions could be considered as building blocks of quantum computing devices as well as for probing into issues related to macroscopic quantum coherence. This is a vast field in its own right and we will be able to provide here only a very rudimentary treatment. Qubits are quantum mechanical systems where a particular degree of freedom can be tuned in such a way that the system can be effectively treated as a TSS (the quantum equivalent of the classical 'on' and 'off' states). In the case of superconducting circuits, there are three degrees of freedom that can be used to make different qubits,

- *Charge* in the superconducting island of a cooper pair box,
- *Flux* enclosed in a superconducting loop,
- *Phase* difference across a Josephson junction.

Corresponding to these degrees of freedom there are three types of superconducting qubits:

  i Charge Qubit,

  ii Flux Qubit,

  iii Phase Qubit.

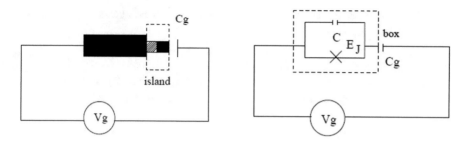

Figure 6.2: Circuit representations of the CPB.

The circuits are tuned in specific ways in order to use these different degrees of freedom; these tuning conditions are specified in terms of the unit energy parameters E_C and E_J, the single electron charging energy and the Josephson coupling energy, respectively. The basic features of these qubits are outlined below with one simple implementation of each qubit.

6.3.1 Charge Qubit

The charge qubit is the first superconducting qubit to have been implemented. The charge qubit is implemented on a Cooper pair box (CPB) circuit, see Fig. (6.2). The qubit Hamiltonian can be arrived at using the quantized version of the classical Hamiltonian. The classical Hamiltonian is given by,

$$H(\delta, p_*) = \frac{1}{2 C_\Sigma} \left(\frac{q}{\hbar}\right)^2 p_*^2 - E_J \cos \delta. \tag{6.19}$$

Here,

$$p_* = \frac{\hbar C_\Sigma}{q} \left(\frac{\hbar}{q}\dot{\delta} - \frac{C_g}{C_\Sigma} V_g\right), \quad \text{and can be rewritten as}$$

$$p_* = \hbar\left(\hat{n} - n_g\right) \text{ ; where the gate charge } n_g = \frac{\hbar}{q} C_g V_g.$$

Further, q is the electron charge, $C_\Sigma = C + C_g$, see Fig. (6.2), δ is the phase difference of the Cooper pair wave function across the junction and $\hat{n}$ is the number operator of excess Cooper pair charges on the island, conjugate to the phase δ. The quantized Hamiltonian can now be written as,

$$\hat{H}_{CPB}(\delta, n) = E_C(\hat{n} - n_g)^2 - E_J \cos \delta. \tag{6.20}$$

Here E_C and E_J are the charging energy of the junction and Josephson energy, respectively. $\hat{n}$ is a number operator and hence can take only integral eigenvalues. Let $|n\rangle$ denote the eigenstate corresponding to the eigenvalue $n : \forall n \in \mathbb{Z}$. Also, for a shift in phase of the state by 2π, the state should not change. The periodicity condition implies that the wave function belongs to the Hilbert space

$\mathcal{L}^2(0, 2\pi)$. Normalizing the wave function over this space allows the eigenstates to be represented as

$$|n\rangle = \frac{1}{\sqrt{2\pi}} e^{in\delta}. \tag{6.21}$$

It can be observed that under the condition that $E_C \gg E_J$, for the value of $n_g = 0.5$, the energy of states $|0\rangle$ and $|1\rangle$ are almost equal, i.e.,

$$\hat{H}_{CPB} |0\rangle = \left(E_C(0 - 0.5)^2 - E_J \cos\delta\right) |0\rangle \approx \frac{1}{4} E_C |0\rangle,$$

$$\hat{H}_{CPB} |1\rangle = \left(E_C(1 - 0.5)^2 - E_J \cos\delta\right) |1\rangle \approx \frac{1}{4} E_C |1\rangle.$$

This implies that $|0\rangle$ and $|1\rangle$ form a stable, degenerate two-state system (TSS) and hence froms a qubit. Therefore, evluating the projected Hamiltonian of the Cooper pair box (CPB) on the space $V = \{|0\rangle, |1\rangle\}$, we find that the projected Hamiltonian $P_{H_{CPB}}$ takes the following form

$$P_{H_{CPB}} = \frac{1}{4} E_C \begin{bmatrix} 1 & 0 \\ 0 & 1 \end{bmatrix} + \begin{bmatrix} E_C(n_g - 0.5) & -\frac{1}{2} E_J \\ -\frac{1}{2} E_J & E_C(n_g - 0.5) \end{bmatrix}.$$

The first term being the degenerate energy of the two states, the effective Hamiltonian can be written in terms of the Pauli spin matrices as,

$$\bar{P}_{H_{CPB}} = E_C(n_g - 0.5)\sigma_z - \frac{1}{2} E_J \sigma_x. \tag{6.22}$$

This is formally equivalent to the two-state Hamiltonian in Eq. (6.2). Here the Pauli matrices are defined as

$$\sigma_x = \begin{pmatrix} 0 & 1 \\ 1 & 0 \end{pmatrix} ; \quad \sigma_y = \begin{pmatrix} 0 & -i \\ i & 0 \end{pmatrix} ; \quad \sigma_z = \begin{pmatrix} 1 & 0 \\ 0 & -1 \end{pmatrix}.$$

The major disadvantage of the charge qubit system is that it is extremely sensitive to charge noise, i.e., the noise of the biasing voltage. The major source of decoherence is the low frequency $1/f$ noise. This problem is overcome by working in the $E_C \geq E_J$ limit, and has been achieved in the *quantronium* qubit.

6.3.2   Flux Qubit

As the name suggests, the flux qubit uses the fact that the flux through a superconducting loop is quantized and the two-state system in this case is in the eigenstates of the flux values. The flux qubit is implemented through an rf(ac)-SQUID circuit, see Fig. (6.3). The quantized version of the rf-SQUID Hamiltonian is

$$\hat{H}_{rfS} = E_C \hat{n}^2 - E_J \cos\delta + E_L \frac{(\delta - \delta_e)^2}{2}. \tag{6.23}$$

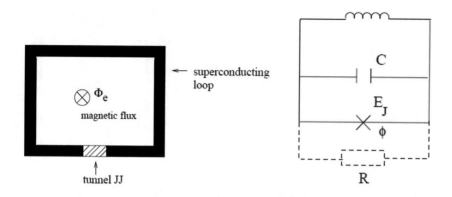

Figure 6.3: Circuit representations of the rf-SQuID.

Here E_L is the energy in the inductor while $(\delta - \delta_e)$ is related to the current through the inductor I_L as $I_L = \frac{\hbar}{q}(\delta - \delta_e)$. Also, $\delta_e = \frac{q}{\hbar}\Phi_e$ is the phase corresponding to the external flux Φ_e through the loop. The flux cutting through the ring should obey the following relation due to the phenomena of flux quantization in a superconducting loop

$$\left(\frac{\Phi_o}{2\pi}\right)\delta + \Phi_e + \Phi_{ind} = m\Phi_o.$$

Here e and ind stand for external and induced, respectively, and $\Phi_o = \frac{2\pi\hbar}{q}$ is the basic quanta of flux through a superconducting loop. In the limit that $E_J \gg E_L$ the effect of the offset charges can be neglected. When the value of δ_e is at π, the effective Hamiltonian (the last two terms of the Hamiltonian) span over two-states of minimum energy that are almost degenerate in energy. These correspond to the two states when the current is flowing in clockwise and counter-clockwise directions.

However, the major difficulty with such a system is that for the system to be close to degeneracy, the value of E_J of the Josephson junction and the value of the self Inductance L should be very high. This will, in turn, make tunnelling between the two states difficult. Also, large inductance implies large magnetic fields which are extremely liable to couple to the environment and hence lead to dephasing of the system. This problem could be overcome by replacing a single rf-SQUID by a loop containing three rf-SQUIDs. Such a structure leads to an energy profile where tunnelling can be made easy, the environmental coupling low and hence to long coherence times.

6.3.3 Phase Qubit

The phase qubit is formed from a simple current biased Josephson junction, see Fig. (6.4). It works in the same limit as that of the flux qubit, $E_J \gg E_C$.

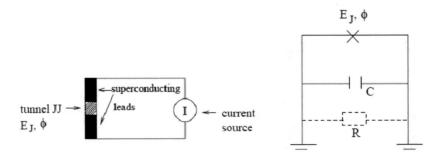

Figure 6.4: Circuit representations of the current-biased Josephson junction.

The quantized version of the corresponding Hamiltonian is

$$\hat{H}_{CBJJ} = E_C\,\hat{n}^2 + E_J\cos\delta - \frac{\hbar}{q}\,I_{bias}\,\delta$$

$$= E_C\,\hat{n}^2 + E_J\left(\cos\delta - \frac{I_{bias}}{I_o}\,\delta\right). \tag{6.24}$$

For $I_{bias} > I_o$, the resistive component of the circuit dominates and the system becomes dissipative with a non-zero value for $\langle\dot{\delta}\rangle$. However, in the sub-critical domain $I_{bias} \approx 0.95 I_o$, the potential follows a *washboard potential* with a few quantum levels at each minima. Along with this if the phase difference is around $\pi/2$, the potential energy of the system can be approximated by a cubic oscillator. Then the effective Hamiltonian reduces to a spin-$\frac{1}{2}$ field in the z- direction.

$$\hat{H}_{PB} = -\frac{\hbar\omega_{01}}{2}\,\sigma_z. \tag{6.25}$$

The most significant advantage of the phase qubit is its simplicity of design and scalability. Though it is insensitive to both charge and flux noise, it is sensitive to low frequency noise in the bias and critical current values.

6.4 Thermodynamics

In the spin-Boson model, the transitions between the two states are sudden. Thus, the *spin* path $x(\tau) = \frac{1}{2}x_0\sigma(\tau)$ for the TSS jumps between $\pm x_0/2$. Therefore, the path $\sigma(\tau)$ jumps between the values ± 1. A very useful representation of the path for $2n$ alternating spin orientations, consisting of n kink-anti-kink pairs, is

$$\sigma^{(n)}(\tau) = 1 + 2\sum_{i=1}^{2n}(-1)^i\Theta(\tau - s_i). \tag{6.26}$$

Here Θ is the step function and the path centers are in chronological order, i.e., $s_{i+1} > s_i$. Substituting the spin path $x(\tau) = \frac{1}{2}x_0\sigma(\tau)$ into the (thermal/imaginary time) Euclidean influence functional $F^E(\sigma)$, see Eq. (23) in Chapter IV, we find

$$F^E(\sigma) = \exp\left\{ -\frac{1}{2} \int_0^{\hbar\beta} d\tau \int_0^\tau d\tau' \mathcal{K}(\tau - \tau')(\sigma(\tau) - \sigma(\tau'))^2/4 \right\}, \qquad (6.27)$$

where

$$\mathcal{K}(\tau) = \frac{x_0^2}{\hbar} K(\tau), \quad \text{and} \quad K(\tau) = \frac{1}{\pi} \int_0^\infty d\omega I(\omega) G_\omega(\tau). \qquad (6.28)$$

Here $I(\omega)$ is the bath spectral density and $G_\omega(\tau)$ is the thermal Green function

$$G_\omega(\tau) = \frac{1}{\hbar\beta} \sum_{n=-\infty}^\infty \frac{2\omega}{\nu_n^2 + \omega^2} e^{i\nu_n\tau}, \qquad (6.29)$$

with $\nu_n = \frac{2\pi n}{\hbar\beta}$ being the Bosonic Matsubara frequency. Substituing the representation of the spin path (6.26) into Eq. (6.27), the time integrals can be done resulting in, for a path with n kink-anti-kink pairs, the Euclidean influence functional,

$$F_n^E = \exp\left\{ \sum_{j=2}^{2n} \sum_{i=1}^{j-1} (-1)^{i+j} \Xi(s_j - s_i) \right\}, \qquad (6.30)$$

where $\Xi(\tau)$ would be the kink interaction term, representing the interaction between two charges,

$$\Xi(\tau) = \int_0^\infty d\omega \frac{I(\omega)}{\omega^2} (G_\omega(0) - G_\omega(\tau)). \qquad (6.31)$$

Here $I(\omega)$ is the spectral density of the spin-Boson model and is given in Eq. (6.10). Eqs. (6.27) and (6.30) are the state and charge representations of the influence functional, respectively. Thus, the influence functional, in the form presented in Eq. (6.30), represents the influence of the sum of all pair interactions of the $2n$ alternating charges.

6.4.1 Partition Function

We have already encountered the partition function in Chapter II, where we saw that having the partition function for a model enables the calculation of a number of useful thermodynamic properties characterizing the system. The partition function of the spin-Boson model is dominated by sequences of separated kinks or instantons. As discussed above, a given path of $2n$ transitions can be grouped into n kink-anti-kink pairs, also known as *bounces*, see for example, the next chapter where we discuss instantons. It is convenient to express the length of the bounces by $\phi_j = s_{2j} - s_{2j-1}$ and intervals between the bounces by $\psi_j = s_{2j+1} - s_{2j}$, where $(j = 1, \ldots, n)$. Also, $s_0 = 0$ and

$s_{2n+1} = \hbar\beta$. In the charge picture, ϕ_j and ψ_j are the intra- and inter-dipole lengths, respectively. As each kink-anti-kink pair or dipole contributes a factor $\Delta^2/4$ to the partition function, the contribution to the partition function from the right or left well for positive bias is

$$\mathcal{Z}_{R/L} = \sum_{n=0}^{\infty} \left(\frac{\Delta}{2}\right)^{2n} \int_0^{\hbar\beta} ds_{2n} \int_0^{s_{2n}} ds_{2n-1} \cdots \int_0^{s_2} ds_1 B_{R/L,n}^{(E)}(\{s_j\}) F_n^{(E)}(\{s_j\}).$$

(6.32)

Here Δ is the renormalized tunnel splitting element [1] and s_j, the kink centers (charge positions), are the collective coordinates. Also,

$$B_{R/L,n}^{(E)}(\{s_j\}) = \exp\left(\pm\frac{1}{2}\hbar\beta\zeta \mp \zeta \sum_{j=1}^n \phi_j\right),$$

(6.33)

and $F_n^{(E)}(\{s_j\})$ is as in Eq. (6.30). The Eq. (6.32) together with Eqs. (6.33) and (6.30) is also known as the *Coulomb gas* representation of the spin-Boson partition function. It is possible to express the time-ordered expression in Eq. (6.32) as a Laplace integral in order to obtain an exact formal solution of the spin-Boson partition function $\mathcal{Z} = \mathcal{Z}_R(\zeta) + \mathcal{Z}_L(\zeta)$ [1].

The partition function can be used to compute useful quantities such as the probability of occupation of the right (left) well in thermal equilibrium as $P_{R/L} = \mathcal{Z}_{R/L}/\mathcal{Z}$, yielding for the thermal average of σ_z

$$\langle\sigma_z\rangle = \frac{1}{\mathcal{Z}}\text{Tr}\{\sigma_z e^{-\beta H}\} = \frac{\mathcal{Z}_R - \mathcal{Z}_L}{\mathcal{Z}}.$$

(6.34)

This in turn is useful for computing thermodynamic quantities such as susceptibility.

6.5 Dynamics

Here we study the dynamics of the spin-Boson model. The principal tool used in this context is the IF approach, developed in Chapter IV. The formal exact series solutions obtained below will be seen to be very cumbersome. This evokes the question whether some suitable approximations can be made. This is indeed so and led to the development of the noninteracting blip approximation (NIBA).

6.5.1 Exact Solution

The approach used in studying the spin-Boson real time dynamics heavily relies on the Influence Functional (IF) techniques discussed in Chapter IV. We will, for simplicity, stick to the case of separable initial conditions, i.e., at the starting point of the evolution, the system, here the TSS, and the reservoir are uncorrelated. Assuming that the system starts out in the right well $|R\rangle$, the initial system-reservoir density matrix is

$$\rho(t=0,\mathcal{C}) = |R\rangle\langle R| \otimes \exp\left\{-\beta\left[H_R - \mathcal{C}\mathcal{P}(t=0)\right]\right\}/\mathcal{Z}_R.$$

(6.35)

Here H_R and $\mathcal{P}$ are as in Eq. (6.7). Also, $\mathcal{C}$ is control parameter determining the shift of the reservoir in the initial state at time $t = 0$. The dynamics of the system starting from Eq. (6.35) can be studied conveniently using the Feynman-Vernon IF, discussed in detail in Chapter IV. Substituting the spin path (6.26) into Eq. (49), Chapter IV, observing that the last term does not contribute here since $\sigma^2(t) = \sigma'^2(t)$, the IF is seen to be

$$F[\sigma, \sigma'; t_0] = \exp\left\{ -\frac{x_0^2}{4\hbar} \int_{t_0}^t dt' \int_{t_0}^{t'} dt''(\sigma(t')\right.$$
$$\left. - \sigma'(t'))(K(t' - t'')\sigma(t'') - K^*(t' - t'')\sigma'(t''))\right\}. \qquad (6.36)$$

Here t_0 is the time at which the dynamics is started and could be zero. The choice of $\mathcal{C} = 0$ or 1 determines the particular type of separable initial condition. They can be combined together in the single relation [1]

$$F_\mathcal{C}[\sigma, \sigma'] = F_{\mathcal{C}=0}[\sigma, \sigma'] \exp\left\{ i\frac{\mathcal{C}}{2} \int_0^t dt'(\sigma(t') - \sigma'(t'))\dot{C}''(t')\right\}. \qquad (6.37)$$

Here $C''(t)$ is the imaginary part of $C(t)$ which is defined as the imaginary time continuation of $\xi(\tau)$, i.e., $C(t) = \Xi(\tau = it)$. The term $\Xi(\tau)$, Eq. (6.31), is in turn defined as $\ddot{\Xi}(\tau) = -\mathcal{K}(\tau)$, where $\mathcal{K}(\tau)$ is as in Eq. (6.28). The explicit form of $C(t)$ can be seen to be

$$C(t) = \int_0^\infty d\omega \frac{I_{SB}(\omega)}{\omega^2} \frac{\cosh(\beta\omega/2) - \cosh[\omega(\beta/2 - it)]}{\sinh(\beta\omega/2)}. \qquad (6.38)$$

Here I_{SB} is the spin-Boson spectral density, Eq. (6.10). Thus, the effect of the state preparation could be thought of as a particular time-dependent bias $\mathcal{C}\dot{C}''(t)$.

Given the IF, we can obtain the propagator and thence the reduced density matrix of the system. All information of the TSS can be obtained from its reduced density matrix with the diagonal elements denoting the population and the off-diagonals, coherence. Thus, for e.g., the average values of the Pauli matrices, the so called *Bloch vectors*, are related to the components of the reduced density matrix as

$$\begin{aligned}
\langle\sigma_x(t)\rangle &= \rho_{1,-1}(t) + \rho_{-1,1}(t), \\
\langle\sigma_y(t)\rangle &= i\rho_{1,-1}(t) - i\rho_{-1,1}(t), \\
\langle\sigma_z(t)\rangle &= \rho_{1,1}(t) - \rho_{-1,-1}(t).
\end{aligned} \qquad (6.39)$$

The quantity $\sigma_z(t)$ describes the population difference of the two localized states and is directly relevant to studies of *macroscopic quantum coherence* (MQC). The spin-Boson dynamics can be expressed in terms of the two-time conditional propagator $J(\xi, t; \xi_0, t_0)$, see Eqs. (7)-(9) in Chapter IV. Here ξ is a two-variable label $\xi = (\eta, \vartheta)$, characterizing the TSS and $\eta(t')$ and $\vartheta(t')$ are symmetric and antisymmetric paths

$$\eta(t') = \frac{1}{2}[\sigma(t') + \sigma'(t')] \quad \text{and} \quad \vartheta(t') = \frac{1}{2}[\sigma(t') - \sigma'(t')]. \tag{6.40}$$

The TSS system can be in one of the four states, denoted by the diagonal terms $RR \equiv \eta = 1$, $LL \equiv \eta = -1$, the so called *sojourn* states [158], and the off-diagonal terms $RL \equiv \vartheta = 1$, $LR \equiv \vartheta = -1$, known as the *blip* states. During a sojourn, $\vartheta(\tau) = 0$, while during a blip, $\eta(\tau) = 0$. Here R, L refer to the right and left wells, respectively. The paths of the TSS, $\sigma(t')$, $\sigma'(t')$ jump between the two discrete values ± 1.

These paths are piecewise constant with sudden jumps in between. Keeping this in mind, the IF in Eq. (6.36) can be put, by performing integration by parts, in the form

$$F[\sigma, \sigma'; t_0] = \exp\left\{ \int_{t_0}^{t} dt' \int_{t_0}^{t'} dt'' \left(\dot{\vartheta}(t') C'(t'-t'') \dot{\vartheta}(t'') + i \dot{\vartheta}(t') C''(t'-t'') \dot{\eta}(t'') \right) \right\}, \tag{6.41}$$

where $C(t) = C'(t) + iC''(t)$ is as defined above.

Let the sojourn length be $s_j = t_{2j+1} - t_{2j}$ and the blip length be $\tau_j = t_{2j} - t_{2j-1}$. Here we assume that the TSS starts from a sojourn state. A general sojourn-to-sojourn path making $2n$ transitions at intermediate times t_j, $j = 1, 2, \ldots, 2n$ is parametrized by

$$\eta^{(n)}(t') = \sum_{j=0}^{n} \eta_j \left[\Theta(t' - t_{2j}) - \Theta(t' - t_{2j+1}) \right],$$

$$\vartheta^{(n)}(t') = \sum_{j=1}^{n} \vartheta_j \left[\Theta(t' - t_{2j-1}) - \Theta(t' - t_{2j}) \right]. \tag{6.42}$$

This path represents a sequence of sojourn and blip dipoles because the system, after every second transition, is again in a sojourn or blip state. Thus, for e.g., $\eta_j = 1$ and $\vartheta_j = 1$ correspond to $(+, -)$ dipoles, while $\eta_j = -1$ and $\vartheta_j = -1$ correspond to $(-, +)$ dipoles. Defining the bath correlations between the blip-pair by $\Lambda_{j,k}$ and that between the sojourn k and a later blip j by $\varkappa_{j,k}$

$$\Lambda_{j,k} = C'_{2j,2k-1} + C'_{2j-1,2k} - C'_{2j,2k} - C'_{2j-1,2k-1},$$

$$\varkappa_{j,k} = C''_{2j,2k+1} + C''_{2j-1,2k} - C''_{2j,2k} - C''_{2j-1,2k+1}, \tag{6.43}$$

the IF, Eq. (6.41), for the path defined in Eq. (6.42) becomes

$$F^{(n)} = A_n B_n,$$

$$A_n = \exp\left\{ -\sum_{j=1}^{n} C'_{2j,2j-1} \right\} \exp\left\{ -\sum_{j=2}^{n} \sum_{k=1}^{j-1} \vartheta_j \vartheta_k \Lambda_{j,k} \right\},$$

$$B_n = \exp\left\{ i \sum_{j=1}^{n} \sum_{k=0}^{j-1} \vartheta_j \varkappa_{j,k} \eta_k \right\}. \tag{6.44}$$

Here $\Lambda_{j,k}$ and $\varkappa_{j,k}$ represents the dipole-dipole interaction between two blips or between a sojourn and succeeding blip dipole, respectively. The blip interactions are in the equation for the real function A_n with the first term reflecting the intrablip or intradipole interactions, while the second term contains the interblip correlations. The form of A_n brings out the point that long blips are suppressed. This in turn implies that TSS is favoured to spend time in the sojourn state. The weight to switch per unit time from a diagonal state η to an off-diagonal state ϑ and *vice versa* is $-i\vartheta\eta\Delta/2$. This could be understood by invoking the spin-Boson Hamiltonian Eq. (6.7) wherein it is clear that the spin flip term is associated with the tunneling element, here we take the renormalized version, $\Delta/2$. Hence, there would be a factor of $-i\Delta/2$ for transitions between (RL), (RR) and between (LL), (LR), while the factor is $i\Delta/2$ for transitions between (RL), (LL) and between (RR), (LR). The weight to stay in the sojourn is unity while that to stay in the jth blip is $\exp(i\zeta\vartheta_j\tau_j)$. The blip factor for n blip states can be written as an overall bias factor

$$K_n = \exp\left(i\zeta \sum_{j=1}^{n} \vartheta_j \tau_j\right). \tag{6.45}$$

From the IF, the propagator J can be obtained, see Chapter IV. The quantities of interest, such as the Bloch vectors, can be obtained from the propagator as

$$
\begin{aligned}
\langle \sigma_z(t) \rangle &= \sum_{\eta=\pm1} \eta J(\eta, t; \eta_0 = 1, 0), \\
\langle \sigma_x(t) \rangle &= \sum_{\vartheta=\pm1} J(\vartheta, t; \eta_0 = 1, 0), \\
\langle \sigma_y(t) \rangle &= i \sum_{\eta=\pm1} \vartheta J(\vartheta, t; \eta_0 = 1, 0).
\end{aligned}
\tag{6.46}
$$

Here we assume that the TSS starts from the diagonal state $\eta_0 = 1$. Collecting the various weight factors, discussed above, the propagator for a sojourn as final state is

$$J(\eta, t; \eta_0, 0) = \delta_{\eta,\eta_0} + \eta\eta_0 \sum_{m=1}^{\infty} \frac{(-1)^m}{2^{2m}} \int_0^t \mathcal{D}_{2m,0}\{t_j\} \sum_{\{\vartheta_j=\pm1\}} A_m K_m \sum_{\{\eta_j=\pm1\}'} B_m, \tag{6.47}$$

while for a blip final state

$$J(\vartheta, t; \eta_0, 0) = -i\vartheta\eta_0 \sum_{m=1}^{\infty} \frac{(-1)^{m-1}}{2^{2m-1}} \int_0^t \mathcal{D}_{2m-1,0}\{t_j\} \sum_{\{\vartheta_j=\pm1\}'} A_m K_m \sum_{\{\eta_j=\pm1\}'} B_m. \tag{6.48}$$

In the above equations $\int_0^t \mathcal{D}$ is a compact notation for the integration over the time-ordered jump times t_j, including the tunneling matrix element Δ. The various summations in the above equations for the propagators denote the sum over all paths, i.e., the sum over the intermediate sojourn and blip states visited by the paths with a given number of transitions, and sum over the possible number of transitions the system may make. Further, the prime in $\{\eta_j = \pm 1\}'$ and $\{\vartheta_j = \pm 1\}'$ indicates that the initial sojourn and the final sojourn or blip states are fixed.

We are now in a position to compute the various Bloch vectors. Substituting Eq. (6.47) into Eq. (6.46) and summing over the intermediate sojourn states $\{\eta_j = \pm 1\}$, the exact series expression of the population $\langle \sigma_z(t) \rangle = P_R(t) - P_L(t)$, where $P_{R/L}(t)$ are the occupation probabilities for the right and left wells, respectively, is obtained as

$$\langle \sigma_z(t) \rangle = 1 + \sum_{m=1}^{\infty} (-1)^m \int_0^t \mathcal{D}_{2m,0}\{t_j\} \frac{1}{2^m} \sum_{\{\vartheta_j = \pm 1\}} \left(M_m^{(+)} K_m^{(s)} - M_m^{(-)} K_m^{(a)} \right).$$

$$(6.49)$$

Here the bias ζ dependence is in

$$K_m^{(s)} = \cos \left(\zeta \sum \vartheta_j \tau_j \right), \quad \text{and} \quad K_m^{(a)} = \sin \left(\zeta \sum \vartheta_j \tau_j \right), \qquad (6.50)$$

while the environmental effect is encoded in

$$M_m^{(+)} = A_m \prod_{k=0}^{m-1} \cos(\phi_{k,m}), \quad \text{and} \quad M_m^{(-)} = A_m \sin(\phi_{0,m}) \prod_{k=1}^{m-1} \cos(\phi_{k,m}). \quad (6.51)$$

Also,

$$\phi_{k,m} = \sum_{j=k+1}^{m} \vartheta_j \varkappa_{j,k}, \qquad (6.52)$$

indicates the environmental correlations between the k th sojourn and the $m-k$ suceeding blips. In Eq. (6.49), $\sum_{\{\vartheta_j = \pm 1\}}$ runs over all intermediate blip states $\{\vartheta_j = \pm 1\}$.

When the damping persists till infinity, i.e., for asymptotic large time, the system is ergodic and reaches thermal equilibrium. Then, the average population $\langle \sigma_z(t \to \infty) \rangle = \langle \sigma_z \rangle_\infty$, using Eq. (6.49), becomes

$$\langle \sigma_z \rangle_\infty = \lim_{t \to \infty} \sum_{m=1}^{\infty} (-1)^{m-1} \int_0^t \mathcal{D}_{2m,0}\{t_j\} \frac{1}{2^m} \sum_{\{\vartheta_j = \pm 1\}} M_m^{(-)} K_m^{(a)}. \qquad (6.53)$$

Note that this equilibrium probability distribution has been obtained using dynamical quantities. It can also be obtained from pure thermodynamic quantities, such as the partition function, and for ergodic systems, the two computations would agree with each other. In a similar fashion, the exact formal series solutions for the coherences $\langle \sigma_x(t) \rangle$ and $\langle \sigma_y(t) \rangle$ can be obtained, by substituting

Eq. (6.48) in Eqs. (6.46), as

$$\langle \sigma_x(t) \rangle = \sum_{m=1}^{\infty} (-1)^{m-1} \int_0^t \mathcal{D}_{2m-1,0}\{t_j\} \frac{1}{2^m} \sum_{\{\vartheta_j = \pm 1\}} \vartheta_m \left(M_m^{(+)} K_m^{(a)} + M_m^{(-)} K_m^{(s)} \right),$$

$$(6.54)$$

$$\langle \sigma_y(t) \rangle = \sum_{m=1}^{\infty} (-1)^{m-1} \int_0^t \mathcal{D}_{2m-1,0}\{t_j\} \frac{1}{2^m} \sum_{\{\vartheta_j = \pm 1\}} \left(M_m^{(+)} K_m^{(s)} - M_m^{(-)} K_m^{(a)} \right).$$

$$(6.55)$$

Keeping in mind that any TSS can be expressed as $\rho(t) = \frac{1}{2}(\mathcal{I} + \sum_{j=1}^{3} \langle \sigma_j(t) \rangle \sigma_j)$, Eqs. (6.54), (6.55) and (6.49) together determine the complete spin-Boson dynamics.

6.5.2 Noninteracting-Blip Approximation

An important approximation made in the study of the dynamics of the spin-Boson model is the noninteracting-blip approximation (NIBA) [158]. The basic assumption made is that the average time spent by the system in a diagonal state is much more than that spent in the off-diagonal state; this could be appreciated from the comment below Eq. (6.44) to the effect that long blips are suppressed. The NIBA involves two technical assumptions:

(a). The sojourn-blip correlation factors $\varkappa_{j,k}$ are set to zero for $j \neq k+1$. Thus, from Eq. (6.43), we see that $\varkappa_{k+1,k} = C''(t_{2k+2} - t_{2k+1})$. The correlations between a sojourn k and the subsequent blips becomes $\vartheta_{k+1} C''(\tau_{k+1})$;

(b). The interblip correlations in A_n, Eq. (6.44), are neglected, i.e., $\Lambda_{j,k}$ are set to zero.

With these approximations, it is easy to see from Eq. (6.44) that the IF reduces to

$$F_{NIBA}^{(n)} = \prod_{j=1}^{n} \exp \left\{ -C'(\tau_j) + i\vartheta_j \eta_{j-1} C''(\tau_j) \right\}. \qquad (6.56)$$

For an Ohmic spectral density, where the upper cutoff frequency of the reservoir $\omega_c \to \infty$, the assumption (a) is exact. Here, NIBA implies neglect of interblip interactions $\Lambda_{j,k}$. The NIBA can be justified in the following limiting cases [1]: (i). *weak coupling and zero bias*, (ii). for bath spectral density $I(\omega) \propto \omega^s$, with $s > 1$ for zero temperature and with $s > 2$ for finite temperatures, and (iii). long blips are suppressed for sub-Ohmic $s < 1$ damping at zero temperature and for $s < 2$ for finite temperatures.

The NIBA corresponds to an expansion in terms of the tunneling matrix element Δ_0, which can also be performed with projection-operator techniques [164]. As a result, the evolution of $\langle \sigma_z(t) \rangle$ becomes

$$\frac{d\langle \sigma_z(t) \rangle}{dt} = -\int_{-\infty}^{t} ds \, f(t-s) \langle \sigma_z(s) \rangle. \qquad (6.57)$$

Here
$$f(s) = \Delta_0^2 \cos(T_1(s)/\pi\hbar) e^{-T_2(s)/\pi\hbar}, \qquad (6.58)$$

with

$$T_1(s) = \int_0^\infty \frac{d\omega}{\omega^2} I(\omega) \sin(\omega s), \qquad (6.59)$$

$$T_2(s) = \int_0^\infty \frac{d\omega}{\omega^2} I(\omega) \coth\left(\frac{\beta\hbar\omega}{2}\right)(1 - \cos(\omega s)). \qquad (6.60)$$

This result was obtained, in a different manner, in [165], by making use of the polaron transformation, discussed above.

6.6   Coupling to Reservoir via an Intermediate Harmonic Oscillator

The spin-Boson model has been put to use in many scenarios. In the case of electron transfer in biomolecules, an electron can travel between two localized sites, say in the same or different molecules. In many situations, the distance between the localized sites is small. This allows for almost free electron transfer between the sites. However, there could be the scenario wherein the sites are separated by larger distances. Then the tunneling process has to be taken into account. A nuclear *reaction* coordinate is coupled to other nuclear or solvent coordinates. This coupling, which leads to friction, if strong enough, has a tendency to slow down motion along the reaction coordinate. This vitiates the assumption of electron transfer being nonadiabatic with respect to the nuclei. A quantum mechanical model for electron transfer using a TSS coupled to an intermediate harmonic oscillator, representing the reaction coordinate, which in turn is coupled to a bath of harmonic oscillators was introduced in [166].

The donor and the acceptor sites of the electron can be considered as a TSS where these two states are identified with the eigenvalues ± 1 of σ_z, respectively. Let the two corresponding positions be $\pm q_0$. The possibility of tunneling is associated with a matrix-element, defined, as above, by $\hbar\Delta/2$, hence the term $\hbar\Delta\sigma_x/2$ in the Hamiltonian. The reaction coordinate is modelled by a harmonic oscillator with the kinetic energy $p_y^2/2M$. The coupling to the TSS is included in the potential term of the oscillator as

$$V(y, \sigma_z) = \frac{1}{2} M\omega^2 (y + y_0\sigma_z)^2 + \frac{1}{2}\zeta\sigma_z. \qquad (6.61)$$

Here ζ is the detuning energy defined before. The reaction coordinate is turn connected in a linear manner to a bath of harmonic oscillators as

$$H_R + H_{yR} = \sum_i \left[\frac{p_i^2}{2m_i} + \frac{1}{2} m_i\omega_i^2\left(x_i + \frac{c_i}{m_i\omega_i^2}y\right)^2\right]. \qquad (6.62)$$

In the above equation, the term $\frac{c_i^2}{m_i\omega_i^2}y^2$ has the effect of setting the minimum of the oscillator bath to zero. We now collect all the above terms to set the electron transfer Hamiltonian as

$$H_{et} = \frac{\hbar\Delta}{2}\sigma_x + \frac{p_y^2}{2M} + V(y,\sigma_z) + \sum_i \left[\frac{p_i^2}{2m_i} + \frac{1}{2}m_i\omega_i^2\left(x_i + \frac{c_i}{m_i\omega_i^2}y\right)^2\right]. \quad (6.63)$$

By transforming to the normal modes, the quadratic part of the Hamiltonian, Eq. (6.63), can be brought to a form similar to the spin-Boson Hamiltonian, Eq. (6.7). Most of the tunneling occurs within a length of order l_{LZ}, the Landau-Zener length, defined as

$$l_{LZ} = \frac{\hbar\Delta}{|F_+ - F_-|}, \quad (6.64)$$

where $F_\pm = -\left(\frac{\partial V(y;\pm)}{\partial y}\right)_{y=y*}$. Here y^* is a point between $\pm y_0$ such that $V(y^*,+) = V(y^*,-)$ and is the crossing point. If the temperature is high enough, the friction is relatively small and the tunneling electron will pass the Landau-Zener region in one time. In this case the probability of tunneling is a small quantity, which could be calculated. If the temperature is low, the friction is relatively large that the electron can stay in the Landau-Zener region and make many transitions. In this case, coherence between the states will get lost in due time.

6.6.1   Effective Spectral Density

We will now introduce a technique which is very useful in the study of the above problem, viz. the effective spectral density [167]. The TSS and the reaction coordinate are influenced by the bath through the spectral density

$$I(\omega) = \frac{\pi}{2}\sum_i \frac{c_i^2}{m_i\omega_i}\delta(\omega - \omega_i). \quad (6.65)$$

For normal velocity dependent friction, this takes the Ohmic form $I(\omega) = \eta\omega e^{-\omega/\omega_c}$, where ω_c is an upper cutoff frequency of the reservoir. Instead of this spectral density we use an effective spectral density $I_{eff}(\omega)$ that describes how the TSS is influenced by the reaction coordinate and bath together. This $I_{eff}(\omega)$ is the same as the one that controls the dynamics of a continuous variable q, of mass μ, moving in a potential $V(q)$ coupled to the reaction coordinate and the bath in the same way as the TSS is coupled. Consider the Hamiltonian

$$H_q = \frac{p_q^2}{2\mu} + V(q) + \frac{P_y^2}{2M} + \frac{1}{2}M\omega^2(y+q)^2 + \sum_i \left[\frac{p_i^2}{2m_i} + \frac{1}{2}m_i\omega_i^2\left(x_i + \frac{c_i}{m_i\omega_i^2}y\right)^2\right]. \quad (6.66)$$

The Hamiltonian equations of motion are

$$\mu \ddot{q} = -V'(q) - M\omega^2(y + q), \tag{6.67}$$

$$M\ddot{y} = -M\omega^2(y + q) - \sum_i c_i x_i - y \sum_i \frac{c_i^2}{m_i \omega_i^2}, \tag{6.68}$$

$$m_i \ddot{x}_i = -m_i \omega_i^2 x_i - c_i y. \tag{6.69}$$

Next, we Fourier transform the Eqs. (6.67) to (6.69):

$$\left(-\mu z^2 + M\omega^2\right) q(z) + M\omega^2 y(z) = -V_z'(q), \tag{6.70}$$

$$\left(-M z^2 + M\omega^2 + \sum_i \frac{c_i^2}{m_i \omega_i^2}\right) y(z) + \sum_i c_i x_i(z) = -M\omega^2 q(z), \tag{6.71}$$

$$x_i(z) = -\frac{c_i}{m_i(\omega_i^2 - z^2)} y(z). \tag{6.72}$$

Here $q(z)$, $y(z)$, $x_i(z)$ and $V_z'(q)$ denote the Fourier transforms of q, y, x_i and $V'(q)$, i.e., $f(z) = \int_{-\infty}^{\infty} dt e^{-izt} f(t)$, with Im(z) < 0 for analytic convergence. Substituting Eq. (6.72) into Eq. (6.71), we obtain

$$y(z) = \frac{-M\omega^2}{M\omega^2 + L(z)} q(z). \tag{6.73}$$

Here

$$L(z) = -z^2 \left[M + \sum_i \frac{c_i^2}{m_i \omega_i^2(\omega_i^2 - z^2)} \right]. \tag{6.74}$$

Next, if we insert Eq. (6.73) into Eq. (6.70), we get, in compact notation

$$K(z)q(z) = -V_z'(q), \tag{6.75}$$

where

$$K(z) = -\mu z^2 + \frac{M\omega^2 L(z)}{M\omega^2 + L(z)}. \tag{6.76}$$

Problem 2: Derive Eqs. (6.73) and (6.75).

In what follows we would need to compute $K(z)$. To do that we start by calculating $L(z)$.

$$
\begin{aligned}
L(z) &= -z^2 \left[M + \sum_i \frac{c_i^2}{m_i \omega_i^2 (\omega_i^2 - z^2)} \right] \\
&= -z^2 \left[M + \frac{2}{\pi} \frac{\pi}{2} \int_0^\infty d\omega' \left(\frac{1}{\omega'(\omega'^2 - z^2)} \sum_i \frac{c_i^2}{m_i \omega_i} \delta(\omega' - \omega_i) \right) \right] \\
&= -z^2 \left[M + \frac{2}{\pi} \int_0^\infty d\omega' \frac{I(\omega')}{\omega'(\omega'^2 - z^2)} \right] \\
&= -z^2 \left[M + \frac{2\eta}{\pi} \int_0^\infty d\omega' \frac{e^{-\omega'/\omega_c}}{(\omega'^2 - z^2)} \right].
\end{aligned}
\tag{6.77}
$$

Here we have made use of the spectral density, Eq. (6.65), and in the last line the Ohmic form of the spectrum. Noting that here $\text{Im}(z) < 0$, the integral in the above equation can be easily computed by the method of residues to yield

$$
L(z) = -Mz^2 + i\eta z.
\tag{6.78}
$$

The effective sepctral density $I_{eff}(\omega)$ can now be found as [167]

$$
I_{eff}(\omega) = \lim_{\epsilon \to 0_+} \text{Im}\left[K(\omega - i\epsilon) \right].
\tag{6.79}
$$

Substituting Eq. (6.78) in Eq. (6.76) and then in Eq. (6.79), we get the following effective spectral density

$$
I_{eff}(\tilde{\omega}) = \frac{\eta \tilde{\omega} \omega^4}{(\omega^2 - \tilde{\omega}^2)^2 + 4\gamma^2 \tilde{\omega}^2}.
\tag{6.80}
$$

Here $\gamma = \eta/2M$ has been used.

Problem 3: Derive Eq. (6.80) and compare it with the Ohmic spectral density.

6.6.2   Application of the Effective Spectrum Method: Asymptotic behavior of the Spin-Boson Model

Here we will apply the effective spectral method to obtain the asymptotic behavior of $P(t) = \langle \sigma_z(t) \rangle$, i.e., $\langle \sigma_z \rangle_\infty$. Even though the formal exact solution is available to us, Eq. (6.53), for practical applications it would be useful to resort to meaningful approximations. The simplest approximation is that of the NIBA, discussed above. Here we will work within this approximation. The desired asymptotic behavior can of course be obtained from the expressions of the IF derived above. However, here we will make use of the intermediate oscillator formalism, enunciated above, to obtain this result [168]. Further, this will serve as a brief introduction to the field of driven *dissipative* quantum tunneling [169].

We consider the problem of a TSS driven by a periodic driving field $-\frac{1}{2}\hbar\hat{\zeta}\cos(\Omega_{dr}t)\sigma_z$. This has the effect of periodically modulating the bias energy of the undriven system. It can be shown [170, 1] that the Laplace transform of $P(t)$, $\hat{P}(\lambda) = \int_0^\infty e^{-\lambda t}P(t)$, obeys the following generalized master equation

$$\lambda\hat{P}(\lambda) = 1 + \int_0^\infty dt e^{-\lambda t}\big[\hat{K}_\lambda^-(t) - \hat{K}_\lambda^+(t)P(t)\big]. \tag{6.81}$$

For periodic driving, as considered here, the kernels $\hat{K}_\lambda^\pm(t)$ can be expanded in a Fourier series as

$$\hat{K}_\lambda^\pm(t) = \sum_{j=-\infty}^\infty k_j^\pm(\lambda)e^{-ij\Omega_{dr}t}. \tag{6.82}$$

This makes it possible to solve Eq. (6.81) recursively. The asymptotic dynamics is periodic with period $2\pi/\Omega_{dr}$ determined by the driving field and

$$\lim_{t\to\infty} P(t) = P_\infty = \sum_{j=-\infty}^\infty p_j e^{-ij\Omega_{dr}t}, \tag{6.83}$$

where

$$p_0 = \frac{k_o^-(0)}{k_o^+(0)} - \sum_{j\neq 0} \frac{k_j^+(0)}{k_o^+(0)}p_j, \tag{6.84}$$

and

$$p_j = \frac{i}{j\Omega_{dr}}\bigg(k_j^-(-ij\Omega_{dr}) - \sum_r k_{j-r}^+(-ij\Omega_{dr})p_r\bigg), \tag{6.85}$$

for $j \neq 0$. Here i stands for the usual complex number $\sqrt{-1}$. Let us take up the case of high frequency driving, i.e., $\Omega_{dr} \gg \{\Gamma, \zeta, \Delta_0\}$, where Γ^{-1} is the mean decay time after which the spin-Boson attains equilibrium and ζ, Δ_0 are the detuning and tunnel splitting terms of the TSS, respectively. This would be appropriate to use in the context of control of chemical reactions. In this case, the asymptotic value P_∞ is given by [169]

$$P_\infty = \frac{k_o^-(0)}{k_o^+(0)}. \tag{6.86}$$

In the above equation,

$$
\begin{aligned}
k_o^-(0) &= \Delta_0^2 \int_0^\infty dt\, h^-(t)\sin(\zeta t)J_0\bigg(\frac{\hat{\zeta}}{\Omega_{dr}}\sin\big(\frac{\Omega_{dr}t}{2}\big)\bigg), \\
k_o^+(0) &= \Delta_0^2 \int_0^\infty dt\, h^+(t)\cos(\zeta t)J_0\bigg(\frac{2\hat{\zeta}}{\Omega_{dr}}\sin\big(\frac{\Omega_{dr}t}{2}\big)\bigg).
\end{aligned} \tag{6.87}
$$

Here

$$
\begin{aligned}
h^-(t) &= e^{-C'(t)}\sin(C''(t)), \\
h^+(t) &= e^{-C'(t)}\cos(C''(t)),
\end{aligned}
\tag{6.88}
$$

and J_0 is the Bessel function of order zero

$$
J_0(x) = \sum_{j=0}^{\infty} \frac{(-1)^j}{(j!)^2}\left(\frac{x}{2}\right)^{2j}.
\tag{6.89}
$$

$C'(t)$ and $C''(t)$ are, as described before, the real and imaginary parts of $C(t)$, Eq. (6.38).

We now use the intermediate oscillator, of frequency Ω, approach. To this effect $C(t)$, Eq. (6.38), is computed with the effective spectral density (6.80)

$$
I_{eff}(\omega) = \frac{\eta\omega\Omega^4}{(\Omega^2 - \omega^2)^2 + 4\gamma^2\omega^2}.
\tag{6.90}
$$

Then $C(t)$ can be expressed as

$$
C(t) = \frac{1}{2}\int_{-\infty}^{\infty} d\omega\, C_a(\omega)C_b(\omega,t),
\tag{6.91}
$$

where

$$
\begin{aligned}
C_a(\omega) &= \frac{\eta\Omega^4}{\omega\left[(\Omega^2 - \omega^2)^2 + 4\gamma^2\omega^2\right]}, \\
C_b(\omega,t) &= \frac{\cosh(\beta\omega/2) - \cosh[\omega(\beta/2 - it)]}{\sinh(\beta\omega/2)}.
\end{aligned}
\tag{6.92}
$$

The integrals can be performed using contour integration [168] to yield

$$
k_0^{\pm}(0) = \sum_{m=-\infty}^{\infty}\sum_{n=-\infty}^{\infty} \Delta^2 \int_0^{\infty} dt\, e^{-C_1'(t)} g_{mn}^{\pm}(t).
\tag{6.93}
$$

Here

$$
\begin{aligned}
g_{mn}^+(t) &= \mathrm{Re}\left[c_{mn}^+(t)\cos(\zeta_{mn}t) + c_{mn}^-(t)\sin(\zeta_{mn}t)\right], \\
g_{mn}^-(t) &= \mathrm{Im}\left[c_{mn}^-(t)\cos(\zeta_{mn}t) - c_{mn}^+(t)\sin(\zeta_{mn}t)\right],
\end{aligned}
\tag{6.94}
$$

$$
\begin{aligned}
c_{mn}^+(t) &= J_n^2\!\left(\frac{\hat{\zeta}}{\Omega}\right) J_m(e^{-\gamma t}w_1)\cos(m\phi)(-i)^m e^{-iA_1}, \\
c_{mn}^-(t) &= J_n^2\!\left(\frac{\hat{\zeta}}{\Omega}\right) J_m(e^{-\gamma t}w_1)\sin(m\phi)(-i)^m e^{-iA_1}.
\end{aligned}
\tag{6.95}
$$

The terms in these equations are

$$
\begin{aligned}
\zeta_{mn} &= \zeta_0 - m\bar{\Omega} - n\Omega_{dr}, \\
\omega_1 &= \sqrt{(A_1 - iB_1)^2 + (A_2 - iB_2)^2}, \\
\tan(\phi) &= -\frac{A_2 - iB_2}{A_1 - iB_1}.
\end{aligned}
\tag{6.96}
$$

Further, $\bar{\Omega}^2 = \Omega^2 - \gamma^2$, and

$$
C_1'(t) = B_1 + \frac{\sinh(\beta\bar{\Omega})/\bar{\Omega} + \sin(\beta\gamma)/\gamma}{\cosh(\beta\bar{\Omega}) + \cos(\beta\gamma)}\frac{\pi\eta\Omega^2 t}{4} - \frac{\pi\eta\Omega^4}{\beta}\sum_{n=1}^{\infty}\frac{(e^{-\nu_n t} - 1)/\nu_n + t}{(\Omega^2 + \nu_n^2)^2 - 4\gamma^2\nu_n^2},
$$

$$
\nu_n = \frac{2\pi in}{\beta},
$$

$$
A_1 = \frac{\pi\eta}{2},
$$

$$
A_2 = -\frac{\pi\eta}{4}\frac{\bar{\Omega}^2 - \gamma^2}{\bar{\Omega}\gamma},
$$

$$
B_1 = \frac{A_2\sinh(\beta\bar{\Omega}) - A_1\sin(\beta\gamma)}{\cosh(\beta\bar{\Omega}) + \cos(\beta\gamma)},
$$

$$
B_2 = \frac{A_1\sinh(\beta\bar{\Omega}) - A_2\sin(\beta\gamma)}{\cosh(\beta\bar{\Omega}) + \cos(\beta\gamma)}.
\tag{6.97}
$$

Also, η, γ are as in Eq. (6.90).

For $\gamma/\Omega \ll 1$ implying $\bar{\Omega} \approx \Omega$ and for not too large temperature T, $\cos(\beta\gamma) \ll \cosh(\beta\Omega)$, $\sin(\beta\gamma) \ll \sinh(\beta\Omega)$. Also, $\tan(m\phi) \approx i\tanh(\frac{m\beta\Omega}{2})$. Using this in Eq. (6.95), we get

$$
i\tanh\left(\frac{m\beta\Omega}{2}\right)c_{mn}^+(t) = c_{mn}^-(t).
\tag{6.98}
$$

If the integrand in Eq. (6.87) is not damped too fast and $\zeta_{mn} = 0$, Eq. (6.96), then the summation in Eq. (6.93) would be dominated by the coefficient of the cosine term. This implies that, for those integers,

$$
\begin{aligned}
g_{mn}^+(t) &\approx \mathrm{Re}\left[c_{mn}^+(t)\right], \\
g_{mn}^-(t) &\approx \tanh\left(\frac{m\beta\Omega}{2}\right)\mathrm{Re}\left[c_{mn}^+(t)\right].
\end{aligned}
\tag{6.99}
$$

It follows that for that specific integer m the asymptotic value of $P(t)$ is

$$
P_\infty = \tanh\left(\frac{m\beta\Omega}{2}\right).
\tag{6.100}
$$

6.7   Guide to Advanced Literature

What has been discussed in this chapter is just the tip of the iceberg. However, just as the tip can identify the iceberg, the tools and techniques laid down here could be used profitably to explore new avenues. We studied the partition function of the spin-Boson model. This can be used to understand the equilibrium properties of open TSS. The formal expressions of the partition function can be used to establish relationships between the Ohmic spin-Boson model with variants of the Kondo model [158] and the Ising model with ferromagnetic $1/r^2$ interaction [171].

We have seen that the exact formal expressions of the spin-Boson dynamics are very cumbersome. Therefore from a practical perspective, it is highly desirable to develop appropriate approximation schemes. The simplest such scheme is the NIBA, which was discussed above. There are approximation schemes that go beyond the NIBA, as would be required for tunneling systems with higher defect concentration. For example, some applications require taking into account the interblip correlations to first order in the coupling strength. In the interacting blip chain approximation the nearest neighbour correlations between blips and phase correlations between neighbouring sojourn-blip pairs is taken into account [1, 172].

Driven TSS and the subsequent driven quantum tunneling, which in its simplest form would be the spin-Boson problem driven by an external laser field, are very useful for understanding processes such as electron transfer. These studies also play an important role in the control of the TLS dynamics via quantum stochastic resonance, a cooperative effect of noise and periodic driving in bistable systems. This is a vast field and is nicely reviewed in [169].

It is also possible to derive a master equation for the spin-Boson model [173] which is valid in the (weak) strong coupling regime. The key ingredient of such a derivation is to consider a variational polaron transformation for the spin-Boson Hamiltonian [174].

Chapter 7

Quantum Tunneling

7.1 Introduction

Tunneling is a *bonafide* quantum mechanical effect [175, 176]. Since it involves barrier penetration, it is also an inherently non-perturbative process. It serves a crucial role in the test of quantum coherence in macroscopic regimes, also known as Macroscopic Quantum Coherence (MQC) [68]. Development in technology has made the concept of tunneling crucial to the development of devices on the nanoscopic, nanometre $10^{-9}m$ range, and mesoscopic, upto a few microns μm, scales. Further, tunneling has important ramifications to almost all branches of physics, such as atomic, molecular, condensed matter physics as well as to quantum field theory and cosmology. A very powerful technique for dealing with tunneling is the semiclassical approximation, which we detail below. Tunneling processes can be broadly classified into two categories: coherent and incoherent tunneling. Coherent tunneling phenomena involve the coherent overlap of wavefunctions located in individual domains, such as, ground states of potential wells, and separated by energy barriers. Incoherent tunneling involves scattering between reservoirs or decay of metastable states into the continuum and hence no overlap of the wavefunction. Tunneling has two perspectives: time independent energy domain considerations and that invoking the time dependent dynamics.

Complex systems usually carry with them the extra baggage of their ambient surroundings whose effect could be dissipation and is the subject of open quantum systems. In dissipative tunneling, numerically exact treatments can be very cumbersome as well as time consuming. Semiclassical approximations provide a good starting point.

We make use of path integral methods throughout this chapter. We begin with an introduction to semiclassical approximation, and the stationary phase approximation, which has become a branch of science on in its own [159]. This implies that the action evaluated at the classical path is much larger than $\hbar$, as a result of which the major contribution to the path integral comes from the classical path and the paths in its neighbourhood. At the same time quantum

© Hindustan Book Agency 2018 and Springer Nature Singapore Pte Ltd. 2018
S. Banerjee, *Open Quantum Systems*, Texts and Readings in Physical Sciences 20,
https://doi.org/10.1007/978-981-13-3182-4_7

effects are taken into consideration. We illustrate the notion of semiclassical approximation by applying it on the harmonic oscillator, where it turns out that the approximation is exact, highlighting the importance and ubiquity of the harmonic oscillator. We make contact with the conventional WKB method.

This sets the scene for a discussion of quantum tunneling by studying the double well potential. It is seen that evaluation of the path integrals in the Euclidean, imaginary time, formulation is best suited for the problem at hand. Here we see that instantons make a natural entry into the scheme of things. We get to grips with the single instanton path integral as well as the multi-instanton one. From this, the concept of tunneling induced splitting, a notion that was used in the previous chapter in the context of the spin-Boson model, comes out naturally. This chapter thus serves to also introduce instantons, which has a huge and diverse literature [159, 177, 43, 160, 1].

Equipped with these tools we move to the topic of dissipative quantum tunneling of a metastable state. Quantum effects are predominant in the regime of low temperatures, i.e., temperatures low compared to the crossover temperature. In these regimes, path integral methods come to the forefront and are perhaps the natural tool for analysis. The notions of semiclassical path integrals and instantons find a natural environment here. Another technique used in this context, and briefly introduced here, is the imaginary free energy method, introduced by Langer [178, 179]. Finally we see that using the language adopted in this chapter, quantum tunneling in the context of open quantum systems is a natural extension, to larger number of degrees of freedom, of the tools developed so far. This should not come as a surprise as the notion of semiclassical path integrals as applied to ordinary quantum mechanics finds a natural extension in quantum field theory [159].

7.2   Semiclassical Approximation

As stated above, tunneling is an inherently non-perturbative process. Hence, it is of importance to develop an understanding of approximation schemes that would help to elucidate some aspects of its non-perturbative character. A very prominent set of tools developed in this context is broadly called semiclassical approximation. Traditional treatments of the semiclassical approximation (SCA) invoke the WKB approximation [23] that necessitates the matching of semiclassical wavefunctions, a cumbersome task. Modern treatments, to which we shall also adhere to, use the path integral formulation of quantum mechanics [159, 43, 41, 180]. Within the framework of the path integral, semiclassical methods find a natural habitat with the approximation corresponding to *orbits* minimizing the action. A further advantage of this is that excursions into open quantum systems by the inclusion of reservoir degrees of freedom is a straightforward extension.

We recall that the Feynman kernel giving the quantum transition amplitude, see Chapter II, Eqs. (99), (103), is

$$\langle x', t' | x, t \rangle = \int Dx_{x,t;x',t'} \exp\left[\frac{i}{\hbar} S(x, \dot{x})\right].\tag{7.1}$$

Here we have absorbed the normalization constant in the measure of the path integral Dx. This expression tells us that the quantum mechanical amplitude to make a transition from x at time t to x' at time t' is given by a path integral over the action and involves a sum over all paths between the same beginning and end points. The path integral formulation of the SCA involves Eq. (7.1) and connects it to the energy levels of the system [159]. Let us denote the eigenfunctions of the Hamiltonian by $|n\rangle$. Now the crucial point is the taking of a trace of the evolution operator $\exp(-iHt/\hbar)$ in two different ways; one over eigenstates of the Hamiltonian $|n\rangle$, with the energy spectrum E_n as the eigenvalue and the other over the position eigenstates $|x_0\rangle$ (recall from Chapter II that the trace can always be performed by using a complete basis set, here the energy or the position basis)

$$
\begin{aligned}
\mathrm{Tr}\left(\exp(-\mathrm{i}Ht/\hbar)\right) &= \sum_n \exp(-iE_n t/\hbar)\\
&= \int_{-\infty}^{\infty} dx_0 \langle x_0 | \exp(-iHt/\hbar) | x_0 \rangle\\
&= \int_{-\infty}^{\infty} dx_0 \int Dx_{x_0, x_0, t} \exp\left[\frac{i}{\hbar} S(x, \dot{x})\right].
\end{aligned}\tag{7.2}
$$

The basic idea is to compute the trace using the path integral, in some approximation, express it in the form of $\sum_n \exp(-iE_n t/\hbar)$ and thus extract the energy levels E_n. The approximation used is the stationary phase approximation [181].

Stationary Phase Approximation:

Consider the multiple integral

$$I = \int_{-\infty}^{\infty} dx_1 \int_{-\infty}^{\infty} dx_2 \cdots \int_{-\infty}^{\infty} dx_N f(\mathbf{x}) \exp\left(-ih(\mathbf{x})\right).\tag{7.3}$$

Here $\mathbf{x}$ is the collective symbol for $\{x_1, x_2, \cdots x_N\}$. Now we assume that the function $h(\mathbf{x})$ has one extremum, stationary point, at $\mathbf{x} = \mathbf{a}$. Taylor expansion of this function about $\mathbf{a}$ yields

$$h(\mathbf{x}) = h(\mathbf{a}) + \frac{1}{2}\zeta_i M_{ij}\zeta_j + O(\zeta^3).\tag{7.4}$$

Here $\zeta_i \equiv x_i - a_i$ is the deviation from the stationary point. Note that the linear term in the expansion is absent because the expansion is around a stationary point. The essence of the stationary phase approximation (SPA) is that if the exponential term $\exp\left(-ih(\mathbf{x})\right)$ oscillates rapidly compared to the other scales in the problem, then the major contribution to the integral, Eq. (7.3), comes from the neighbourhood of the stationary point of $h(\mathbf{x})$, i.e., around $\mathbf{x} = \mathbf{a}$. Then, to the leading order, the terms that are of the third order and higher in the deviation from the stationary point can be neglected. This implies that the terms $O(\zeta^3)$ in Eq. (7.4) can be neglected in this approximation. In that case I, Eq. (7.3), becomes

$$
\begin{aligned}
I &\approx f(\mathbf{a})\exp\left(-ih(\mathbf{a})\right)\int d\zeta_1\int d\zeta_2\cdots\int d\zeta_N \exp\left(-i\frac{1}{2}\zeta_i M_{ij}\zeta_j\right) \\
&= f(\mathbf{a})\exp\left(-ih(\mathbf{a})\right)\left(\frac{2\pi}{i}\right)^{N/2}(\text{DetM})^{-1/2}.
\end{aligned}
\tag{7.5}
$$

The last step involves a standard Gaussian integration and DetM corresponds to the determinant of the matrix M.

Problem 1: Do the Gaussian integration needed to achieve the last step in Eq. (7.5).

The path integral being a multiple integral, the SPA applied to the integral in Eq. (7.3) has a straightforward extension to it. Now the role played by $h(\mathbf{x})$ is taken over by the action $S[x(t)]$. Consider the Lagrangian $L = \frac{1}{2}(\dot{x})^2 - V(x)$. The functional Taylor expansion of the action, about the classical path $x_{cl}(t)$, gives

$$
S[x(t)] = S[x_{cl}(t)] + \frac{1}{2}\int_0^t d\tau\zeta(\tau)O(\tau)\zeta(\tau) + O(\zeta^3).
\tag{7.6}
$$

Here $\zeta(\tau) \equiv x(\tau) - x_{cl}(\tau)$, $\zeta(0) = \zeta(t) = 0$ and

$$
O(\tau) \equiv -\frac{\partial^2}{\partial\tau^2} - \left(\frac{\partial^2 V}{\partial x^2}\right)_{x_{cl}(\tau)}.
\tag{7.7}
$$

Note that the role played by the matrix M_{ij} in Eq. (7.5) is subsumed here by the operator O. Applying this to the path integral in Eq. (7.1) we have

$$
\begin{aligned}
\langle x',t|x,0\rangle &= \int Dx_{x,0;x',t}\exp\left[\frac{i}{\hbar}S(x,\dot{x})\right] \\
&\approx \mathcal{N}'(t)\exp\left[\frac{i}{\hbar}S[x_{cl}(t)]\right](\text{DetO}(t))^{-1/2}.
\end{aligned}
\tag{7.8}
$$

Here all the $2\pi i\hbar$ factors have been absorbed in the measure $\mathcal{N}$, see Eq. (103), Chapter 2. It goes without further ado that Eq. (7.8) is an approximation which involves neglecting cubic and higher order terms in $\zeta(t)$, the quantum fluctuations. It implies that the action $S[x_{cl}(t)]$ is large compared with respect to $\hbar$. Then the exponential oscillates rapidly such that the major contribution to the integral, in the path integral, comes from the neighbourhood of the stationary point. The leading term of the contribution comes from the classical path $x_{cl}(t)$ and the paths in its neighbourhood, hence the notation semiclassical approximation. Quantum effects are included through the terms $\mathcal{N}'(t)(\text{DetO}(t))^{-1/2}$.

The method breaks down if $\text{DetO}(t) = 0$. This happens due some symmetry in the problem or its spontaneous breakdown. We will have occasion to discuss this later. If the integrand has more than one stationary point then, provided these points are widely separated, SPA implies that each stationary path will make a separate additive contribution.

An Illustrative Example:

We now illustrate the SPA with the ubiquitous harmonic oscillator with $V(x) = \frac{1}{2}\omega^2 x^2$. Starting from Eq. (7.2) we have

$$\text{Tr}\big(\exp(-\text{iHt}/\hbar)\big) = \sum_n \exp(-iE_n t/\hbar)$$

$$= \int_{-\infty}^{\infty} dx_0 \int Dx_{x_0,x_0,t} \exp\left[\frac{i}{\hbar}\int_0^t dt\big(\frac{1}{2}(\dot{x})^2 - \frac{1}{2}\omega^2 x^2\big)\right]. \tag{7.9}$$

The classical solution $x_{cl}(t)$ for the harmonic oscillator from $x_{cl}(0) = x_0$ to $x_{cl}(t) = x_0$ is well known and its action is [43]

$$S[x_{cl}(t)] = -2\omega x_0^2 \frac{\sin^2(\omega t/2)}{\sin(\omega t)}. \tag{7.10}$$

We implement SPA by expanding the action, about the classical path $x_{cl}(t)$, as in Eq. (7.6) to get

$$S[x(t)] = S_{cl} + \frac{1}{2}\int_0^t d\tau \zeta(\tau)\left(-\frac{\partial^2}{\partial \tau^2} - \omega^2\right)\zeta(\tau). \tag{7.11}$$

Here $\zeta(t)$ is, as defined before, the deviation from the classical path, the quantum fluctuation and is zero at the end points, i.e., $\zeta(0) = \zeta(t) = 0$. Since the harmonic oscillator is quadratic in $x(t)$, there are no cubic or higher order terms in $\zeta(t)$ in Eq. (7.11). Thus the SPA coincides, in this case, with the exact result. This is one of the reasons for the proliferation of the use of the harmonic oscillator in the literature. Using Eq. (7.8) in Eq. (7.9) and making use of Eq. (7.11) we have

$$\sum_n \exp(-iE_n t/\hbar) = \int_{-\infty}^{\infty} dx_0 \exp\left[\frac{i}{\hbar}S[x_{cl}(t)]\right]\int D[\zeta(\tau)] \tag{7.12}$$

$$\times \exp\left(\frac{i}{\hbar}\int_0^t d\tau\big(\frac{1}{2}\zeta\mathcal{O}(\tau)\zeta\big)\right)$$

$$= \int_{-\infty}^{\infty} dx_0 \exp\left[\frac{i}{\hbar}S[x_{cl}(t)]\right]\big(\text{Det}\mathcal{O}(t)\big)^{-1/2}\mathcal{N}'(t), \tag{7.13}$$

where $\mathcal{O}(\tau) = -\frac{\partial^2}{\partial \tau^2} - \omega^2$. The next task is the determination of the determinant of the operator $\mathcal{O}(t)$. This can be achieved by considering its eigenvalue equation. Note that since $\zeta(0) = \zeta(t) = 0$, a suitable eigenfunction would be $\sin\left(\frac{n\pi\tau}{t}\right)$, and the eigenvalue equation is

$$\left(-\frac{\partial^2}{\partial\tau^2} - \omega^2\right)\sin\left(\frac{n\pi\tau}{t}\right) = \left(\frac{n^2\pi^2}{t^2} - \omega^2\right)\sin\left(\frac{n\pi\tau}{t}\right). \tag{7.14}$$

Here n are positive integers. From this, using the eigenvalues, the determinant can be obtained as

$$\begin{aligned}
\left(\text{Det}\mathcal{O}(t)\right)^{-1/2} &= \prod_{n=1}^{\infty}\left(\frac{n^2\pi^2}{t^2} - \omega^2\right)^{-1/2} \\
&\equiv \alpha(t)\prod_{n=1}^{\infty}\left(1 - \frac{\omega^2 t^2}{n^2\pi^2}\right)^{-1/2}. \tag{7.15}
\end{aligned}$$

It can be shown that the infinite product on the *RHS* of the above equation is equal to $\left(\frac{\sin(\omega t)}{\omega t}\right)^{-1/2}$ [182]. Using Eq. (7.15) and Eq. (7.10), Eq. (7.13) becomes

$$\begin{aligned}
\sum_n \exp(-iE_n t/\hbar) &= \alpha(t)\mathcal{N}'(t)\int dx_0 \exp\left(-2i\omega x_0^2\frac{\sin^2(\omega t/2)}{\sin(\omega t)}\right)\left(\frac{\omega t}{\sin(\omega t)}\right)^{1/2} \\
&= \alpha(t)\mathcal{N}'(t)\frac{1}{2i\sin(\omega t/2)}(2\pi i\hbar t)^{1/2} \\
&= \alpha(t)\mathcal{N}'(t)(2\pi i\hbar t)^{1/2}\sum_{n=0}^{\infty}\exp\left[-i(n+\frac{1}{2})\omega t\right]. \tag{7.16}
\end{aligned}$$

In the *RHS* of the above equation, the second line is a result of a simple Gaussian integration and the third line is due to the following geometric series

$$\sum_{n=0}^{\infty}\exp\left[-i(n+\frac{1}{2})\omega t\right] = \frac{1}{2i\sin(\omega t/2)}. \tag{7.17}$$

Problem 2: Prove the above statements.

In Eq. (7.16), we adjust the constants such that $\alpha(t)\mathcal{N}'(t) = (2\pi i\hbar t)^{-1/2}$. Then a comparison of the *LHS* with the *RHS* yields the familiar result

$$E_n = (n+\frac{1}{2})\hbar\omega. \tag{7.18}$$

Contact with WKB:
    We now bring out the connection between the semi-classical path integral methods, discussed above, with the traditional WKB methods. We begin with the transition amplitude in Eq. (7.8). As seen there, the determinant in Eq.

(7.7) needs to be evaluated. From a careful analysis, see for e.g. [177], it can be shown that

$$\langle x',t|x,0\rangle \approx \mathcal{N}'(t)\exp\left[\frac{i}{\hbar}S[x_{cl}(t)]\right]\left(\mathrm{Det}O(t)\right)^{-1/2}$$

$$= \frac{\mathcal{N}'(t)}{\sqrt{\phi_{V''}^{(0)}(t)}}\exp\left[\frac{i}{\hbar}S[x_{cl}(t)]\right]. \tag{7.19}$$

Here $V'' = \frac{\partial^2}{\partial x^2}$ and $\phi_{V''}^{(0)}(t)$ satisfies the equation

$$\left(m\frac{d^2}{dt^2}+V''(x_{cl})\right)\phi_{V''}^{(0)}(t) = 0. \tag{7.20}$$

Since the classical equations of motion, the Euler-Lagrange equations, for the Lagrangian $L = \frac{1}{2}(\dot{x})^2 - V(x)$ is

$$m\frac{d^2 x_{cl}}{dt^2}+V'(x_{cl}) = 0, \tag{7.21}$$

the Eq. (7.20) can be recast in the following convenient form

$$\left(m\frac{d^2 x_{cl}}{dt^2}+V''(x_{cl})\right)\frac{dx_{cl}}{dt} = 0. \tag{7.22}$$

A comparison of Eqs. (7.20) and (7.22) yields

$$\phi_{V''}^{(0)}(t) \propto \frac{dx_{cl}}{dt} = p(x_{cl}). \tag{7.23}$$

Here $p(x_{cl})$ is the momentum associated with $x_{cl}(t)$. Using Eq. (7.23) in Eq. (7.19) we can see that the transition amplitude becomes

$$\langle x',t|x,0\rangle = \frac{\mathcal{N}'(t)}{\sqrt{p(x')}}\exp\left[\frac{i}{\hbar}S[x_{cl}(t)]\right]. \tag{7.24}$$

This clearly brings out the desired connection between the semi-classical path integral methods, with the traditional WKB methods [183]. As a side note, let us revisit the wavefunction of a particle $\psi(x)$ in a potential $V(x)$, in the WKB picture. It is given by

$$\psi(x) = \frac{\mathcal{N}}{\sqrt{p(x)}}\exp\left[\frac{i}{\hbar}S[x_{cl}(t)]\right]. \tag{7.25}$$

Here $\mathcal{N}$ is a normalization factor. From this, the probability density

$$\psi^*(x)\psi(x) \propto \frac{1}{p(x)}. \tag{7.26}$$

The probability density is seen to be inversely proportional to the momentum. This is consistent with our classical intuition in that the slower the particle, i.e., lesser velocity, the more likely it is to be found. Thus, the WKB wavefunction is consistent with our classical understanding and hence the terminology *semiclassical* coined to it and to its counterparts discussed here.

7.3   Double Well Potential

The prototype example of quantum tunneling is the double well potential. We have been acquainted with this in the last chapter in the context of the Spin-Boson model. Tunneling in a double well has found myriad applications in the literature, ranging from condensed matter [1, 176] to cosmology [177, 184]. We will give a sketch of the model, and discuss the effect of tunneling in it via WKB, briefly, and by the method of instantons. This will also serve as an introduction to the notion of the instanton [159, 43].

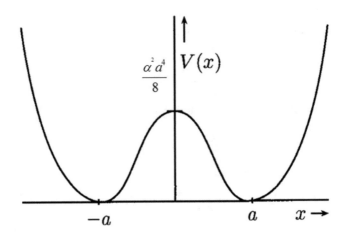

Figure 7.1: Double well potential with local minima at $\mp a$. Figure adapted from [43].

Consider the motion of a particle of mass m, in one dimension, moving in an anharmonic potential

$$V(x) = \frac{\alpha^2}{8}\left(x^2 - a^2\right)^2. \qquad (7.27)$$

Here α and a are constants. The potential $V(x)$ has the shape of a double well with two minima at $x = \pm a$, see Fig. (7.1). Due to its form it is sometimes also referred to as the Mexican Hat Potential . Note that the potential has a local maximum at the origin $x = 0$, with the potential cutting the ordinate at $V(x) = \frac{\alpha^2 a^4}{8}$. Thus, the picture emerges of two wells centered around $x = \pm a$ and separated by a wall or barrier of height $\frac{\alpha^2 a^4}{8}$. This highlights the role of the parameter α as the coupling between the two wells. For infinite large coupling α, the potential separates into two separate symmetrical wells. A particle in one well stays put there. From the reflection symmetry of the potential, both the wells will have degenerate energy levels. However, when the coupling α is finite, there exists the possibility of tunneling between the two wells. This breaks the energy degeneracy and results in a splitting of the energy levels. The task then is to compute this splitting. It is quite straightforward to compute it using standard WKB techniques leading to the splitting $\Delta E = E_2 - E_1$ between the erstwhile degenerate energy levels E_1, E_2, corresponding to the energy of the symmetric and anti-symmetric combinations of the ground state wavefunction $\phi_0(x)$, respectively [183]

$$\Delta E = \frac{4e}{\pi}\sqrt{\hbar m}\omega^{3/2}ae^{-\frac{1}{\hbar}S_0}. \tag{7.28}$$

The parameter ω is identified with the frequency of harmonic oscillations near the well minima, i.e., $V''(x = \pm a) = \alpha^2 a^2 = m\omega^2$. Also, $S_0 = \frac{2m\omega a^2}{3} = \frac{2m^2\omega^3}{3\alpha^2}$. The damped exponential in the *RHS* of Eq. (7.28) is a typical signature of tunneling.

Next we attempt the same problem using path integration. This will, as seen in the sequel, make use of the Euclidean formulation of path integrals, typically of use in discussions related to tunneling and have been introduced earlier, see for e.g., Eq. (18) in Chapter 4. Further, this exercise will serve to give a brief introduction of the concept of instantons and its uses. We begin with the transition amplitude, Eq. (7.1), but with the initial and final times changed to $\mp\frac{t}{2}$, respectively, for notational convenience

$$\langle x_f, t/2|x_i, -t/2\rangle = \langle x_f|e^{-\frac{i}{\hbar}Ht}|x_i\rangle = \int Dx_{x_i,-t/2;x_f,t/2}\exp\left[\frac{i}{\hbar}S(x,\dot{x})\right]. \tag{7.29}$$

We will be interested in the asymptotics of the dynamics, i.e., in the limit $t \to \infty$. The action for the double well potential, Eq. (7.27), is

$$S[x] = \int_{-t/2}^{t/2} d\tau\left(\frac{1}{2}m\dot{x}^2 - \frac{\alpha^2}{8}(x^2 - a^2)^2\right). \tag{7.30}$$

A very convenient way to evaluate path integrals is to use the Euclidean formulation which implies rotation to imaginary time $\tau \to -i\tau$ [185]. The appearance of imaginary time can also be motivated from the following physical grounds. Tunneling entails particle motion under a barrier; the total energy is less than the potential energy. If we interpret the difference as a negative kinetic energy, it would not be difficult to imagine imaginary velocity and hence imaginary time. Going to Euclidean space the transition amplitude assumes the form

$$\langle x_f|e^{-\frac{1}{\hbar}Ht}|x_i\rangle = \int Dx\exp\left[-\frac{1}{\hbar}S_E(x)\right], \tag{7.31}$$

where the subscript E denotes Euclidean. It is important to note that the Euclidean action has the form

$$S_E[x] = \int_{-t/2}^{t/2} d\tau\left(\frac{1}{2}m\dot{x}^2 + \frac{\alpha^2}{8}(x^2 - a^2)^2\right). \tag{7.32}$$

Problem 3: Derive Eq. (7.32) from Eq. (7.30).

Note that time τ in Eq. (7.32) is imaginary and $\dot{x}$ denotes a derivative $w.r.t$ imaginary time. We now use the semiclassical path integral method, invoking SPA, to this path integral. The classical equation of motion can be obtained from the action, Eq. (7.32), as

$$m\ddot{x} - V'(x) = 0. \tag{7.33}$$

Here, as usual, dot denotes derivative $w.r.t$ time and dash, in the superscript, denotes a spatial derivative. Also, $V(x)$ is as in Eq. (7.27). From Eq. (7.33), it emerges that the Euclidean equations correspond to particle motion in an inverted potential $-V(x)$. The corresponding Euclidean energy is

$$E = \frac{1}{2}m\dot{x}^2 - V(x). \tag{7.34}$$

The energy landscape now indicates hills at $x = \pm a$ and a valley in between at $x = 0$. $x(t) = \pm a$ is an obvious solution of Eq. (7.33). This would correspond to the situation, in real time, of small harmonic oscillations about the bottom of either well. In the asymptotic, large time limit, there are further nontrivial solutions $x_{cl}(t)$ of Eq. (7.33), which are in the context of quantum tunneling, as will be seen shortly, called *bounce*. They can be seen to be

$$x_{cl} = \pm a \tanh\left(\frac{\omega(t - \tau_c)}{2}\right). \tag{7.35}$$

Here τ_c is a constant denoting the time when the solution reaches the valley of the Euclidean potential and is arbitrary, due to time translational invariance. Further, ω is, as discussed above, identified with the harmonic oscillations frequency near the well minima, i.e., $V''(x = \pm a) = \alpha^2 a^2 = m\omega^2$.

Problem 4: Verify that Eq. (7.35) is indeed a solution of Eq. (7.33).

It can be easily seen that the asymptotic behavior of the solutions depicted in Eq. (7.35) are

$$\begin{aligned} x_{cl}(t \to \infty) &= \pm a, \\ x_{cl}(t \to -\infty) &= \mp a. \end{aligned} \tag{7.36}$$

These are depicted in Figs. (7.2). The solutions thus correspond to, in the Euclidean picture, to have started from the top of one hill at $t \to -\infty$ and going over to the other top at $t \to \infty$. They carry the topological index [159]

$$Q = \frac{1}{2a}(x_{cl}(\infty) - x_{cl}(-\infty)) = 1. \tag{7.37}$$

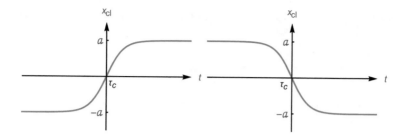

Figure 7.2: The two instanton solutions depicted in Eqs. (7.35) and (7.36).

Noting that for Eq. (7.35)

$$
\begin{aligned}
\dot{x}_{cl} &= \pm \frac{a\omega}{2} \operatorname{sech}^2\left(\frac{\omega(t - \tau_c)}{2} \right) \\
&= \mp \frac{\omega}{2a}\left(x_{cl}^2 - a^2 \right) \\
&= \mp \sqrt{2V(x_{cl})m},
\end{aligned}
\tag{7.38}
$$

we see that the Euclidean energy, Eq. (7.34), for such solutions are

$$
E = \frac{1}{2}m\dot{x}_{cl}^2 - V(x_{cl}) = 0.
\tag{7.39}
$$

Hence, the nontrivial solutions, Eq. (7.35), correspond to minimum energy solutions. The corresponding action is seen to be

$$
\begin{aligned}
S_E[x_{cl}] &= \int_{-\infty}^{\infty} dt \left(\frac{1}{2}m\dot{x}_{cl}^2 + V(x_{cl}) \right) \\
&= \int_{-\infty}^{\infty} dt\, m\dot{x}_{cl}^2 = m \int_{\mp a}^{\pm a} dx_{cl}\, \dot{x}_{cl} \\
&= \frac{2m^2\omega^3}{3\alpha^2}.
\end{aligned}
\tag{7.40}
$$

The solutions in Eq. (7.35) are thus finite action classical solutions in Euclidean space and are known as the *instanton*, starting from $-\infty$ and ending up in ∞ and hence would be the analogue of *kinks* discussed in the last chapter in the context of the partition function of the spin-Boson model, or the *anti-instanton*, starting from ∞ and ending up in $-\infty$, being analogous to *anti-kinks*. There the kink-anti-kink pair was termed as *bounce*, which as will be noticed in the discussions to follow, play a crucial role in quantum mechanical tunneling. The corresponding Euclidean Lagrangian, cf. Eq. (7.32) is

$$
\begin{aligned}
L_E &= \frac{1}{2}m\dot{x}_{cl}^2 + V(x_{cl}) \\
&= \frac{ma^2\omega^2}{4} \operatorname{sech}^4\left(\frac{\omega(t - \tau_c)}{2} \right).
\end{aligned}
\tag{7.41}
$$

The Lagrangian is thus seen to have a narrow spread around $t = \tau_c$ with the width at half maximum being

$$\Delta t \propto \frac{1}{\omega} = \frac{\sqrt{m}}{\alpha a}. \tag{7.42}$$

Thus, the instantons could be thought of as finite action Euclidean solutions with a finite size of about $\frac{1}{\omega}$.

Single Instanton Path Integral:

We now proceed to evaluate the (Euclidean) path integral, which is basically Eq. (7.31) with x_i and x_f being replaced by $\mp a$, using the semiclassical methods discussed above. As a result, we have

$$\langle a | e^{-\frac{1}{\hbar} H t} | - a \rangle_{1\mathrm{I}} = \int Dx \exp \left[-\frac{1}{\hbar} S_E(x) \right]$$

$$= \mathcal{N'}_E(t) \exp \left[-\frac{1}{\hbar} S_E[x_{cl}(t)] \right] \left(\mathrm{Det}\mathrm{O}(\mathrm{t}) \right)^{-1/2}. \tag{7.43}$$

The subscript 1I stands for one instanton while the subscript E stands, as usual, for Euclidean. Here we have made use of Eq. (7.8), in the Euclidean domain. The path integral is obtained by performing the Gaussian integration over the quantum fluctuations $\zeta(\tau) \equiv x(\tau) - x_{cl}(\tau)$ with $\zeta(-a) = \zeta(a) = 0$, with $x_{cl} = a \tanh \left(\frac{\omega(t-\tau_c)}{2} \right)$ and

$$O(\tau) \equiv -\frac{\partial^2}{\partial \tau^2} + \left(\frac{\partial^2 V}{\partial x^2} \right)_{x_{cl}(\tau)}. \tag{7.44}$$

When the range of the parameter τ tends to infinity, the operator $O(\tau)$ will have a zero eigenvalue. This is a consequence of the translational invariance of the action in the τ variable. This causes a technical problem, i.e., it leads to the divergence of Eq. (7.43). Also, the translational invariance ensures that given a solution $x_{cl}(\tau)$, $x_{cl}(\tau - \tau_c)$, for any τ_c, say, the center of the instanton and called the *collective coordinate*, will also be an instanton solution and hence will contribute to Eq. (7.43). This cumulative effect is taken care of by multiplying Eq. (7.43) by $\int_{-t/2}^{t/2} d\tau_c$. The way to proceed is to replace the zero mode coordinate by the collective coordinate by a suitable change of variables. Such an analysis on Eq. (7.43) leads to [159, 43, 177]

$$\langle a | e^{-\frac{1}{\hbar} H t} | - a \rangle_{1\mathrm{I}} = \mathcal{N'}_E(t) J \exp \left[-\frac{1}{\hbar} S_0 \right] \left(\mathrm{Det}\mathrm{O'}(\mathrm{t}) \right)^{-1/2} \int_{-t/2}^{t/2} d\tau_c. \tag{7.45}$$

Here $J = \left(\frac{S_0}{m} \right)^{1/2}$ is the Jacobian involved in the transformation used to take care of the collective coordinate. Also, S_0 stands for $S_E[x_{cl}(t)]$ and $\mathrm{Det}\mathrm{O'}(t)$ denotes the determinant of the operator $O(t)$, Eq. (7.44), with the zero mode excluded. Since the range is from $-t$ to t, it can be broken up into sub-intervals and the above path integral can be evaluated by evaluating the operator $e^{-\frac{1}{\hbar}Ht}$ in these sub-intervals. Baring the intervals where the instanton size differs substantially from zero, see the discussion on its finite size above, the operator $O'(t)$ can be approximated by a harmonic oscillator of frequency ω, i.e., $-\frac{\partial^2}{\partial\tau^2} + \frac{\partial^2 V}{\partial x^2} \approx -\frac{\partial^2}{\partial\tau^2} + \omega^2$. Thus,

$$\left(\text{Det}O'(t)\right)^{-1/2} = \left[\text{Det}\left(-\frac{\partial^2}{\partial\tau^2} + \omega^2\right)\right]^{-1/2}I. \tag{7.46}$$

Here I is a constant independent of t in the limit of t going to infinity and can be shown to be [43, 177, 160]

$$I = \sqrt{\frac{2m\omega}{\hbar}}\,2a\omega\left(\frac{S_0}{m}\right)^{-1/2}. \tag{7.47}$$

Using this, Eq. (7.45) becomes

$$\langle a|e^{-\frac{1}{\hbar}Ht}|-a\rangle_{11} = \mathcal{N}'_E(t)JI\exp\left[-\frac{1}{\hbar}S_0\right]\left[\text{Det}\left(-\frac{\partial^2}{\partial\tau^2} + \omega^2\right)\right]^{-1/2}\int_{-t/2}^{t/2}d\tau_c. \tag{7.48}$$

Note that $\mathcal{N}'_E(t)\left[\text{Det}\left(-\frac{\partial^2}{\partial\tau^2} + \omega^2\right)\right]^{-1/2}$ is the result of the Gaussian functional integral encountered before in the context of the harmonic oscillator path integral, now analytically continued to imaginary time, i.e., $t \to -it$. Its value, in real time, is $\left(\frac{1}{2\pi i\hbar t}\frac{m\omega t}{\sin(\omega t)}\right)^{1/2}$, see Eqs. (121) and (122) in Chapter 2. By analytical continuation to imaginary time we see that

$$\mathcal{N}'_E(t)\left[\text{Det}\left(-\frac{\partial^2}{\partial\tau^2} + \omega^2\right)\right]^{-1/2} = \left(\frac{1}{2\pi\hbar}\frac{m\omega}{\sinh(\omega t)}\right)^{1/2}. \tag{7.49}$$

Note that time t on the *RHS* of the above equation denotes imaginary time. In the limit of time going to infinity, we get

$$\lim_{t\to\infty}\left(\frac{1}{2\pi\hbar}\frac{m\omega}{\sinh(\omega t)}\right)^{1/2} \to \left(\frac{m\omega}{\pi\hbar}\right)^{1/2}e^{-\omega t/2}. \tag{7.50}$$

Collecting these terms we have the one instanton contribution to be

$$\lim_{t\to\infty}\langle a|e^{-\frac{1}{\hbar}Ht}|-a\rangle_{11} = JI\exp\left[-\frac{1}{\hbar}S_0\right]\left(\frac{m\omega}{\pi\hbar}\right)^{1/2}e^{-\omega t/2}\int_{-t/2}^{t/2}d\tau_c. \tag{7.51}$$

Since the above path integral is in the limit of infinite time t, configurations of such instantons with arbitrarily large separations, separations much larger than the instanton size, are favourable. The single anti-instanton, i.e., where the solution starts from $t = \infty$ and goes over to $t = -\infty$, path integral can be seen to be the same.

Multi-instanton Contributions and Tunneling induced Splitting:

A string of widely separated instantons and anti-instantons also satisfy the Euclidean equations of motion, Eq. (7.33). They form what is sometimes referred to as the dilute instanton gas. Thus their contributions to the transition amplitudes will be multiplicative. In such a scenario we can see that the n instanton solution, using Eq. (7.51), is

$$\left(\frac{m\omega}{\pi\hbar}\right)^{1/2} e^{-\omega t/2} \frac{(\alpha t)^n}{n!}. \tag{7.52}$$

Here $\alpha = 2\sqrt{\frac{2m}{\hbar}}(\omega)^{3/2} a e^{-\frac{1}{\hbar}S_0}$. The factor t^n is obtained by integrals, such as $\int_{-t/2}^{t/2} d\tau_c$, over the centers of the n instantons while $n!$ takes care of the indistinguishability of the n instantons. Only an even number of instantons and anti-instantons will contribute to transition amplitudes like $\langle a|e^{-\frac{1}{\hbar}Ht}|a\rangle$ and $\langle -a|e^{-\frac{1}{\hbar}Ht}|-a\rangle$, while an odd number will contribute to terms like $\langle a|e^{-\frac{1}{\hbar}Ht}|-a\rangle$ and $\langle -a|e^{-\frac{1}{\hbar}Ht}|a\rangle$. Adding the contributions we get, for example,

$$\begin{aligned}
\langle -a|e^{-\frac{1}{\hbar}Ht}|-a\rangle &= \sum_n \left(\frac{m\omega}{\pi\hbar}\right)^{1/2} e^{-\omega t/2} \frac{(\alpha t)^{2n}}{2n!} \\
&= \frac{1}{2}\left(\frac{m\omega}{\pi\hbar}\right)^{1/2} \left(e^{-(\frac{\omega}{2}-\alpha)t} + e^{-(\frac{\omega}{2}+\alpha)t}\right). \tag{7.53}
\end{aligned}$$

Similarly,

$$\begin{aligned}
\langle a|e^{-\frac{1}{\hbar}Ht}|-a\rangle &= \sum_n \left(\frac{m\omega}{\pi\hbar}\right)^{1/2} e^{-\omega t/2} \frac{(\alpha t)^{2n+1}}{(2n+1)!} \\
&= \frac{1}{2}\left(\frac{m\omega}{\pi\hbar}\right)^{1/2} \left(e^{-(\frac{\omega}{2}-\alpha)t} - e^{-(\frac{\omega}{2}+\alpha)t}\right). \tag{7.54}
\end{aligned}$$

The transition amplitude can also be expressed by inserting a complete basis of energy eigenstates. Identifying the two low lying states, low energy levels, of the Hamiltonian H with energy eigenvalues $E_{\pm}$ by $|\pm\rangle$, we have for t tending to infinity

$$\begin{aligned}
\langle -a|e^{-\frac{1}{\hbar}Ht}|-a\rangle &\approx \langle -a|e^{-\frac{1}{\hbar}Ht}|-\rangle\langle -|-a\rangle + \langle -a|e^{-\frac{1}{\hbar}Ht}|+\rangle\langle +|-a\rangle \\
&= e^{-\frac{1}{\hbar}E_-t}\langle -a|-\rangle\langle -|-a\rangle + e^{-\frac{1}{\hbar}E_+t}\langle -a|+\rangle\langle +|-a\rangle. \tag{7.55}
\end{aligned}$$

Comparing Eq. (7.55) with Eq. (7.53) we see that

$$E_{\pm} = \hbar\left(\frac{\omega}{2} \pm \alpha\right). \tag{7.56}$$

The splitting of the low lying energy levels $\Delta E = E_+ - E_-$ is thus

$$\begin{aligned}
\Delta E &= 2\hbar\alpha \\
&= 4\sqrt{2m\hbar}\, \omega^{3/2} a \exp\left[-\frac{1}{\hbar}S_0\right]. \tag{7.57}
\end{aligned}$$

It is instructive to compare the splitting obtained by the instanton technique, Eq. (7.57), with that using the standard WKB method, Eq. (7.28). It is worth noting that even though the standard WKB method yields the correct exponential factor, the instanton method is more useful for estimating the coefficient of the exponential and is also simpler to extend to the case of higher dimensions, as could be envisaged in a quantum field theoretic or open quantum system application [186].

7.4   Quantum Tunneling

The problem of transition by a system from one state, say a metastable state, to another by getting over a barrier is a theme that is recurrent in many areas of physics, ranging from nuclear [187], low temperature physics [1, 188], chemical kinetics to transport in biomolecules [189, 176]. At high temperatures the principal escape mechanism is thermal activation, while at low temperatures quantum effects start playing an important role. Thus, for example, at zero temperature the system, localized in the metastable well ground state, can only rely on quantum tunneling to effect an escape! The modern era of thermal driven state transition theory could be said to have its beginning in Kramer's work [190] where the classical aspects of the problem in the presence of weak and moderate to strong damping were considered. The quantum mechanical aspects of the problem taking into account the effect of dissipation on tunneling was initiated in [188, 191] from the perspective of macroscopic quantum coherence [68]. This was generalized in [192].

Let us visualize the following setup: a particle of mass m is moving in an external potential $V(x) = \frac{1}{2}m\omega_0^2 x^2\left(1 - \frac{2x}{3x_b}\right)$, a quadratic plus cubic potential with a single metastable minimum zero at $x = 0$. The potential is such that it is negative for $x > x_{ext}$, referred to as the *exit point* of the barrier. The barrier height is $V(b)$, see Fig. (7.3). A particle that moves out from the exit point is assumed to not return in finite time. The coordinate x would be the tunneling degree of freedom which for chemical reactions would be the reaction coordinate [189]. Metastable nature of the state becomes pertinent when the barrier, here $V(b)$, is large enough such that the decay time of the metastable state is much longer compared to the other characteristic time scales in the problem, such as τ_{rel}, the relaxation time in the locally stable well, thermal time $\hbar\beta$, correlation time scale of the noise τ_n, and the time scales related to the potential curvature at the minima and at the barrier top, i.e., $\tau_0 = \omega_0^{-1}$ and $\tau_b = \omega_b^{-1}$, respectively. Here $\omega_0 = \left(V''(0)/m\right)^{1/2}$ and $\omega_b = \left(-V''(x_b)/m\right)^{1/2}$ are the frequencies of small oscillations about the well minima and unstable maxima, respectively. Note the minus sign in ω_b, which is indicative of the unstable nature of the barrier frequency. This hinders the decay process. Weak metastability implies $V_b \gg k_B T$ and $V_b \gg \hbar\omega_0$. A convenient parametrization of the escape rate α_e from the metastable well is

$$\alpha_e = \Theta e^{-\xi}. \tag{7.58}$$

The quantity Θ is called the *attempt* frequency of the particle in the well towards the barrier and ξ is a measure of the barrier size that needs to be overcome.

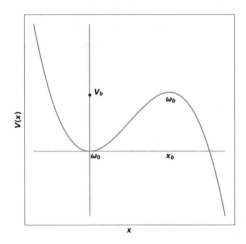

Figure 7.3: Quadratic plus cubic potential $V(x) = \frac{1}{2}m\omega_0^2 x^2 \left(1 - \frac{2x}{3x_b}\right)$ with a stable local minima at ω_0 at $x = 0$, unstable local maxima at ω_b at $x = x_b$ and a barrier height of V_b.

As mentioned above, a careful study of the problem of escape over a barrier in the presence of damping, in the classical regime, was made by Kramers [190]. He modelled the problem as a nonlinear Brownian motion and studied the evolution of the probability density $p(x, v, t)$ using a Fokker-Planck equation, see Chapter 3. The result obtained for the escape rate α_e is [189]

$$\alpha_e = \Theta_{cl}\, e^{-V_b/k_B T}. \tag{7.59}$$

Here the classical attempt frequency $\Theta_{cl} = \rho \frac{\omega_0}{2\pi}$, where ρ is a constant depending upon the frequency ω_b and the damping term. For large damping, the problem can be studied in a similar manner, now making use of the Smoluchowski equation, see Chapter 3. Note the similarity between the forms of Eqs. (7.58) and (7.59). The maximum rate of escape is achieved at a damping value intermediate between the moderate to strong damping regimes [1]. For moderate to strong damping, thermal equilibrium is established inside the well; however this is not so for weak damping. Here the motion of the particle inside the well is oscillatory with damping causing a gradual change in the distribution of energy.

Quantum tunneling has a long history [193, 194]. In [187, 195] tunneling was used to explain the radioactive decay of nuclei, in particular application of tunneling was made to understand alpha decay. An early attempt at providing a quantum mechanical escape rate involved the Wigner function ρ^W [196], a well known quasiprobability distribution, see for e.g. [197] and references therein,

$$\alpha_e = \frac{1}{Z_0} \frac{1}{2\pi\hbar} \int dx dp \; \frac{p}{m} \Theta(p)\delta(x - x_b)\rho^W(x,p). \qquad (7.60)$$

Here $\Theta(p)$ and $\delta(x - x_b)$ are the usual step and delta functions, respectively. Also, Z_0 is the partition function of the particle inside the well. Another form of the quantum rate expression, making use of inputs from quantum scattering, is [198]

$$\alpha_e = \frac{1}{Z_0} \mathrm{Re} \left\{ \mathrm{Tr}\left(e^{-\beta H} \mathcal{F} \mathcal{P} \right) \right\}, \qquad (7.61)$$

where H is the Hamiltonian and the flux operator $\mathcal{F}$ is

$$\mathcal{F} = \delta(x - x_b)p/m, \qquad (7.62)$$
$$\mathcal{P} = \Theta(p). \qquad (7.63)$$

It would be pertinent here to note that in quantum mechanics we are dealing with operators, hence the expressions above are operators. Using the identity

$$e^{-\beta H} = \frac{1}{\pi\hbar} \lim_{\epsilon \to 0^+} \mathrm{Im} \int_0^\infty dE \; e^{-\beta E} \int_0^\infty d\tau e^{(E+i\epsilon-H)\tau/\hbar}, \qquad (7.64)$$

the quantum rate expression Eq. (7.61) becomes

$$\alpha_e = \frac{1}{Z_0} \frac{1}{2\pi\hbar} \int_0^\infty dE \; p(E)e^{-\beta E}, \qquad (7.65)$$

where $p(E)$ is the transmission probability at energy E and is given by

$$p(E) = \lim_{\epsilon \to 0^+} \mathrm{Im} \int_0^\infty d\tau e^{(E+i\epsilon)\tau/\hbar} \int dx \delta(x - x_b)|\dot{x}|_{x=x_b} \langle x|e^{-H\tau/\hbar}|x\rangle. \qquad (7.66)$$

Eq. (7.65) with Eq. (7.66) is the quantum mechanical rate function for a one-dimensional system.

   Next we use *semiclassical methods* to tackle Eq. (7.65). To this end repeated use is made of the SPA, as developed above, to the matrix element involving $e^{-H\tau/\hbar}$ and then to the time integral $\int_0^\infty d\tau$ [159]. The SPA over $e^{-H\tau/\hbar}$ gives apart from the determinant factors, the exponential $e^{-S_{cl}(\tau)/\hbar}$ and a phase factor which is a multiple of $\pi/2$. This factor is due to the singularity of the velocity $\dot{x}$ whenever the path goes through a turning point [199, 183], also known as conjugate points [200, 201, 202]. The phase per turning point is $e^{i\pi/2}$. For a path which traverses the same one-dimensional orbit n times the phase factor is $e^{in\pi}$, as there are two turning points per traversal. Here $S_{cl}(\tau)$ is the action of the extremal path in the upside down potential $-V$; note that we are using Euclidean time, i.e.,

$$S_{cl}(\tau) = \int_0^\tau d\tau' \left(\frac{1}{2}m\dot{x}^2(\tau') + V(x(\tau')) \right). \qquad (7.67)$$

The SPA applied to the time integral leads to the exponential $\exp - [S_{cl}(\tau) - E\tau]/\hbar$. The stationary points are *periodic orbits* with energy E in the upside-down potential $-V(x)$. Expanding about the stationary points and retaining terms upto the Gaussian approximation, i.e., upto the second term in the Taylor series, we have

$$S_{cl}(\tau) - E\tau = W(E) + \frac{1}{2} \frac{\partial^2 S_{cl}}{\partial \tau^2}\bigg|_{\tau=\tau_{st}} (\tau - \tau_{st})^2. \tag{7.68}$$

Here τ_{st} is the Euclidean time for which the energy integral is stationary. Also,

$$W(E) = 2 \int_{x_1}^{x_2} dx \ \sqrt{2m(V(x) - E)}. \tag{7.69}$$

The limits x_1 and x_2 in the above equation are the turning points, represented by the zeros of the integrand, of the classical motion in $-V(x)$ with total energy

$$E = V(x(\tau)) - \frac{1}{2} m\dot{x}^2(\tau). \tag{7.70}$$

The contribution of the periodic orbit with one cycle to the transmission probability is

$$p_1(E) = -e^{-i\pi} e^{-W(E)/\hbar}. \tag{7.71}$$

For energy $E \ll V_b$, the path stays inside the well and bounces back from the outer turning point x_2, hence the nomenclature *bounce* for the periodic path satisfying Eq. (7.70). The factor $e^{-i\pi}$ in Eq. (7.71) is due to the conjugate points, as discussed above. For a path with n cycles this phase factor is $e^{-i\pi n} = (-1)^n$. Proceeding in this fashion, and assuming that the stationary points are sufficiently separated from each other, the transmission probability becomes

$$p(E) = \sum_{n=1}^{\infty} (-1)^{n-1} e^{-nW(E)/\hbar} = \frac{1}{1 + e^{W(E)/\hbar}}. \tag{7.72}$$

In the semiclassical limit $\hbar \to 0$, the Boltzmann average in the Eq. (7.65) is dominated by the periodic orbit with one cycle of period $\tau = -\partial W/\partial E = \hbar\beta$, and the quantum rate expression becomes

$$\alpha_e = \frac{1}{Z_0} \frac{1}{2\pi\hbar} \int_0^{\infty} dE \ e^{-(\beta E + W(E)/\hbar}, \tag{7.73}$$

which upon using SPA yields [203]

$$\alpha_e = (Z_0)^{-1} (2\pi\hbar|\tau_B'|)^{-1/2} e^{-S_B(\hbar\beta)/\hbar}, \quad T < T_0, \tag{7.74}$$

where

$$|\tau_B'| = \left| \frac{\partial^2 W(E)}{\partial E^2} \right|_{E=E_{\hbar\beta}}, \tag{7.75}$$

and T_0 is defined as $k_B T_0 = \frac{\hbar \omega_b}{2\pi}$, with ω_b being the frequency of small oscillations around the barrier top, see Fig. (7.3). T_0 is, roughly speaking, the *crossover* temperature, where the transition from the classical to quantum decay occurs. Also, $S_B(\hbar\beta)$ is the action, Eq. (7.67), for the bounce trajectory with period $\hbar\beta$ and total conserved energy $E = E_{\hbar\beta}$. Thus,

$$S_B(\hbar\beta) = W(E_{\hbar\beta}) + E_{\hbar\beta}\hbar\beta. \tag{7.76}$$

The Eqs. (7.74) to (7.76) provide us with the semiclassical quantum rate in the regime $0 \leq k_B T < \hbar\omega_0$, where ω_0 is the frequency of small oscillations around the potential minimum, see Fig. (7.3).

In the regime $E \ll V_b$, i.e., for $T \ll T_0$, the deep quantum tunneling regime with the thermal energy very small as compared to $\hbar\omega_0$, the quantum rate expression can be shown to be reduced to [1]

$$\alpha_e = \frac{e^{-\hbar\omega_0/2k_B T}}{Z_0} \exp\left(\alpha_0^2 \frac{S_0}{\hbar} e^{-\hbar\omega_0/k_B T}\right)\psi_0, \tag{7.77}$$

where $Z_0 = (2\sinh(\hbar\beta\omega_0/2))^{-1}$, α_0 is a constant depending upon the barrier shape, S_0 is the zero energy bounce action

$$S_0 = W(0) = 2\int_0^{x_0} dx \ \sqrt{2mV(x)}. \tag{7.78}$$

A comparison of Eq. (7.78) with Eq. (7.69) makes it clear that 0 and x_0 in the above equation are the turning points, representing zeros of the quadratic plus cubic potential $V(x)$, Fig. (7.3), with x_0 being the point where the potential again crosses the x axis in the direction taken by the particle exiting the potential, here the right quadrature. For this potential $S_0 = \frac{36V_b}{5\omega_0}$ and

$$\psi_0 = \alpha_0\omega_0 \left(S_0/2\pi\hbar\right)^{1/2} e^{-S_0/\hbar}, \tag{7.79}$$

is the probability per unit time that a particle localized in the ground state of the metastable potential well escapes by quantum tunneling. Note that Eq. (7.77) is obtained by taking a thermal average of the decay out of the different energy levels $E_n = \hbar\omega_0(n+1/2)$ over a canonical distribution $e^{-\beta E_n}$. Thus, for example, the rate of decay out of the nth excited state is [204]

$$\psi_n = \frac{1}{n!}\left(\alpha_0^2 \frac{S_0}{\hbar}\right)^n \psi_0, \tag{7.80}$$

and its average with the canonical distribution

$$\alpha_e = \frac{1}{\sum_{j=0}^{\infty} e^{-\beta E_j}} \sum_{i=0}^{\infty} \psi_i e^{-\beta E_i}. \tag{7.81}$$

Substituting Eq. (7.80) in Eq. (7.81) we obtain the quantum rate expression in Eq. (7.77).

Imaginary Free Energy Method: A thermodynamic equilibrium approch to the problem of quantum rate can now be discussed. This has its precussor in the work of Langer [178, 179]. This involves the free energy, see Chapter 2, of the metastable system. The free energy can be obtained from the partition function which in turn is defined, for the case of metastable potentials, by an analytical continuation from a stable to the metastable potential of interest, here the potential depicted in Fig. (7.3). Analytical continuation leads to an imaginary part of the free energy of the metastable state which in turn is connected to the decay rate of the system.

Thus, for a metastable system the partition function is

$$\mathcal{Z} = \sum_j e^{-\beta(E_j - i\hbar\psi_j/2)}. \tag{7.82}$$

Here ψ_n is the decay rate out of the nth excited state, see Eq. (7.80). For weak metastability $V_b \gg \hbar\omega_0$, it can be readily seen that $\hbar\psi_n \ll E_n$ for all n, where the notations have the same implications as discussed above. Hence

$$\mathcal{Z} = \mathrm{Re}\mathcal{Z} + i\mathrm{Im}\mathcal{Z} \approx \sum_j e^{-\beta E_j} + i(\hbar\beta/2) \sum_j \psi_j e^{-\beta E_j}. \tag{7.83}$$

Since the free energy $\mathcal{F}$ is connected to the partition function $\mathcal{Z}$ as $\mathcal{F} = -\frac{1}{\beta} \ln \mathcal{Z}$, the imaginary part of the free energy, generally small compared to the real part, is

$$\mathrm{Im}\mathcal{F} = -\frac{1}{\beta} \frac{\mathrm{Im}\mathcal{Z}_b}{\mathcal{Z}_0} = -\frac{\hbar}{2} \frac{\sum_j \psi_j e^{-\beta E_j}}{\sum_n e^{-\beta E_n}}. \tag{7.84}$$

$\mathrm{Im}\mathcal{Z}_b$, the imaginary part of $\mathcal{Z}_b$, is determined by the properties of the barrier, while $\mathcal{Z}_0$ is real and determined by the properties of the potential well. It should be kept in mind that the dominant stationary point for the computation of $\mathcal{Z}_b$ is the periodic bounce path $x_B(\tau)$, in the upside down potential $-V(x)$, with period $\hbar\beta$ in the regime $T < T_0$, i.e., for temperatures less than the crossover temperature T_0, which is basically the quantum tunneling regime.

Problem 5: Using Eq. (7.83) convince yourself that Eq. (7.84) is correct.

A comparison with Eq. (7.81) yields

$$\alpha_e = -\frac{2}{\hbar}\mathrm{Im}\mathcal{F}. \tag{7.85}$$

It is worth noting that Eq. (7.85) generalizes the decay rate of the ground state at $T = 0$ to finite temperature T and is valid in the regime $T \leq T_0$. For $T > T_0$, the rate can be shown to be [203]

$$\alpha_e = -\frac{2}{\hbar} \frac{\beta}{\beta_0} \mathrm{Im}\mathcal{F}. \tag{7.86}$$

Here β and β_0 correspond to the temperatures T and T_0, respectively. The physical picture that emerges is that for conditions of weak metastability, as discussed above, the partition function is dominated by stationary points of the action. There are three solutions, two trivial constant solutions, viz. $\bar{x}(\tau) = x_b$ and $\bar{x}(\tau) = 0$, corresponding to the particle sitting at the barrier top or well minimum, respectively and one nontrivial bounce solution $x_B(\tau)$, about which we will have more to talk on shortly. This situation is analogous to the one discussed for the solutions of Eq. (7.33). For temperatures less than the crossover temperature T_0, there always exists a periodic *bounce* solution, in the upside down potential $-V(x)$ with a period $\hbar\beta$. The shortest period is that controlled by the frequency of small oscillations over the barrier top, i.e., $2\pi/\omega_b$. For temperatures below T_0, the bounce $x_B(\tau)$ is usually smaller than the action of the constant path $\bar{x}(\tau) = x_b$. Since the constant path action has an exponential contribution, the path integral below T_0 is dominated by the bounce. At $T = T_0$ the periodic bounce goes over to the harmonic oscillation. The respective actions are equal and also coincide with the action of the constant path $x = x_b$. The crossover temperature T_0 varies from system to system, being quite large for electrons and very small for tunneling in Josephson junctions [1]. It should be noted that the imaginary free energy method is applicable when thermal equilibrium in established inside the well. This in turn implies that the relaxation time scale inside the well should be short compared to the average escape time from the well.

7.5   Transition to Open Systems

We now generalize our discussion to $N + 1$ degrees of freedom, N for the reservoir and one for the system of interest. This, as should be appreciable is towards making the transition to Open Quantum Systems. The semiclassical path integral method, discussed in this chapter, is well suited to make this transition. This together with the thermodynamic imaginary free energy approach, introduced above, is particularly suited to study tunneling in the wider context of open systems. As a consequence of the connection of the free energy $\mathcal{F}$ to the partition function $\mathcal{Z}$, we will rely on the path integral of the partition function of the *reduced* damped system. The path integral representation of the partition function has been introduced earlier, see Eq. (126) in conjugation with Eq. (108), Chapter 2. Here the path integral would yield an integral of $\exp\{-S_{open}^E\}$ over all periodic paths $x(\tau)$ with period $\tau = \hbar\beta$. The superscript E denotes Euclidean and hence the time τ is imaginary. Further, S_{open}^E is the Euclidean action of the dissipative open system of interest of mass m, velocity

$\dot{x}$, moving in a potential $V(x)$ and is

$$S_{open}^{E}[x] = \int_{0}^{\hbar\beta} d\tau \left(\frac{1}{2} m\dot{x}^2 + V(x(\tau)) \right)$$
$$+ \frac{1}{2} \int_{0}^{\hbar\beta} d\tau \int_{0}^{\tau} d\tau' K(\tau - \tau')(x(\tau) - x(\tau'))^2,$$
$$= \int_{0}^{\hbar\beta} d\tau \left(\frac{1}{2} m\dot{x}^2 + V(x(\tau)) \right) + \int_{0}^{\hbar\beta} d\tau \int_{0}^{\tau} d\tau' k(\tau - \tau')x(\tau)x(\tau').$$
(7.87)

The kernel $K(\tau)$ is

$$K(\tau) = \frac{1}{\pi\hbar\beta} \sum_{n=-\infty}^{\infty} \int_{0}^{\infty} d\omega I(\omega) \frac{2\omega}{\nu_n^2 + \omega^2} e^{i\nu_n\tau}, \qquad (7.88)$$

cf., Eqs. (28) and (29), Chapter 6. The kernel $k(\tau)$ is related to $K(\tau)$ by

$$k(\tau) = \mu : \delta(\tau) : -K(\tau), \qquad (7.89)$$

where

$$\mu = \sum_{n=1}^{N} \frac{c_n^2}{m_n\omega_n^2} = \frac{2}{\pi} \int_{0}^{\infty} d\omega \frac{I(\omega)}{\omega}, \qquad (7.90)$$

and $: \delta(\tau) :$ is the periodically continued, in imaginary Matsubara time, δ function

$$: \delta(\tau) := \frac{1}{\hbar\beta} \sum_{n=-\infty}^{\infty} e^{i\nu_n\tau} = \sum_{n=-\infty}^{\infty} \delta(\tau - n\hbar\beta). \qquad (7.91)$$

With the help of Eqs. (7.89), (7.90) and (7.91), $k(\tau)$ can be expressed as

$$k(\tau) = \frac{1}{\pi\hbar\beta} \sum_{n=-\infty}^{\infty} \int_{0}^{\infty} d\omega \frac{I(\omega)}{\omega} \frac{2\nu_n^2}{\nu_n^2 + \omega^2} e^{i\nu_n\tau}. \qquad (7.92)$$

Here $I(\omega)$ is the spectral density of the reservoir which phenomenologically models the reservoir action and $\nu_n = \frac{2\pi n}{\hbar\beta}$ is the Bosonic Matsubara frequency. It is worth pointing out here that Eq. (7.87) is the Euclidean action for a general linear dissipation, i.e., where the system-reservoir interaction is linear in the respective coordinates. As shown before, Eq. (7.33), motion in Euclidean space corresponds to motion in an upside down potential. The $N + 1$ dimensional upside down potential landscape $-V(x, \mathbf{q})$, cf. Eqs. (26) to (28), Chapter 4, is

$$V(x, \mathbf{q}) = V(x) + \frac{1}{2} \sum_{n=1}^{N} m_i \omega_n^2 \left(q_n - \frac{c_n}{m_n\omega_n^2} x \right)^2. \qquad (7.93)$$

The potential landscape $-V(x, \mathbf{q})$ is concave up and hence stable in one direction but concave down and hence unstable in N directions. The periodic orbit, in the partition function, is thus characterized by N stability angles.

The tunneling contribution to the periodic orbit is dominated by the periodic orbit with period $\hbar\beta$. At $T = 0$, the total energy is zero resulting in an infinite periodicity. At finite T the period becomes $\hbar\beta$. The periodic orbit is referred to as the most probable escape path [1]. The turning points x_1 and x_2, cf. Eq. (7.69), are now appropriate points on the potential landscape of the inverted potential $-V(x, \mathbf{q})$. The most probable escape path leaves the surface containing the point corresponding to the turning point x_1 at time $-\hbar\beta/2$, passes through the well of $-V(x, \mathbf{q})$, hits the surface containing the turning point x_2, perpendicularly, at time zero and returns back to the starting point at time $\hbar\beta/2$. The projection of the most probable escape path on the x axis is the *bounce* path $x_B(\tau)$. This is a stationary point of the action, Eq. (7.87). In the spirit of the semiclassical approximation, discussed in this chapter, the quantum fluctuations about the bounce path $x_B(\tau)$ will be made in the Gaussian approximation. Needless to say, the zero mode, encountered before would have to be dealt with care.

The stationary points for a stable potential are a minima of the action. However, for a metastable potential, the action, for the stationary path, has a saddle point in function space with an unstable direction, where the action further reduces. This in turn implies that the potential has another local minimum corresponding to negative eigenvalue. The mode with negative eigenvalue has to be dealt with using analytical continuation, by deforming the integration contour from the stable to the metastable situation. As a result, the partition function acquires an exponentially small imaginary part which in turn is picked up by the free energy, leading to the decay rate by the imaginary free energy method. This is the essence of the approach, adopted here, to quantum tunneling for open quantum systems; making use of semiclassical methods along with the imaginary free energy technique.

We now briefly go through the steps needed to understand quantum tunneling in open quantum systems. At temperatures below the crossover temperature T_0, the stationary point of interest is the bounce trajectory $x_B(\tau)$. The action evaluated upto the second order (Gaussian approximation) about the path $x(\tau)$ obtained by adding the quantum fluctuations $\chi(\tau)$ to the bounce, i.e., $x(\tau) = x_B(\tau) + \chi(\tau)$ is, in the spirit of Eq. (7.11),

$$S[x] = S_B + \frac{m}{2} \int_0^{\hbar\beta} d\tau \, \chi(\tau) \mathbf{O}[x_B(\tau)]\chi(\tau), \qquad (7.94)$$

with the fluctuation operator $\mathcal{O}[x_B(\tau)]$

$$\mathbf{O}[x_B(\tau)]\chi(\tau) = \left(-\frac{\partial^2}{\partial \tau^2} + \frac{1}{m} V''[x_B(\tau)] \right) \chi(\tau) + \frac{1}{m} \int_0^{\hbar\beta} d\tau' \, k(\tau - \tau')\chi(\tau'). \quad (7.95)$$

Here S_B is basically $S_{open}^E[x_B]$, Eq. (7.87). Note that a comparison between Eqs. (7.95) and (7.11), see also Eq. (7.44), reveals the similar structure of the fluctuation operator modulo the fact that the former, Eq. (7.95), modelling an open system has a nonlocal term, basically the kernel $k(\tau - \tau')$, which is absent in the later, as that represented a single harmonic oscillator. This also serves to illustrate the difference between a single, closed and an $N + 1$ dimensional, open system.

Expanding the quantum fluctuations $\chi(\tau)$ in the complete basis of the modes $\phi_n(\tau)$, normalized in the interval $(0, \hbar\beta)$, i.e.,

$$\chi(\tau) = \sum_n \lambda_n \phi_n(\tau), \tag{7.96}$$

the action $S[x]$ becomes

$$S[x] = S_B + \frac{m}{2} \sum_n \mathcal{O}_n[x_B]\lambda_n^2. \tag{7.97}$$

Here, $\mathcal{O}_n[x_B]$ are the eigenvalues of $\mathbf{O}[x_B(\tau)]$ for the periodic boundary conditions. The determinant of $\mathbf{O}[x_B(\tau)]$ in the diagonal basis is

$$\mathbf{D}[x_B] = \prod_n \mathcal{O}_n[x_B]. \tag{7.98}$$

From the equation of motion of the bounce $x_B(\tau)$, which is basically the stationary solution of the Euclidean action of linear quantum dissipation, in Eq. (7.87), it can be shown that $\dot{x}_B(\tau)$ is an eigenmode of the operator $\mathbf{O}[x_B(\tau)]$ with eigenvalue zero. The corresponding normalized zero mode is

$$\phi_1(\tau) = \sqrt{\frac{m}{S_0}} \dot{x}_B(\tau). \tag{7.99}$$

Here S_0 is the zero mode normalization factor and is

$$S_0 = m \int_0^{\hbar\beta} d\tau \, \dot{x}_B^2(\tau). \tag{7.100}$$

The mode $\phi_1(\tau)$ defines a time translation of the bounce, and is akin to the time translational invariance of the instanton solution described earlier. As a matter of fact, what we have here is the generalization of our earlier discussion on instantons to open quantum systems. We have encountered the zero mode solutions before, in our discussions of the single instanton solutions in a quartic double well potential. These solutions lead to a divergence and need to be carefully handled as, for e.g., by a suitable change of variables. In the present context this leads to

$$\frac{1}{\sqrt{\mathcal{O}_1[x_B]}} \rightarrow \sqrt{\frac{S_0}{2\pi\hbar}} \int_0^{\hbar\beta} d\tau_0 = \sqrt{\frac{S_0}{2\pi\hbar}} \hbar\beta. \tag{7.101}$$

The zero mode $\phi_1(\tau)$ has one node. This implies the existence of a nodeless eigenmode of $\mathbf{O}[x_B(\tau)]$ with a negative eigenvalue. This negative eigenvalue indicates an unstable mode, a trademark signature of tunneling. This happens because the action for the constant path $x(\tau) = x_b$, at the barrier top, is a saddle point. The functional integral is now performed by deforming the integration contour of the variable corresponding to the unstable mode into the upper half of the complex plane along the direction of steepest descent [178]. This analytical continuation gives an imaginary contribution to the partition function which in turn contributes to the imaginary part of the free energy, Eq. (7.84), and hence to the decay rate

$$\alpha_e = \Theta_{qm} e^{-S_B/\hbar}. \tag{7.102}$$

Note the formal similarity of Eq. (7.102) with Eq. (7.58). Thus, Θ_{qm} would be the attempt frequency of the particle in the well towards the barrier and $S_B/\hbar$ is the barrier dependent part of tunneling, which here is the Euclidean action evaluated at the bounce. The attempt frequency Θ_{qm} turns out to be [189]

$$\Theta_{qm} = \left(\frac{S_0}{2\pi\hbar}\right)^{1/2} \left(\frac{D_0}{|D'[x_B]|}\right)^{1/2}. \tag{7.103}$$

Here S_0 is as in Eq. (7.100), D_0 and $D'[x_B]$ are connected with the Gaussian fluctuations about the constant paths $x(\tau) = 0$ and $x(\tau) = x_b$, respectively. The prime on $D'[x_B]$ indicates that the zero eigenvalue is omitted. The source of the origin of the prefactor term Θ_{qm} is not difficult to fathom. We have to keep in mind that in the problem at hand, we are dealing with the path integral about two constant paths: (a). about the local potential mimimum x_0 and (b). about the unstable local maximum at $x = x_b$, see Figure (7.3) where $x_0 = 0$. If we now apply semiclassical analysis to the path integral, c.f., Eq. (7.43), we get our desired result, Eq. (7.102) with Eq. (7.103).

Dissipative quantum tunneling from the ground state in a metastable potential was initiated in [188] and extended to finite temperatures in [205]. We end this section with a brief discussion of the thermal enhancement of macroscopic quantum tunneling. The tunneling rate is enhanced by thermal effects as finite T opens up the further avenue that the particle can now tunnel from an excited state in the well. The leading thermal enhancement at low T comes from the temperature dependence of the bounce action. For a spectral density $I(\omega) \propto \omega^s$, the thermal enhancement for an undamped system is exponentially weak, i.e., it goes as $e^{-\hbar\omega_0/k_B T}$, while interestingly for a damped system, the enhancement goes like a power law, i.e., proportional to T^{1+s}. The power law exponent is a distinctive feature of the particular damping model employed and is independent of the form of the metastable potential [206]. Thus a precise measurement of the power of the thermally induced algebraic enhancement of the tunneling probability would provide valuable information regarding the nature of the coupling to the reservoir.

7.6   Guide to further reading

Tunneling is a vast subject and has applications to many areas, besides the ones discussed here, such as those in nuclear, atomic and molecular physics.

As a matter of fact, there are books exclusively devoted to it [175, 176, 207]. We have, in this chapter, indicated the way to attack the problem of quantum tunneling from a metastable well using semiclassical methods. The regime discussed is typically at temperatures much lesser than the crossover temperature. As the crossover temperature is approached and as one goes beyond it, thermal effects need to be carefully taken into consideration. This requires quite a bit of analysis which, though technically not more advanced than the ones undertaken here, are nevertheless very demanding. The reader who wishes to pursue this further can start by going through the book by Weiss [1]. In the high temperature regime, i.e., for temperatures more than the crossover temperature, the theory is seen to naturally agree with the original analysis of Kramers.

An interesting avenue in this context is the analysis of tunneling using a real-time description, in contrast to the imaginary time method promulgated here [208]. To use this approach in the context of open quantum systems is one of the problems of current research in this field. Another area worth venturing into is the study of tunneling in the presence of external driving.

Chapter 8

Open Quantum System at Interface with Quantum Information

8.1 Introduction

This chapter is devoted to the interface between open system ideas and the burgeoning field of quantum information. Quantum information [38, 209] is, as the name suggests, the broad name given to information tasks that make use of the laws of quantum mechanics. It encompasses within its purview, communication, computation and foundational information theoretical tasks. Information theoretic ideas pervade the whole of physics and make inroads beyond it. As it involves encoding, transmission and decoding of information as *bits* or *qubits*, all of which are very sensitive to their ambient environment, they provide a fruitful ground for the application of open system ideas. In fact, this chapter and the next one bear testimony to this.

This chapter attempts to provide a succinct introduction to the use of the formalism of open quantum systems to various facets of quantum information. This involves some tools that are very suitable for the task at hand. Among these are the notions of quantum operations, Lindblad evolutions (to which we have been introduced before in Chapter 3), and channel-state duality. After a discussion of these concepts, we make a foray into their application to concrete quantum information problems. The examples chosen range from the computational to the algorithmic and purely information theoretic aspects of quantum information. Thus, for example, evolution of quantum mechanical correlations, including the well known entanglement, are studied along with quantum cryptography. The role of open quantum systems is illustrated by applying a very useful quantum noisy channel, the squeezed generalized amplitude damping (SGAD) channel [140, 210], which is discussed in some detail in the early part of the chapter, to the chosen applications. Interaction with the ambient environment causes noise and it is desirable, for the efficient implementation of the quantum informational task, to control and correct it. Error correction is

© Hindustan Book Agency 2018 and Springer Nature Singapore Pte Ltd. 2018
S. Banerjee, *Open Quantum Systems*, Texts and Readings in Physical Sciences 20,
https://doi.org/10.1007/978-981-13-3182-4_8

the name coined for efforts in this direction. We conclude the chapter by an introduction to ideas of quantum error correction.

8.2 Role of Noise in Quantum Information: Introduction to Tools and Techniques

We briefly review some concepts of quantum mechanics, from the perspective of quantum information. It may be fruitful for the reader at this stage to take another look at Chapter 2. The fundamental unit of quantum information is the *qubit*, which could be any two-level quantum system with levels $|0\rangle$ and $|1\rangle$. A classical bit string can be described as a vector over the Galois field Z_2. Analogously, string of n qubits is a vector in the Hilbert space $\mathcal{H} = \mathcal{H}_1 \otimes \mathcal{H}_2 \otimes \cdots \otimes \mathcal{H}_n$ of dimension $d = 2^n$ over a field of complex numbers $\mathcal{C}$, where $\mathcal{H}_1, \ldots, \mathcal{H}_n$ represent the Hilbert space of individual qubits and $\otimes$ is the tensor product. Accordingly, if $|0\rangle$ and $|1\rangle$ are the two possible states of a quantum system, then the linear combination

$$|\psi\rangle = \alpha|0\rangle + \beta|1\rangle, \tag{8.1}$$

is also a bonafide state with $\alpha, \beta \in \mathcal{C}$. This is known as principle of *superposition* in quantum mechanics. In general, pure state in an n-qubit system can be written as the superposition

$$|\Psi\rangle = \sum_{i=0}^{2^n-1} = \alpha_i|i_n\rangle, \tag{8.2}$$

where $|i_n\rangle$ is basis element for $\mathcal{H}$ and $\sum_{i=0}^{2^n-1} |\alpha_i|^2 = 1$.

Quantum entanglement   A multi-particle superposition in which the state is not factorizable as a product of states of individual particles, is referred to as entanglement [211, 212]. Entanglement can be harnessed to create quantum *parallelism* in a quantum computational task, exploiting a superposition of evaluation of all possible inputs y to a function $g(y)$, as

$$2^{-n/2}\sum_{y} |y\rangle|0\rangle \rightarrow 2^{-n/2}\sum_{i} |y\rangle|g(y)\rangle. \tag{8.3}$$

Entanglement is a ubiquitous feature of quantum mechanics, and is considered to be a powerhouse of quantum information [38]. Let $|\psi\rangle \in \mathcal{H}_1 \otimes \mathcal{H}_2$ be a pure state $\mathcal{H}_1$ and $\mathcal{H}_2$. Then $|\psi\rangle$ is said to be separable if there exist states $|\psi_1\rangle \in \mathcal{H}_1$ and $|\psi_2\rangle \in \mathcal{H}_2$ such that $|\psi\rangle = |\psi_1\rangle \otimes |\psi_2\rangle$; else it is entangled. There are two kinds of evolution in quantum mechanics: a continuous, norm-preserving (i.e., unitary) Schrödinger evolution and a discontinuous, probabilistic evolution following a measurement.

Continuous evolution. The time evolution of a quantum state is determined by the Shrödinger equation

$$-i\hbar\frac{d}{dt}|\psi(t)\rangle = H|\psi(t)\rangle, \tag{8.4}$$

where H is the Hamiltonian of the system.

Measurement. Observables in quantum mechanics are represented by Hermitian operators. A measurement operator

$$\mathcal{P} = \sum_{i=0}^{2^n-1} \lambda_i |i_n\rangle\langle i_n|, \tag{8.5}$$

updates the state $|\Psi\rangle$ to $|i_n\rangle$ with probability $|\langle i_n|\Psi\rangle|^2$ with the corresponding measurement outcome being λ_i. Conservation of probability implies the completeness condition $\sum_{i=0}^{2^n-1} |i_n\rangle\langle i_n| = \mathbb{I}$. For example, after measurement in the $\{|0\rangle, |1\rangle\}$ basis, $|\psi\rangle$ collapses to either $|0\rangle$ or $|1\rangle$ with probability $|\alpha|^2 = |\langle 0|\psi\rangle|^2$ or $|\beta|^2 = |\langle 1|\psi\rangle|^2$, respectively, with $|\alpha|^2 + |\beta|^2 = 1$. $|\alpha|^2$ and $|\beta|^2$ can be accessed by performing measurement on an ensemble of $|\psi\rangle$.

More generally, measurement can be represented by any partition of unit operator. This constitutes a positive operator valued measurement (POVM). Unlike projectors, POVM elements need not form an idempotent matrix. A projector corresponding to a degenerate eigenvalue is sometimes called incomplete, mainly in quantum error correction where syndrome measurements are such measurements.

Density matrix representation

When dealing with mixed states, as is often the case in the context of open quantum systems, the object of interest is the density matrix ρ and has been introduced earlier in Chapter 2.

Geometrically, the pure state of a qubit can be represented as a point on the three-dimensional sphere known as *Bloch* sphere, while the mixed states are points within the sphere. This can be seen by parameterizing the Eq. (8.1) with θ and ϕ as

$$|\psi\rangle = \cos\frac{\theta}{2}|0\rangle + e^{-i\phi}\sin\frac{\theta}{2}|1\rangle, \tag{8.6}$$

where the norm, $\cos^2\frac{\theta}{2} + \sin^2\frac{\theta}{2} = 1$, represents the surface of a sphere with unit radius. The density matrix of a qubit in the Pauli basis is

$$\rho = \frac{I_2 + \hat{n}\cdot\sigma}{2}, \tag{8.7}$$

where $\hat{n} = \{\hat{n}_x, \hat{n}_y, \hat{n}_z\}$, $\sigma = \{\sigma_x, \sigma_y, \sigma_z\}$ and I_2 is the 2×2 identity matrix. The Bloch vector $\hat{n}$ is the expectation value of ρ in the Pauli basis, i.e., $\hat{n} \equiv$

$(\langle\sigma_x\rangle, \langle\sigma_y\rangle, \langle\sigma_z\rangle)$. The Pauli operators are traceless and have the following well known representation in the computational basis

$$\sigma_x = \begin{pmatrix} 0 & 1 \\ 1 & 0 \end{pmatrix}; \quad \sigma_y = \begin{pmatrix} 0 & -i \\ i & 0 \end{pmatrix}; \quad \sigma_z = \begin{pmatrix} 1 & 0 \\ 0 & -1 \end{pmatrix}. \tag{8.8}$$

The computational basis vectors $\{|0\rangle, |1\rangle\}$ are eigenvectors of σ_z. From Eq. (8.7) $\mathrm{Tr}(\rho^2) = 1$ corresponds to $\hat{n}_x^2 + \hat{n}_y^2 + \hat{n}_z^2 = 1$ and represents the surface of the Bloch sphere. Further, it can be verified that for mixed states $\mathrm{Tr}(\rho^2) = \hat{n}_x^2 + \hat{n}_y^2 + \hat{n}_z^2 < 1$, representing the interior of the Bloch sphere. The *completely mixed state* $\rho = I_2/2$ is the center of the Bloch sphere which represents an equal mixture of $|0\rangle$ and $|1\rangle$.

Operations on quantum states

Operations required for quantum information processing are known as gates, which are unitary operations, mathematically represented by unitary matrices U, that satisfy $UU^\dagger = U^\dagger U = I$. The U transforms initial state to the required final state $|\psi'\rangle$ as,

$$|\psi'\rangle = U|\psi\rangle, \tag{8.9}$$

which can be read by performing a measurement. It can be observed that $\langle\psi'|\psi'\rangle = 1$, implying that U is an isometry of the evolution, i.e., a trace preserving operation. This evolution corresponds to that described by the Schrödinger equation, Eq. (8.4), where $U = e^{-iHt/\hbar}$. The Pauli operators are unitary and can perform various required operations in quantum computation. The basic gates used in quantum computation are bit-flip (σ_x), phase-flip (σ_z), combination of bit and phase flips ($\sigma_y = \sigma_x\sigma_z$), Hadamard and controlled NOT (CNOT). A Hadamard gate does a $\frac{\pi}{4}$-rotation on the qubit space, while a CNOT is a two-qubit gate that flips the second qubit conditioned on the state of the first qubit, i.e., $|0\rangle\langle 0| \otimes I + |1\rangle\langle 1| \otimes X$. Using a CNOT and $SU(2)$ (single qubit gate) gate, any unitary quantum operation on qubits can be simulated and hence they form a universal set of gates for quantum computation.

In general, the gates on qubits and qubits themselves would be noisy due their unavoidable interaction with the surrounding, leading to non-unitary evolution. We now turn to this issue.

Problem 1: Show that, using Eq. (8.9), $\langle\psi'|\psi'\rangle = \langle\psi|\psi\rangle$.

Problem 2: Let $\langle x|\psi\rangle = \psi(x) = \left(\frac{\pi}{a}\right)^{-1/4} e^{-ax^2/2}$. Show that $\Delta\hat{X}\Delta\hat{P} = \frac{\hbar}{2}$; where $\Delta\hat{X} = \sqrt{\langle(\hat{X})^2\rangle - (\langle\hat{X}\rangle)^2}$ and $\hat{P} = -i\hbar\frac{\partial}{\partial\hat{X}}$. Note that $\hat{X}$ and $\hat{P}$ correspond to the position and momentum operators, respectively.

Problem 3: A quantum system is in the state:

$$|\psi\rangle = 2i|u_1\rangle - 3|u_2\rangle + i|u_3\rangle,$$

where the $|u_i\rangle$, $i = 1, 2, 3$, constitute an orthonormal basis. Write down the column vector representing this vector in the given basis; do the same for the row vector.

Problem 4: A quantum system is in the state:

$$|\psi\rangle = \frac{\sqrt{2}}{3}|\phi_1\rangle + \frac{\sqrt{3}}{3}|\phi_2\rangle + \frac{2}{3}|\phi_3\rangle,$$

where the $|\phi_i\rangle$, $i = 1, 2, 3$, constitute an orthonormal basis. These states are eigenvestors of a Hamiltonian operator such that: $H|\phi_1\rangle = E|\phi_1\rangle$; $H|\phi_2\rangle = 2E|\phi_2\rangle$; $H|\phi_3\rangle = 3E|\phi_3\rangle$. Is $|\psi\rangle$ normalized? If energy is measured, what are the probabilities of obtaining E, $2E$ and $3E$?

Problem 5: In some orthonormal basis $\{|u_1\rangle, |u_2\rangle, |u_3\rangle\}$ an operator A acts as:

$$\begin{aligned}
A|u_1\rangle &= 2|u_1\rangle, \\
A|u_2\rangle &= 3|u_1\rangle - i|u_3\rangle, \\
A|u_3\rangle &= -|u_2\rangle.
\end{aligned}$$

Write the matrix representation of the operator.

Problem 6: In some orthonormal basis $\{|u_1\rangle, |u_2\rangle, |u_3\rangle\}$ an operator T is: $T = -|u_1\rangle\langle u_1| + |u_2\rangle\langle u_2| + 2|u_3\rangle\langle u_3| - i|u_1\rangle\langle u_2| + |u_2\rangle\langle u_1|$. Calculate Tr(T).

8.2.1   Quantum noise

The chief barrier in realizing quantum computation, apart from the difficulty to scale-up, is that a quantum system is rarely truly isolated; it tends to interact with its environment (alternatively referred to as bath or reservoir) and is thus usually *open*. These interactions are often unwanted (though exceptions exist [213]) and difficult-to-eliminate. They show up as *noise* in quantum information processing systems and *decohere* the system. Decoherence causes decay of the quantum information about the coherence in the system in a basis determined by the interaction Hamiltonian [214], which could be the position basis or the eigen basis of the system Hamiltonian, leading to familiar classical behavior. One of the first testing grounds for open system ideas was in quantum optics [6]. Its application to other areas gained momentum from the works of Caldeira and Leggett [215], and Zurek [216, 217], among others.

Quantum operations

Any evolution consistent with the general rules of quantum mechanics can be described by a linear, completely positive map, called quantum operation ($\mathcal{E}$) [38].

A useful notion in this context is that of complete positivity. Consider any positive map $\mathcal{E}$ on the system Q_1; if an extra system R of arbitrary dimensionality is introduced, and $(\mathcal{I} \otimes \mathcal{E})(A)$ is positive on any positive operator A on the combined system RQ_1, where $\mathcal{I}$ denotes the identity map on system R, then $\mathcal{E}$ is completely positive.

A unitary evolution is a special case of a quantum operation; general quantum operations can describe non-unitary evolutions, due to coupling with environment. Any such quantum operation can be composed from elementary operations:

- unitary transformations: $\mathcal{E}_1(\rho) = U\rho U^\dagger$, effected by the unitary operator U;

- addition of an auxiliary system: $\mathcal{E}_2(\rho) = \rho \otimes \sigma$: here ρ is the original system and σ is the auxiliary one;

- partial traces: $\mathcal{E}_3(\rho) = \mathrm{Tr}_B(\rho)$. In the context of open quantum systems, the partial trace is usually performed over the reservoir degrees of freedom;

- projective measurements: $\mathcal{E}_4(\rho) = P_k \rho P_k / \mathrm{Tr}(P_k \rho)$, with $P_k^2 = P_k$. It should be emphasized that though we talk about projective measurements here, in general more general measurements are possible [38].

We now wish to interpret the above results in terms of familiar noisy channels : How can these environmental effects affect quantum computing? In operator-sum representation, the action of a superoperator $\mathcal{E}$ due to environmental interaction is

$$\rho \longrightarrow \mathcal{E}(\rho) = \sum_k \langle e_k | U(\rho \otimes |f_0\rangle\langle f_0|) U^\dagger | e_k \rangle = \sum_j E_j \rho E_j^\dagger,$$

where the unitary operator U represents free evolution of system, environment, as well as the interaction between the two; $|f_0\rangle$ is the environment's initial state and $\{|e_k\rangle\}$ is a basis for the environment. We assume, here, that the environment-system starts in a separable state.

$E_j \equiv \langle e_k | U | f_0 \rangle$ implements the dynamical mapping and a partition of unity, i.e., $\sum_j E_j^\dagger E_j = \mathcal{I}$. The E_js are known as Kraus operators, the extension of unitary operators to non-unitary evolution. Any transformation that can be represented as an operator-sum is a completely positive (CP) map [38]. This formulation is known as the quantum operations formalism [218, 219]. The superoperator $\mathcal{E}$ is a map and is also interchangeably called a quantum channel. Some familiar noise channels are the depolarizing channel, the dephasing channel, the amplitude damping channel and generalized amplitude damping channel [38]. A generalization of the latter, in which the thermal bath is squeezed, known as squeezed generalized amplitude damping channel was introduced in [220].

Lindblad Evolutions

An important class of noisy evolutions, used extensively in quantum optics and quantum information, is the Lindbladian evolutions. They have been introduced and discussed earlier in Chapter 3. Here we provide a simple, intuitive sketch to the Lindbladian evolution.

Let us look at the dynamics of the system on a timescale δt. It should satisfy two conditions:

(A). $\delta t \ll \tau_S$: the timescale is small compared to the characteristic timescale of the system τ_S, determined by, say, the natural frequency of the system. The system density matrix evolves only a little in this time interval;

(B). $\delta t \gg \tau_B$: At the same time δt is long compared to the time over which the environment/bath forgets its information about the system τ_B. This time scale would be the reservoir memory timescale and would be typically associated with the high-frequency cutoff in the reservoir spectral density and the time scale associated with the reservoir temperature, which measures the relative importance of quantum to thermal effects.

Since we look for dynamics beyond time τ_B, the evolution through time δt should be described by a quantum operation on the current system density matrix. Hence,

$$\rho_S(\delta t) = \mathcal{E}(\rho_S(0)) = \sum_k E_k \rho_S(0) E_k^\dagger = \rho_S(0) + O(\delta t). \tag{8.10}$$

It follows that the Kraus operators should be of the form

$$\begin{aligned} E_0 &= \mathcal{I}_S + (K - \frac{i}{\hbar}H)\delta t, \\ E_k &= \sqrt{\delta t}L_k, \quad k \geq 1. \end{aligned}$$

K, H are arbitrary Hermitian operators; L_k are also arbitrary and are called the Lindblad operators. The normalization of Kraus operators gives

$$\mathcal{I}_S = \mathcal{I}_S + (2K + \sum_k L_k^\dagger L_k)\delta t + O((\delta t)^2), \tag{8.11}$$

implying that $K = -\frac{1}{2}\sum_k L_k^\dagger L_k$. Therefore

$$\rho_S(\delta t) = \rho_S(0) - \left\{ \frac{i}{\hbar}[H, \rho_S] - \sum_k \left[L_k \rho_S(0) L_k^\dagger - \frac{1}{2}\{\rho_S(0), L_k^\dagger L_k\} \right] \right\}\delta t + O((\delta t)^2). \tag{8.12}$$

Here, $\{A, B\} = AB + BA$. Taking the limit $\delta t \longrightarrow 0$, the Lindblad master equation is obtained as

$$\frac{d\rho_S}{dt} = \frac{1}{i\hbar}[H, \rho_S] + \sum_k \left[L_k \rho_S(0) L_k^\dagger - \frac{1}{2}\{\rho_S(0), L_k^\dagger L_k\} \right]. \tag{8.13}$$

If the evolution were unitary, there are no Lindblad operators, then the above master equation reduces to $\frac{d\rho_S}{dt} = \frac{1}{i\hbar}[H, \rho_S]$, the usual Schrödinger-von Neumann equation. This derivation gives no clue to the microscopic origins of the Lindbladians. That would require a more detailed derivation, of the kind presented earlier in Chapter 3.

Illustrative Example:

We reconsider the example of the decay of a two-level system interacting with a radiation field (bath) in the weak Born-Markov, rotating wave approximation. It follows from Eq. (104), Chapter 3, that the Lindbladian evolution can be expressed compactly as

$$\frac{d}{dt}\rho^s(t) = \sum_{j=1}^{2}\left(2R_j\rho^s R_j^\dagger - R_j^\dagger R_j\rho^s - \rho^s R_j^\dagger R_j\right), \qquad (8.14)$$

where $R_1 = (\gamma_0(N_{th} + 1)/2)^{1/2}\sigma_-$, $R_2 = (\gamma_0 N_{th}/2)^{1/2}\sigma_+$. Note that if $T = 0$, a single Lindblad operator suffices. All the other terms are as described before, in Chapter 3. A useful tool for the solution of the above Lindblad equation is by invoking the representation of the two-level density matrix in terms of Pauli operators as

$$\begin{aligned}
\rho^S(t) &= \frac{1}{2}(I + \langle\vec{\sigma}(t)\rangle.\vec{\sigma}) \\
&= \begin{pmatrix} \left(\frac{1}{2}\right)(1 + \langle\sigma_z(t)\rangle) & \langle\sigma_-(t)\rangle \\ \langle\sigma_+(t)\rangle & \left(\frac{1}{2}\right)(1 - \langle\sigma_z(t)\rangle) \end{pmatrix}.
\end{aligned}$$

Using, for e.g.,

$$\frac{d}{dt}\langle\sigma_z(t)\rangle = \text{Tr}\left(\sigma_z\frac{d}{dt}\rho^S(t)\right), \qquad (8.15)$$

and likewise for the other two Pauli operators, we get three linear differential equations which can be easily solved to yield the Bloch vectors

$$\langle\sigma_x(t)\rangle = e^{-\frac{\gamma_0}{2}(2N_{th}+1)t}\langle\sigma_x(0)\rangle,$$

$$\langle\sigma_y(t)\rangle = e^{-\frac{\gamma_0}{2}(2N_{th}+1)t}\langle\sigma_y(0)\rangle,$$

$$\langle\sigma_z(t)\rangle = e^{-\gamma_0(2N_{th}+1)t}\langle\sigma_z(0)\rangle - \frac{1}{(2N_{th}+1)}\left(1 - e^{-\gamma_0(2N_{th}+1)t}\right). \qquad (8.16)$$

Problem 7: Derive the linear differential equations for the Bloch vectors and solve them to get Eq. (8.16).

Problem 8: The density matrix for a given state is:

$$\rho = \begin{pmatrix} 3/4 & -i/4 \\ i/4 & 1/4 \end{pmatrix}.$$

Find out the Bloch vector for this state. Is it pure or mixed? If the σ_z operator is measured, find the probability of finding $|1\rangle$.

Problem 9: A density operator for some system is given by

$$\rho = \begin{pmatrix} 2/3 & 1/6 - i/3 \\ 1/6 + i/3 & 1/3 \end{pmatrix}.$$

Find out the Bloch vector for this state. Is it pure or mixed? A measurement of spin is made in the z-direction. What is the probability that the measurement result is spin-down? What is the probability that the measurement is spin-up?

We will next introduce and discuss a very useful technique that makes use of the duality between the state and the corresponding channel, the Choi-Jamiolwski isomorphism [221, 222, 223].

8.2.2   Channel-State Duality

Channel-state duality refers to the correspondence between quantum channels and bipartite states, and finds many uses in quantum information theory. It is sometimes referred to as the Choi-Jamiolwski isomorphism. It refers to the statement that any channel (i.e., quantum operation, or equivalently, any linear, completely positive, trace-preserving map) from the state space of an input quantum system to that of an output system corresponds to a bipartite state of the tensor product of the two relevant systems. This correspondence links dynamics to kinematics, and is not merely mathematical but also has a fundamental physical meaning, profound consequences, and a plethora of applications. Thus, for example, it allows for the characterization of the channel based on the correlations between the output system and the ancilla, used for purifying the system [38].

Let $\mathcal{H}_1$ be a finite-dimensional Hilbert space and $\mathcal{B}(\mathcal{H})_1$ be the Hilbert space of all bounded linear operators on $\mathcal{H}_1$ equipped with a finite Hilbert-Schmidt inner product. Let $\mathcal{H}_2$ be another finite-dimensional Hilbert space (which may or may not be identical to $\mathcal{H}_1$). For any linear map $\mathcal{X} : \mathcal{B}(\mathcal{H})_1 \to \mathcal{B}(\mathcal{H})_2$ sending operators on $\mathcal{H}_1$ to operators on $\mathcal{H}_2$, consider the following correspondence

$$\chi \to \rho_\chi = \mathcal{I} \otimes \chi(|\phi\rangle\langle\phi|) = \sum_{ij} e_{ij} \otimes \chi(e_{ij}). \tag{8.17}$$

Here $\mathcal{I}$ is the identity operation, $|\phi\rangle = \sum_i |i\rangle \otimes |i\rangle$ is the canonical maximally entangled (unnormalized) state in $\mathcal{H}_1 \otimes \mathcal{H}_1$ and $e_{ij} = |i\rangle\langle j|$, where $|i\rangle$ is an orthonormal basis in $\mathcal{H}_1$. The correspondence in Eq. (8.17) is an isomorphism between the spaces of linear maps from $\mathcal{B}(\mathcal{H})_1 \to \mathcal{B}(\mathcal{H})_2$ and the space of bipartite operators on $\mathcal{H}_1 \otimes \mathcal{H}_2$. Here $\mathcal{B}(\mathcal{H})_1$, $\mathcal{B}(\mathcal{H})_2$ stand for the set of bounded linear operators on the Hilbert spaces $\mathcal{H}_1$, $\mathcal{H}_2$, respectively. In the physically relevant cases, these maps are restricted to subsets of the space of general linear maps, for example, positive maps, completely positive maps, channels. It can be used to establish the channel-state duality. Thus, for example, if χ is a completely positive map, then ρ_χ is a positive operator. This sets up a duality between the channel, here the map χ, and state ρ_χ.

Applications of Channel-State Duality:

(a). *Factorization law of entanglement decay*

The content of this is that the evolution of entanglement in a bipartite entangled state under a local one-sided channel can be fully characterized by its action on a maximally entangled state [224]. The amount of entanglement at any time t in a given initially entangled two-qubit pure state $|\Phi\rangle$, under the action of a one-sided quantum channel χ, is equal to the product of the initial entanglement in the given state and the entanglement in the state obtained by applying the channel on one side of a two-qubit maximally entangled state. Thus,

$$C\big(\mathcal{I} \otimes \chi\big)\big(|\Phi\rangle\langle\Phi|\big) = C\big(|\Phi\rangle\langle\Phi|\big) C\big((\mathcal{I} \otimes \chi)(|\phi^+\rangle\langle\phi^+|)\big). \qquad (8.18)$$

Here $|\phi^+\rangle = \frac{1}{\sqrt{d}} \sum_{i=1}^{d} |ii\rangle$ is a maximally entangled state in a $d \otimes d$ Hilbert space and C is concurrence, a useful quantitative measure of entanglement [225]. We will have more to say on entanglement later on in this chapter. The usefulness of the factorization law is that, using it, a study of entanglement of a general state evolution is reduced to the concurrence in the state obtained after the evolution of the maximally entangled state $|\phi^+\rangle$ [226]. For a derivation of the factorization law of entanglement decay using channel-state duality, we refer the reader to [224].

(b). *Kraus operators of the Squeezed Generalized Amplitude Damping Channel*

The Squeezed Generalized Amplitude Damping (SGAD) channel [220], generalizes the notion of the well known Amplitude Damping (AD) and Generalized Amplitude Damping (GAD) channels to the case of finite reservoir squeezing. As squeezing is a quantum resource, the SGAD channel finds a number of uses in quantum information processing, some of which will be discussed later on in this chapter. In the spirit of Eq. (8.14), the master equation depicting the evolution of the reduced density matrix of the qubit interacting with a squeezed thermal bath expressed in a manifestly Lindblad form has the form [2, 220, 227]

$$\frac{d}{dt}\rho(t) = \sum_{j=1}^{2} \left(2R_j \rho R_j^\dagger - R_j^\dagger R_j \rho - \rho R_j^\dagger R_j\right), \qquad (8.19)$$

where $R_1 = [\gamma_0(N_{\text{th}} + 1)/2]^{1/2} R$, $R_2 = [\gamma_0 N_{\text{th}}/2]^{1/2} R^\dagger$ and $R = \sigma_- \cosh(s) + e^{i\Phi}\sigma_+ \sinh(s)$. $\gamma_0 = (4\omega^3 |\vec{d}|^2)/(3\hbar c^3)$, is the spontaneous emission rate and σ_+, σ_- are the standard Pauli raising and lowering operators. Also $N = N_{\text{th}}[\cosh^2(s) + \sinh^2(s)] + \sinh^2(s)$, where $N_{\text{th}} = 1/(e^{\hbar\omega/k_B T} - 1)$ is the Planck distribution giving the number of thermal photons at the frequency ω; s and Φ are bath squeezing parameters. Eq. (8.19) guarantees that the evolution of the density operator is completely positive (CP). If $T = 0$, then R_2 vanishes, and a single Lindblad operator suffices. The general map generated by the Eq. (8.19) is the SGAD channel [220], which generalizes the notion of the AD and GAD channels [38]. These amplitude damping class of channels are non-unital and contractive, mapping any initial state to a unique asymptotic state; a consequence of the fluctuation-dissipation theorem.

We now present a useful technique for the calculation of the Kraus operators of the SGAD channel. This can be used to construct similar Kraus operators. The technique makes essential use of the Choi-Jamiolwski isomorphism. Proceeding as in Eq. (8.16), the Bloch vectors for the SGAD channel, Eq. (8.19), are [220]:

$$\begin{aligned}
\langle\sigma_x(t)\rangle &= A\langle\sigma_x(0)\rangle - B\langle\sigma_y(0)\rangle, \\
\langle\sigma_y(t)\rangle &= G\langle\sigma_x(0)\rangle - B\langle\sigma_y(0)\rangle, \\
\langle\sigma_z(t)\rangle &= H\langle\sigma_z(0)\rangle - Y,
\end{aligned} \qquad (8.20)$$

where

$$A = [1 + \frac{1}{2}(e^{\gamma_0 a t} - 1)(1 + \cos(\Phi))]e^{\frac{-\gamma_0(2N+1+a)t}{2}},$$

$$B = \sin(\Phi)\sinh(\frac{\gamma_0 a t}{2})e^{\frac{-\gamma_0(2N+1)t}{2}},$$

$$G = [1 + \frac{1}{2}(e^{\gamma_0 a t} - 1)(1 - \cos(\Phi))]e^{\frac{-\gamma_0(2N+1+a)t}{2}},$$

$$H = e^{-\gamma_0(2N+1)t},$$

$$Y = \frac{(1 - e^{-\gamma_0(2N+1)t})}{2N+1}. \qquad (8.21)$$

In Eq. (8.21), $a = \sinh(2s)[2N_{th} + 1]$ and all the other terms are as defined above.

Problem 10: Derive the Eqs. (8.20).

Consider the maximally entangled (unnormalized) state $|\tilde{\psi}\rangle = |00\rangle + |11\rangle$. We construct the Choi matrix of the SGAD channel from its action on ρ, the state reconstructed from the Bloch vectors in Eq. (8.20), as

$$\chi_{\mathcal{E}} \equiv (I \otimes \mathcal{E})|\tilde{\psi}\rangle\langle\tilde{\psi}| = \begin{pmatrix} \frac{1+H-Y}{2} & 0 & 0 & \frac{A+G}{2} \\ 0 & \frac{1-H+Y}{2} & \frac{A-G}{2} - iB & 0 \\ 0 & \frac{A-G}{2} + iB & \frac{1-H-Y}{2} & 0 \\ \frac{A+G}{2} & 0 & 0 & \frac{1+H+Y}{2} \end{pmatrix}.$$
$$(8.22)$$

Let the spectral decomposition of this yield,

$$\chi_\mathcal{E} = \sum_{j=0}^{3} |\xi_j\rangle\langle\xi_j|, \tag{8.23}$$

where $|\xi_j\rangle$ are the eigenvectors normalized to the value of the eigenvalue. By Choi's theorem [221, 228], each $|\xi_j\rangle$ yields a Kraus operator obtained by folding the d^2 (here, 4) entries of the eigenvector into $d \times d$ (2×2) matrix, essentially by taking each sequential d-element segment of $|\xi_j\rangle$, writing it as a column, and then juxtaposing these columns to form the matrix.

The eigenvectors corresponding to non-vanishing eigenvalues are found to be

$$|\xi_0\rangle = (0, i\frac{(\sqrt{1-H-\Psi})(\Psi+Y)}{2B+i(G-A)}, \sqrt{1-H-\Psi}, 0),$$

$$|\xi_1\rangle = (0, i\frac{(\sqrt{1-H+\Psi})(\Psi-Y)}{2B+i(G-A)}, \sqrt{1-H+\Psi}, 0),$$

$$|\xi_3\rangle = (\frac{-\sqrt{1+H-\eta}(Y+\eta)}{A+G}, 0, 0, \sqrt{1+H-\eta}),$$

$$|\xi_4\rangle = (\frac{-\sqrt{1+H+\eta}(Y-\eta)}{A+G}, 0, 0, \sqrt{1+H+\eta}). \tag{8.24}$$

From above eigenvectors we obtain the Kraus representation for the SGAD channel [210]

$$J_\pm = \frac{1}{M_\pm}\begin{pmatrix} 0 & \sqrt{1-H\mp\Psi} \\ i\frac{(\sqrt{1-H\mp\Psi})(\Psi\pm Y)}{2B+i(G-A)} & 0 \end{pmatrix},$$

$$K_\pm = \frac{1}{N_\pm}\begin{pmatrix} \frac{-\sqrt{1+H\mp\eta}(Y\pm\eta)}{A+G} & 0 \\ 0 & \sqrt{1+H\mp\eta} \end{pmatrix}, \tag{8.25}$$

where $\Psi = \sqrt{(A-G)^2 + 4B^2 + Y^2}$, $\eta = \sqrt{(A+G)^2 + Y^2}$, and $M_\pm = \sqrt{2}\sqrt{1 + \left|\frac{\mp Y+\Psi}{2B+i(G-A)}\right|^2}$,

$N_\pm = \sqrt{2}\sqrt{1 + \left|\frac{Y\pm\eta}{A+G}\right|^2}$. It should be noted here that there are infinitely many Kraus operator representations even within the same representation basis of the system, depending on the choice of tracing basis $\{|e_k\rangle\}$ of the environment. The different Kraus representations are connected by an appropriate unitary transformation. This is so for the SGAD channel as well and can be seen in [210].

Problem 11: *For the bravehearts!* Sketch the steps from Eq. (8.22) to finally arive at the Kraus representation for the SGAD channel, Eq. (8.25).

(c). *Qubit Channels: Some Remarks*

The channel-state duality can be used to study various quantum channels geometrically. This provides a powerful as well as elegant tool to develop a deeper understanding of the channels and also helps in their classification. Given a map $\mathcal{E}$ that maps the algebra of $m \times m$ complex matrices to another matrix algebra, we may define the rank of the channel as that of the matrix associated with $\mathcal{E}$ [229]. Here, by virtue of the Choi isomorphism, one may associate a rank with the channel, identified with that of the corresponding Choi matrix. From the above analysis of the Choi matrix corresponding to the SGAD channel, it can be shown that the rank of the $SGAD$ channel for a qubit is either 2 or 4. This is not a general quantum feature, and there do exist noise channels for qubits with odd rank greater than 1. A prominent example of a rank 3 channel is the Pauli channel with Kraus operator elements I, σ_x and σ_y with weights p, q and r, such that $p + q + r = 1$.

Using the Choi-Jamiolwski isomorphism, the set of unitaries on a qudit maps to pure states in V, the set of two-qudit states isomorphic to CP maps on a single qudit. The general state of a two-qubit density operator is given by:

$$\rho = \frac{1}{4} \left(I \otimes I + \sum_j r_j \sigma_j \otimes I_2 + s_j I_2 \otimes \sigma_j + \sum_{j,k} t_{j,k} \sigma_j \otimes \sigma_k \right), \qquad (8.26)$$

where r_j, s_j are the analogues of the Bloch vectors in the single qubit case and the tensor $t_{j,k}$, also called the tensor polarization [230], are generally complex numbers subject to requirement $\rho = \rho^\dagger$ and $Tr(\rho) = 1$. Letting $|\psi\rangle = \frac{1}{\sqrt{2}}(|00\rangle + |11\rangle$, we have

$$|\psi\rangle\langle\psi| = \frac{1}{4} \left(I \otimes I + \sigma_x \otimes \sigma_x - \sigma_y \otimes \sigma_y + \sigma_z \otimes \sigma_z \right) = (\mathcal{E}_\mathcal{I} \otimes I)|\psi\rangle\psi| \equiv \mathcal{I}, \qquad (8.27)$$

where $\mathcal{E}_\mathcal{I}$ is the trivial noise, corresponding to the identity operator. Under the Choi isomorphism, the corresponding state is therefore $\mathcal{I}$.

The *phase flip* channel is represented by the set of Kraus operators $[\sqrt{\alpha}I, \sqrt{(1-\alpha)}Z]$, where Z stands for the Pauli operator σ_z and α is a real positive number such that $0 \leq \alpha \leq 1$. It is closely related to the *phase damping* channel, characterized by the Kraus-operators $\left[\sqrt{\beta}I, \sqrt{(1-\beta)}P_0, \sqrt{(1-\beta)}P_1 \right]$, where $P_0 = |0\rangle\langle 0|$ and $P_1 = |1\rangle\langle 1|$ are projectors and β is a real positive number such that $0 \leq \beta \leq 1$. The phase damping channel is strictly a subset of the phase flip channel. The *generalized depolarizing or Pauli* channels have Pauli operators (apart from a factor) as their Kraus operators, i.e., $\left[\sqrt{\alpha}I, \sqrt{\beta}\sigma_x, \sqrt{\gamma}\sigma_y, \sqrt{\delta}\sigma_z \right]$, where $\alpha, \beta, \gamma, \delta \geq 0$ are real numbers satisfying $\alpha + \beta + \gamma + \delta = 1$. Every Pauli channel is a member of the polytope given by four pure points. Note that the convex hull of a given finite number of pure points is a *convex polytope*. Geometrically, a polytope can be visualized as an object or tile with flat sides. Using Choi isomorphism, it can be shown [210] that the set of all Pauli channels, the polytope $\hat{\mathcal{P}}$, is a 3-simplex (a tetrahedron) embedded within V. The phase flip channel $\hat{\mathcal{F}}$ corresponds to a proper subset of $\hat{\mathcal{P}}$, and the volume of phase damping channels in this set is $\frac{1}{2}$. This structure has been studied using affine maps on Bloch sphere in [231], where it was shown that the fraction of the channels that can be simulated with a one-qubit environment is $\frac{3}{8}$. The *depolarizing* channel has a Kraus representation $\left[\sqrt{\frac{1+3p}{4}}I, \sqrt{\frac{3(1-p)}{4}}\sigma_x, \sqrt{\frac{3(1-p)}{4}}\sigma_y, \sqrt{\frac{3(1-p)}{4}}\sigma_z\right]$. The Choi matrix for this process has a convex structure which is just the two-qubit Werner state $pI \otimes I + (1-p)|\psi\rangle\langle\psi|$. The set $\hat{\mathcal{D}}$ of all depolarizing channels forms a 1-simplex embedded within $\hat{\mathcal{P}}$.

In contrast to these Pauli class of channels, the geometry of the SGAD channel turns out to be quite different, illustrating their more complicated nature. The entire family of amplitude damping channels, culminating in the SGAD channel, lack a convex structure, giving a clue to why they are inherently different from the Pauli channels [210].

8.3   Selected Applications of Open Quantum Systems to Quantum Information Processing

Now we elucidate the role played by open system ideas to various facets of quantum information processing by discussing selected applications.

8.3.1   Environment-Mediated Quantum Deleter

Quantum computation is well known to solve certain types of problems more efficiently than classical computation [38]. Although quantum mechanical linearity endows a quantum computer with greater-than-classical power [232], it also imposes certain restrictions, such as the prohibition on cloning [233] and on deleting [234]. The latter result means that quantum mechanics does not allow us to delete a copy of an arbitrary quantum state perfectly.

Requirement of Open Quantum System

A quantum computational task can be broadly divided into three stages:

- (A). Initializing the quantum computer, by putting all qubits into a standard 'blank state';

- (B). Executing the unitary operation that performs the actual computation. This is the area where "decoherence" is an obstacle. A variety of techniques, including quantum error correction [235], dynamic decoupling [236], fault tolerant quantum computation, decoherence-free subspaces [237], among others exist to combat decoherence.

- (C). Performing measurements to read off results.

In step (A), we must be able to erase quantum memory at the end of a computational task, in order to prepare the state of a quantum computer for a subsequent task. What is required is a quantum mechanism that with high probability allows us to prepare standard 'blank states'. It is clear that no unitary process can achieve this, since true deletion would be irreversible, and hence non-unitary. Further, the no-deleting theorem implies that no qubit state can be erased against a copy [234]. A direct method for initializing the quantum computer would be to measure all qubits in the computational basis. This results in a statistical mixture of $|0\rangle$'s and $|1\rangle$'s, and there is no unitary way in a closed system to convert the $|1\rangle$'s while retaining the $|0\rangle$'s. However, open quantum systems, in particular a decohering environment, can affect non-unitary evolution of a sub-system of interest. We are thus led to conclude that decoherence is in fact *necessary* for step (A), since there would be no other way to delete quantum information.

Here this insight is used to argue that decoherence can be useful to quantum computation [213]. In particular, it is shown that a dissipative environment, the *amplitude-damping channel* in the parlance of quantum information theory, can serve as an effective deleter of quantum information. Note that the amplitude-damping channel belongs to the family of the SGAD channel discussed above and can be deduced from it by setting the reservoir temperature T and squeezing parameters, such as s, to zero.

To illustrate our idea we discuss the fidelity as a function of temperature. Fidelity is a measure of the closeness of the state of interest, here the reduced density matrix of the system $\rho^s(t)$, obtained by solving Eq. (8.16) and setting $T = 0$, from the target state $|0\rangle$. It is defined as

$$
\begin{aligned}
f(t) \quad &= \quad \sqrt{\langle 0|\rho^s(t)|0\rangle} = \sqrt{\frac{1 - \langle \sigma_3(t)\rangle}{2}} \\
&= \quad \frac{1}{\sqrt{2}}\left[(1 - e^{-\Gamma t}\langle \sigma_3(0)\rangle) + \frac{(1 - e^{-\Gamma t})}{2N + 1}\right]^{1/2},
\end{aligned}
\qquad (8.28)
$$

where $\Gamma \equiv \gamma_0(2N + 1)$ and $\langle \sigma_3(0)\rangle$ is the expectation value of σ_3 at time $t = 0$. The plot of fidelity as a function of the temperature T in Fig. (8.1) brings out the point that under the action of the amplitude damping channel, operating at $T = 0$, the final state reduces to the desired target state $|0\rangle$, corresponding to unit fidelity.

Problem 12: Derive Eq. (8.28).

8.3.2   Geometric Phase (GP) in Open Quantum Systems

Let us begin with a brief history of GP. Pancharatnam [238] defined a phase characterizing the interference of classical light in distinct states of polarization.

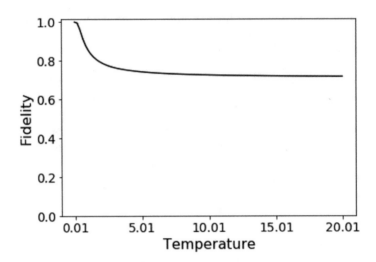

Figure 8.1: Fidelity $(f(t))$ falls as a function of temperature $(T$, in units where $\hbar \equiv k_B \equiv 1)$ until it reaches the value $1/\sqrt{2}$ corresponding to a maximally mixed state. The case shown here corresponds to $\theta_0 = 0$, $\gamma_0 = 0.6$, $\omega = 1.0$ and time $t = 12$. Here we set the squeezing parameters s and Φ to zero. Figure adapted from [213].

Berry [239] discovered that under cyclic adiabatic evolution, a system acquires an extra phase over the dynamical phase, which is popularly called the Berry phase, and which in its general perspective is referred to as the GP. Simon [240] established the geometric nature of GP, linked to notion of parallel transport, i.e., GP depends only on the area covered by the motion, independent of how the motion is executed. This is a consequence of the holonomy in a line bundle over parameter space. Generalization of GP to non-adiabatic and to non-cyclic evolutions were made in [241], [242], respectively. That GP is a consequence of quantum kinematics was shown in [243]. GP was defined for nondegenerate density operators undergoing unitary evolution in [244]. A kinematic approach to define GP in mixed states undergoing nonunitary evolution was developed in [245], which we make use of in our study of GP in an open system model.

The geometric nature of GP implies an inherent fault tolerance and would be useful for quantum computers. There have been proposals to observe GP in superconducting nanocircuits [246]. Here the effect of the environment is never negligible [247]. The above points provide a strong motivation for studying GP in the context of Open Quantum Systems [248, 249, 250, 251]. Here we discuss, following [249], a prototype model of GP in open quantum systems, i.e., the GP of a two-level system or a qubit.

Geometric Phase (GP) in a Two-Level Open System

Here we provide an example of GP, in the context of open quantum systems, by applying the formalism of [245] to a two-level system, undergoing quantum non-demolition (QND), see Chapter 3, Section 8, or dissipative evolution [249].

The system Hamiltonian is

$$H_S = \frac{\hbar\omega}{2}\sigma_z. \tag{8.29}$$

We assume that the system interacts with a squeezed-thermal bath via a QND or a dissipative interaction. We have [245]

$$\Phi_{\mathrm{GP}} = \arg\left(\sum_{k=1}^{N}\sqrt{\lambda_k(0)\lambda_k(\tau)}\langle\Psi_k(0)|\Psi_k(\tau)\rangle e^{-\int_0^\tau dt\langle\Psi_k(t)|\dot\Psi_k(t)\rangle}\right). \tag{8.30}$$

Here $\lambda_k(\tau)$, $\Psi_k(\tau)$ are the eigenvalues and eigenvectors, respectively, of the reduced density matrix of the system. We now sketch a calculation for the evolution of the GP of a two-level system evolving under the influence of an SGAD (dissipative) channel, Eq. (8.20). The nomenclature of the various terms is as defined there. This illustrates, in a simple fashion, the influence of dissipation on the GP of a qubit.

The reduced density matrix of the system can be written as

$$\rho^s(t) = \begin{pmatrix} \frac{1}{2}(1+A) & Be^{-i\omega t} \\ B^*e^{i\omega t} & \frac{1}{2}(1-A) \end{pmatrix}, \tag{8.31}$$

where, invoking the Bloch vector representation of the density matrix of a two-level system,

$$A \equiv \langle\sigma_3(t)\rangle = e^{-\gamma_0(2N+1)t}\langle\sigma_3(0)\rangle - \frac{1}{(2N+1)}\left(1 - e^{-\gamma_0(2N+1)t}\right), \tag{8.32}$$

$$\begin{aligned} B &= \left[1 + \frac{1}{2}\left(e^{\gamma_0 a t} - 1\right)\right]e^{-\frac{\gamma_0}{2}(2N+1+a)t}\langle\sigma_-(0)\rangle \\ &+ \sinh\left(\frac{\gamma_0 a t}{2}\right)e^{i\Phi - \frac{\gamma_0}{2}(2N+1)t}\langle\sigma_+(0)\rangle. \end{aligned} \tag{8.33}$$

For the determination of GP we need the eigenvalues and eigenvectors of the Eq. (8.31). The eigenvalues are

$$\lambda_\pm(t) = \frac{1}{2}(1+\epsilon_\pm), \quad \text{where} \quad \epsilon_\pm = \pm\sqrt{A^2 + 4R^2}. \tag{8.34}$$

Here R is given by

$$
\begin{aligned}
R^2 &= \frac{1}{4}\Big[\{1 + \frac{1}{2}(1 + \cos(\Phi))\left(e^{\gamma_0 a t} - 1\right)\}e^{-\frac{\gamma_0}{2}(2N+1+a)t}\langle\sigma_1(0)\rangle \\
&\quad - \sin(\Phi)\sinh(\frac{\gamma_0 a t}{2})e^{-\frac{\gamma_0}{2}(2N+1)t}\langle\sigma_2(0)\rangle\Big]^2 \\
&\quad + \frac{1}{4}\Big[\{1 + \frac{1}{2}(1 - \cos(\Phi))\left(e^{\gamma_0 a t} - 1\right)\}e^{-\frac{\gamma_0}{2}(2N+1+a)t}\langle\sigma_2(0)\rangle \\
&\quad - \sin(\Phi)\sinh(\frac{\gamma_0 a t}{2})e^{-\frac{\gamma_0}{2}(2N+1)t}\langle\sigma_1(0)\rangle\Big]^2 .
\end{aligned} \tag{8.35}
$$

At time $t = 0$, $\lambda_+(0) = 1$ and $\lambda_-(0) = 0$, hence for the purpose of GP we need only the eigenvalue $\lambda_+(t)$, and its corresponding normalized eigenvector

$$
|\Psi_+(t)\rangle = \sin\left(\frac{\theta_t}{2}\right)|1\rangle + e^{i(\chi(t)+\omega t)}\cos\left(\frac{\theta_t}{2}\right)|0\rangle, \tag{8.36}
$$

where

$$
\sin\left(\frac{\theta_t}{2}\right) = \frac{2R}{\sqrt{4R^2 + (\epsilon_+ - A)^2}} = \sqrt{\frac{\epsilon_+ + A}{2\epsilon_+}}, \tag{8.37}
$$

and $\chi(t)$ is given by

$$
\begin{aligned}
\tan(\chi) &= \Big[\{1 + \frac{1}{2}(1 - \cos(\Phi))\left(e^{\gamma_0 a t} - 1\right)\}e^{-\frac{\gamma_0}{2}a t}\langle\sigma_2(0)\rangle \\
&\quad - \sin(\Phi)\sinh(\frac{\gamma_0 a t}{2})\langle\sigma_1(0)\rangle\Big] \\
&\quad \div \Big[\{1 + \frac{1}{2}(1 + \cos(\Phi))\left(e^{\gamma_0 a t} - 1\right)\}e^{-\frac{\gamma_0}{2}a t}\langle\sigma_1(0)\rangle \\
&\quad - \sin(\Phi)\sinh(\frac{\gamma_0 a t}{2})\langle\sigma_2(0)\rangle\Big].
\end{aligned} \tag{8.38}
$$

It can be seen that for $t = 0$, $\chi(0) = \phi_0$, $\sin\left(\frac{\theta_t}{2}\right) = \sqrt{\frac{1+\langle\sigma_3(0)\rangle}{2}} \equiv \cos\left(\frac{\theta_0}{2}\right)$ and $\cos\left(\frac{\theta_t}{2}\right) = \sqrt{\frac{1-\langle\sigma_3(0)\rangle}{2}} \equiv \sin\left(\frac{\theta_0}{2}\right)$, as expected. Here θ_0 and ϕ_0 are the polar and azimuthal angles of the Bloch sphere representing the two-level system, respectively. Use of Eqs. (8.34), (8.36) in Eq. (8.30) we obtain the desired GP as

$$
\begin{aligned}
\Phi_{\text{GP}} &= \arg\Big[\{\frac{1}{2}\left(1 + \sqrt{A^2(\tau) + 4R^2(\tau)}\right)\}^{\frac{1}{2}} \\
&\quad \times \Big\{\cos\left(\frac{\theta_0}{2}\right)\sin\left(\frac{\theta_\tau}{2}\right) \\
&\quad + e^{i(\chi(\tau)-\chi(0)+\omega\tau)}\sin\left(\frac{\theta_0}{2}\right)\cos\left(\frac{\theta_\tau}{2}\right)\Big\} \\
&\quad \times e^{-i\int_0^\tau dt(\dot{\chi}(t)+\omega)\cos^2(\frac{\theta_t}{2})}\Big].
\end{aligned} \tag{8.39}
$$

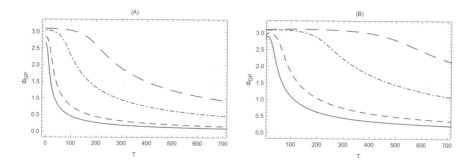

Figure 8.2: GP (Φ_{GP} in radians) as function of temperature (T) for dissipative interaction with a bath of harmonic oscillators. Here $\omega = 1$, $\theta_0 = \pi/2$, the large-dashed, dot-dashed, small-dashed and solid curves, represent, $\gamma_0 = 0.005, 0.01, 0.03$ and 0.05, respectively. Fig. (A) zero squeezing, Fig. (B): squeezing non–vanishing, with $s = 0.4$ and $\Phi = 0$. GP falls with T. Effect of squeezing: GP varies more slowly with T, by broadening peaks and flattening tails. Counteractive action of squeezing on influence of T on GP useful for practical implementation of GP phase gates. Figure adapted from [249].

It can be easily seen from Eq. (8.39) that if the terms determining the influence of the environment on the two-level system, encapsulated here by the terms γ_0, a and Φ, are set to zero, we obtain for $\tau = \frac{2\pi}{\omega}$, $\Phi_{GP} = -\pi(1 - \cos(\theta_0))$, the expected result for the unitary evolution of an intial pure state [252]. In a similar fashion, the case of GP under QND evolution can be handled.

Using these, we can illustrate the effect of open systems, on the evolution of GP. It can be shown that temperature suppresses the GP in the pure dephasing case. However, in the dissipative case, Fig. (8.2), interestingly, reservoir squeezing makes the GP vary more slowly with T, by broadening the peaks and flattening the tails. Counteractive action of squeezing on influence of T on GP would be useful for practical implementation of GP phase gates. Of course, asymptotically the GP falls with temperature, a signature of quantum to classical transition.

8.3.3   Classical Capacity of a Squeezed Generalized Amplitude Damping Channel

A quantum communication channel can be used to perform a number of tasks, such as, transmitting classical or quantum information. A natural question to ask is how information communicated over a squeezed generalized amplitude damping channel gets degraded [213].

Consider the following situation: there is a sender A and receiver B; A has a classical information source producing symbols $X = 0, \cdots, n$ with probabilities $p_0, \cdots, p_n$ which are encoded as quantum states ρ_j ($0 \le j \le n$) and communicated to B, whose optimal measurement strategy maximizes the accessible information, which is bounded above by the *Holevo bound*

$$\chi = S(\rho) - \sum_j p_j S(\rho_j), \tag{8.40}$$

where $\rho = \sum_j p_j \rho_j$; ρ_j are various initial states and $S(\rho)$ is the von Neumann entropy. Assume A encodes binary symbols of 0 and 1 in terms of pure, orthogonal states of the form $|\psi(0)\rangle = \cos(\frac{\theta_0}{2})|1\rangle + e^{i\phi_0}\sin(\frac{\theta_0}{2})|0\rangle$, and transmits them over the squeezed generalized amplitude damping channel ($\mathcal{E}$). Further assume that A transmits messages as product states, i.e., without entangling them across multiple channel use. Then, the (product state) classical capacity $\mathcal{C}$ of the quantum channel is defined as the maximum of $\chi(\mathcal{E})$ over all ensembles $\{p_j, \rho_j\}$ of possible input states ρ_j .

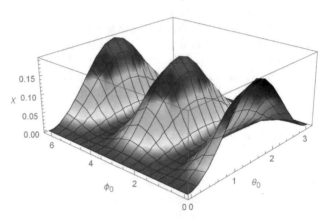

Figure 8.3: Holevo bound χ for a squeezed generalized amplitude damping channel with $\Phi = 0$, over the set $\{\theta_0, \phi_0\}$, which parametrizes the ensemble of input states $\{(\theta_0, \phi_0), (\theta_0 + \pi, \phi_0)\}$, corresponding to the symbols 0 and 1, respectively, with probability of the input symbol 0 being $f = 0.5$. Here temperature $T = 5$, $\gamma_0 = 0.1$, time $t = 5.0$ and bath squeezing parameter $s = 1$. The figure is adapted from [220].

The channel capacity $\mathcal{C}$ is seen, in Fig. (8.3), to correspond to the optimal value of $\theta_0 = \pi/2$, i.e., the input states $\frac{1}{\sqrt{2}}(|0\rangle \pm |1\rangle)$ for $\phi_0 = 0$. From Fig. (8.4), it emerges that χ is maximized for states $\frac{1}{\sqrt{2}}(|0\rangle \pm |1\rangle)$, when the pair of input states are given by $(\theta_0 = \frac{\pi}{2}, \phi_0 = 0)$ and $(\theta_0 = \frac{\pi}{2} + \pi, \phi_0 = 0)$. A comparison of the solid and small-dashed (small-dashed and large-dashed) curves demonstrates the expected degrading effect on the accessible information, of increasing the bath exposure time t (increasing T). A comparison of the large-dashed and dot-dashed curves demonstrates the dramatic effect of including squeezing. In particular, whereas squeezing improves the accessible information for the pair of input states $\frac{1}{\sqrt{2}}(|0\rangle \pm |1\rangle)$, it is detrimental for input states (θ_0, ϕ_0) given by $(0, 0)$ (i.e., $|1\rangle$) and $(\pi, 0)$ (i.e., $|0\rangle$).

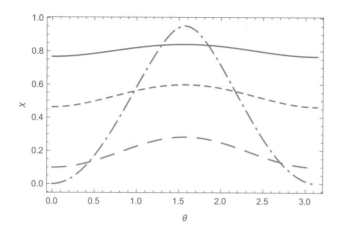

Figure 8.4: Optimal source coding for the squeezed amplitude damping channel, with χ plotted against θ_0 corresponding to the "0" symbol. Here $\Phi = 0$, $\gamma_0 = 0.05$ and $f = 0.5$. The solid and small-dashed curves represent temperature $T = 0$ and bath squeezing parameter $s = 0$, but $t = 1$ and 4, respectively. The large-dashed and dot-dashed curves represent $T = 2.5$ and $t = 2$, but with $s = 0$ and 2, respectively. The figure is adapted from [220].

A comparison between the dotted and solid curves, in Fig. (8.5), shows that squeezing can improve $\mathcal{C}$. This highlights the possible usefulness of squeezing to noisy quantum communication.

8.3.4   Application to Quantum Cryptography

The field of quantum cryptography could be said to have started with the Bennett-Brassard protocol [253], usually called BB84, for quantum key distribution (QKD). In QKD two remote legitimate users (Alice and Bob) can establish an unconditionally secure key through transmission of qubits. Since the pioneering work of Bennett and Brassard several protocols for different cryptographic tasks have been proposed. While most of the initial works on quantum cryptography [253, 254, 255] were concentrated on QKD, eventually explorations were begun in other directions [256]. A protocol for direct secure quantum communication using entangled photon pairs was proposed in [257]. Protocols for deterministic secure quantum communication (DSQC) were later proposed [258, 259, 260], in which the receiver can read out the secret message only after the transmission of at least one bit of additional classical information for each qubit. A set of protocols exist which do not require exchange of classical information. Such protocols are generally referred to as protocols for "quantum secure direct communication" (QSDC) [261].

DSQC and QSDC protocols are reducible to secure QKD protocols in the sense that the former equipped with a source of quantum randomness, yield the

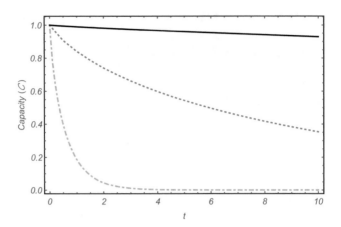

Figure 8.5: Interplay of squeezing and temperature on the classical capacity $\mathcal{C}$ of the squeezed amplitude damping channel (with input states $\frac{1}{\sqrt{2}}(|0\rangle \pm |1\rangle)$), and $f = 1/2$, corresponding to the optimal coding). Here $\Phi = 0$ and $\gamma_0 = 0.1$. The dotted and dot-dashed curves correspond to zero squeezing s, and temperature $T = 0$ and 7, respectively. The solid curve corresponds to $T = 7$ and $s = 3$. The figure is adapted from [220].

latter. A conventional QKD protocol generates an unconditionally secure key by quantum means but then uses classical cryptographic resources to encode the message. No such classical means are required in DSQC and QSDC.

Let us consider a variant of a DSQC protocol based on rearrangement of the orders of particles and dense coding [262]. It should be noted that dense coding refers to the transmission of two bits of information using one qubit [38]. To be precise, we study a DSQC protocol for a specific task, which may be visualized as involving three parties, *viz.*, Charlie as the boss who controls the information, and Alice and Bob as his subordinates. This can also be visualized as an application of *controlled dense coding*, in which the controller is Charlie, who determines how much classical information is delivered to Bob after Alice sends him all her dense coding qubits. Only if the 2-bit information corresponding to choice of Bell state is made available by Charlie to Bob can the latter recover Alices information. By varying the information he gives, Charlie can continuously vary the information recovered by Bob. We study the performance of the protocol by considering the channel of transmission of Alice to Bob to be noisy [263], subjected to the squeezed generalized amplitude damping channel.

The model can be considered as a three-party quantum secret sharing scenario, in which Charlie prepares a Bell state, of which he transmits one half to Alice and the other half to Bob. From this perspective, the latter two receive an ensemble of Bell-states:

$$\rho_{\text{AB}} \equiv \sum_{j,k} a_{j,k} |B_{j,k}\rangle\langle B_{j,k}|, \tag{8.41}$$

where j denotes the parity bit and k the phase bit. Also, $|B_{0,0/1}\rangle \equiv \frac{1}{2}(|00\rangle\pm|11\rangle)$ and $|B_{1,0/1}\rangle = \frac{1}{2}(|01\rangle\pm|10\rangle)$. Alice encodes two bits on her qubit using the four Pauli operators of the superdense coding protocol [262], and sends the qubit to Bob via a possibly insecure channel.

In the noiseless case, Bob measures the two qubits in his possession to obtain the state that corresponds to Alice's encoding. However, Bob can decode the full information only if Charlie shares the full classical *key* information κ that would make the initial entangled state pure. More generally (as detailed below), Bob recovers Alice's transmitted bits depending on the key information obtained from Charlie. Thus, Charlie acts as a *cryptographic switch* who can determine the level of information Alice sends to Bob *after* the full transmission of her qubit.

In this manner, we can consider a family of protocols in which the key information κ varies continuously as $0 \leq \kappa \leq \kappa_{max} = 2$. There are a number of ways to implement κ, and it is assumed that Alice, Bob and Charlie agree upon one such convention at the start of the protocol. One such method would be to start with the assumption that the mixed state provided by Charlie is a Werner state [38, 264], and we parametrize the amount of key information Charlie reveals by means of a single variable ψ. The joint state of Alice and Bob, given by Eq. (8.41), is assumed to have the form

$$\rho_{AB}^{(0,0)}(\psi) = a\Pi_{0,0} + b \sum_{j,k\neq 00} \Pi_{j,k}, \tag{8.42}$$

where $\Pi_{j,k}$ is projector to the Bell state $B_{j,k}$, $a = 0.25 + 0.75\sin(\psi)$, $b = 0.25(1 - \sin(\psi))$ and $b \equiv \frac{1-a}{3}$. After transmitting the qubits to Alice and Bob, Charlie announces the value of angle ψ over a public channel. In so doing, the amount of information provided by him is $\kappa = 2 - H(a, b, b, b)$ bits, where $H(\cdot)$ is Shannon entropy. If $\kappa = 2$ ($a = 1$), then Bob knows Charlie had sent out a $|B_{0,0}\rangle$ state, and can work out Alice's encoded information. Similarly other Werner states are possible. The maximum information Bob can extract from this ensemble is the Holevo quantity χ for the ensemble (8.42). Figure (8.6) shows how Bob's information increases with key information in the noiseless case.

As stated above, we now consider noise to be a squeezed generalized amplitude damping channel acting on Alice's qubit transmission. In Fig. (8.7), the variation of Bob's recovered information, quantified by the Holevo quantity χ, as a function of bath squeezing s, and Charlie's information κ, is depicted. The Holevo quantity χ increases with κ, but not as much as in the noiseless case (Fig. (8.6)), because of the randomness introduced by the noise. Fig. (8.8) depicts the effect, on the Holevo quantity, of squeezing as a function of time. Interestingly, it is seen that squeezing can have favorable influence in some regimes. For sufficiently early times, squeezing fights thermal effects to cause an increase in the recovered information.

The significance of noise is that Alice and Bob may consume some of the Bell pairs to determine the noise level, and decide whether it is too high to

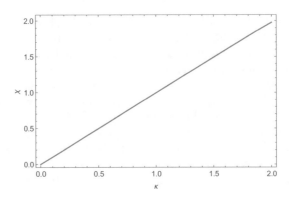

Figure 8.6: Information recovered by Bob, quantified by the Holevo quantity χ, as a function of the key information κ communicated by Charlie, in the noiseless case. The figure is adapted from [263].

permit secure information transfer, assuming conservatively that all the noise is due to Eve. Further information on the influence of noise on general crypto-graphic protocols can be had from [264, 265, 266].

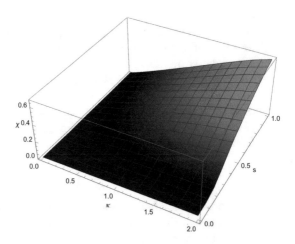

Figure 8.7: Information recovered by Bob, quantified by the Holevo quantity χ, as a function of the squeezing parameter s, coming from the SGAD Channel, and key information κ communicated by Charlie. The time of evolution $t = 0.8$, while temperature $T = 0.2$ (in units where $\hbar \equiv k_B \equiv 1$). The figure is adapted from [263].

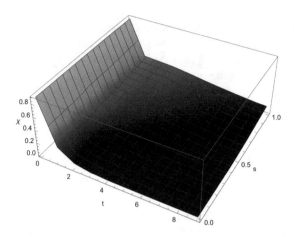

Figure 8.8: Information recovered by Bob, quantified by the Holevo quantity χ, as a function of the SGAD channel parameters s (squeezing) and t (time of evolution), assuming Charlie communicates one bit of information. We note that, for sufficiently early times, squeezing fights thermal effects (here $T = 0.2$) to cause an increase in the recovered information. The figure is adapted from [263].

8.3.5   Dynamics of Quantum Correlations under Open Quantum System evolutions

A capsule on various facets of quantum correlations

Quantum correlations is the terminology developed to address various forms of correlations, in the quantum mechanical regime. Till the turn of the last century, they were thought to be synonymous with entanglement, a notion that was soon seen to be incorrect. At present, there are a number of quantum correlations, characterized based upon some foundational or operational criterion. We first provide a brief summary of some of the well known aspects of quantum correlations. This is followed by illustrating them, dynamically, on an open system model.

Bell inequalities

As stated above, quantum correlations are a *many-faceted* entity. Bell inequalities were one of the first tools to detect entanglement. Given a pair of qubits in the state ρ, the elements of correlation matrix T are $T_{mn} = Tr\left[\rho(\sigma_m \otimes \sigma_n)\right]$. If u_i $(i = 1, 2, 3)$ are the eigenvalues of the matrix $T^\dagger T$ then the Bell-CHSH inequality can be written $M(\rho) < 1$ [267, 268], where $M(\rho) = \max(u_i + u_j)$ $(i \neq j)$. Violations of this inequality, i.e., $M(\rho) > 1$ is a witness of the fact that the system is entangled. However, as pointed by Werner [264], states that satisfy the inequality could also be entangled. This lead to the development of the teleportation fidelity.

Teleportation Fidelity

Teleportation provides an operational meaning to entanglement; whenever $F_{\max} > 2/3$, teleportation is possible. $F_{\max}$ is computed in terms of the eigenvalues $\{u_i\}$ of $T^\dagger T$. $F_{\max} = \frac{1}{2}\left(1 + \frac{1}{3}N(\rho)\right)$, where $N(\rho) = \sqrt{u_1} + \sqrt{u_2} + \sqrt{u_3}$ [268]. A useful inequality involving $M(\rho)$ and $F_{\max}$

$$F_{max} \geq \frac{1}{2}\left(1 + \frac{1}{3}M(\rho)\right) \geq \frac{2}{3} \text{ if } M(\rho) > 1. \tag{8.43}$$

Concurrence

For a mixed state ρ of two qubits, the concurrence, which is a measure of entanglement, is

$$C = \max(\lambda_1 - \lambda_2 - \lambda_3 - \lambda_4, 0). \tag{8.44}$$

Here λ_i are the square root of the eigenvalues, in decreasing order, of the matrix $\rho^{\frac{1}{2}}(\sigma_y \otimes \sigma_y)\rho^*(\sigma_y \otimes \sigma_y)\rho^{\frac{1}{2}}$ where ρ is computed in the computational basis $\{|00\rangle, |01\rangle, |10\rangle, |11\rangle\}$ [225]. For a two-qubit system, concurrence is equivalent to the entanglement of formation which can then be expressed as a monotonic function of concurrence C as

$$E_F = -\frac{1 + \sqrt{1 - C^2}}{2}\log_2\left(\frac{1 + \sqrt{1 - C^2}}{2}\right) - \frac{1 - \sqrt{1 - C^2}}{2}\log_2\left(\frac{1 - \sqrt{1 - C^2}}{2}\right). \tag{8.45}$$

Discord

Quantum discord is the difference between two classically equivalent expressions for mutual information [35] when extended to the quantum regime [269, 270]. Given the density matrix of the joint state of the systems a and b, ρ_{ab}, the two expressions of mutual information are

$$\begin{aligned} I(\rho_{ab}) &\equiv S(\rho_a) + S(\rho_b) - S(\rho_{ab}), \\ J(\rho_{ab}) &\equiv S(\rho_a) - S(\rho_{a|b}). \end{aligned} \tag{8.46}$$

In the above equation, $S(\rho_{ab})$ is the usual von Neumann entropy, $I(\rho_{ab})$ is the quantum mutual information, while $S(\rho_{a|b})$ is the quantum conditional entropy. Classically, these two expressions are equal, but in the context of quantum mechanics, $S(\rho_{a|b})$ depends upon the measurement procedure and is given by optimization over all measurements, for example, the set of projective measurements $\{P_i\}$ as

$$S(\rho_{a|b}) = \min_{P_i} \sum_i p_i S(\rho_{ab}^i), \tag{8.47}$$

where p_i is the probability that the measurement $\{P_i\}$ is performed on the state ρ_{ab} resulting in the state ρ_{ab}^i. $J(\rho_{ab})$ then becomes

$$J(\rho_{ab}) = S(\rho_a) - \min_{P_i} \sum_i p_i S(\rho_{ab}^i). \tag{8.48}$$

Using Eq. (8.48), quantum discord $QD(\rho_{ab})$ is defined as

$$QD(\rho_{ab}) = I(\rho_{ab}) - J(\rho_{ab}). \tag{8.49}$$

There are states that are not entangled, but have discord.

Geometric discord

Since quantum discord involves optimization, it is generally difficult to compute. To ease the computational complexity, for the case of two qubits, geometric discord was proposed [271, 272] and is $D_G(\rho) = \frac{1}{3}[\|\vec{x}\|^2 + \|T\|^2 - \lambda_{max}(\vec{x}\vec{x}^\dagger + TT^\dagger)]$ where T is the correlation matrix, $\vec{x}$ is the vector whose components are $x_m = \text{Tr}(\rho(\sigma_m \otimes \mathbb{I}_2))$, and $\lambda_{max}(K)$ is the maximum eigenvalue of the matrix K.

Measurement induced disturbance (QMID)

QMID ($\mathcal{QM}$) quantifies the quantumness of the correlation between the quantum bipartite states shared between two parties, using popular quantum information parlance, Alice and Rob. For the given $\rho'_{A,R}$, if ρ'_A and ρ'_R are the reduced density matrices, then the mutual information that quantifies the correlation between Alice and Rob is

$$I = S(\rho'_A) + S(\rho'_R) - S(\rho'_{A,R}), \tag{8.50}$$

where $S(\rho)$ is the von Neumann entropy. If $\rho'_A = \sum_i \lambda^i_A \Pi^i_A$ and $\rho'_R = \sum_j \lambda^j_R \Pi^j_R$ denotes the spectral decomposition of ρ'_A and ρ'_R, respectively, then the state $\rho'_{A,R}$ after measuring in the joint basis $\{\Pi_A, \Pi_R\}$ is

$$\Pi(\rho'_{A,R}) = \sum_{i,j} (\Pi^i_A \otimes \Pi^j_R)\rho'_{A,R}(\Pi^i_A \otimes \Pi^j_R). \tag{8.51}$$

QMID [273] is

$$\mathcal{QM}(\rho'_{A,R}) = I(\rho'_{A,R}) - I(\Pi(\rho'_{A,R})), \tag{8.52}$$

is a measure of quantumness of the correlation.

Dynamics of the Reduced Density Matrix for two-qubit Dissipative system

Consider the Hamiltonian, describing the dissipative, position dependent, interaction of two qubits with bath (modelled as a 3-D electromagnetic field (EMF)) via dipole interaction as [274, 275, 276]

$$H = H_S + H_R + H_{SR}$$
$$= \sum_{n=1}^{N=2} \hbar\omega_n S^z_n + \sum_{\vec{k}s} \hbar\omega_k (b^\dagger_{\vec{k}s} b_{\vec{k}s} + 1/2)$$
$$- i\hbar \sum_{\vec{k}s} \sum_{n=1}^{N} [\vec{\mu}_n \cdot \vec{g}_{\vec{k}s}(\vec{r}_n)(S^+_n + S^-_n)b_{\vec{k}s} - h.c.].$$

Here, $\vec{\mu}_n$ represents the transition dipole moments, dependent on the different atomic positions $\vec{r}_n$. Also,

$$S^+_n = |e_n\rangle\langle g_n|, \ S^-_n = |g_n\rangle\langle e_n|, \tag{8.53}$$

are the dipole raising and lowering operators satisfying the usual commutation relations and

$$S_n^z = \frac{1}{2}(|e_n\rangle\langle e_n| - |g_n\rangle\langle g_n|), \qquad (8.54)$$

is the energy operator of the nth atom. The creation and annihilation operators, of the field mode (bath) $\vec{k}s$ with the wave vector $\vec{k}$, frequency ω_k and polarization index $s = 1, 2$, are $b_{\vec{k}s}^\dagger$, $b_{\vec{k}s}$, respectively. The system-reservoir (S-R) coupling constant

$$\vec{g}_{\vec{k}s}(\vec{r}_n) = (\frac{\omega_k}{2\varepsilon\hbar V})^{1/2}\vec{e}_{\vec{k}s}\, e^{i\vec{k}\cdot r_n}, \qquad (8.55)$$

is seen to be dependent on the atomic position r_n. Here, V denotesthe normalization volume and $\vec{e}_{\vec{k}s}$, unit polarization vector of the field. This leads to a number of interesting dynamical effects. Assuming separable initial conditions, and taking a trace over the bath, the reduced density matrix of the qubit system in the interaction picture and in the usual Born-Markov, rotating wave approximation (RWA) is obtained as

$$
\begin{aligned}
\frac{d\rho}{dt} &= -\frac{i}{\hbar}[H_{\tilde{S}}, \rho] - \frac{1}{2}\sum_{i,j=1}^{2}\Gamma_{ij}[1+\tilde{N}](\rho S_i^+ S_j^- + S_i^+ S_j^- \rho - 2S_j^- \rho S_i^+) \\
&\quad - \frac{1}{2}\sum_{i,j=1}^{2}\Gamma_{ij}\tilde{N}(\rho S_i^- S_j^+ + S_i^- S_j^+ \rho - 2S_j^+ \rho S_i^-) \\
&\quad + \frac{1}{2}\sum_{i,j=1}^{2}\Gamma_{ij}\tilde{M}(\rho S_i^+ S_j^+ + S_i^+ S_j^+ \rho - 2S_j^+ \rho S_i^+) \\
&\quad + \frac{1}{2}\sum_{i,j=1}^{2}\Gamma_{ij}\tilde{M}^*(\rho S_i^- S_j^- + S_i^- S_j^- \rho - 2S_j^- \rho S_i^-).
\end{aligned}
$$

Here,

$$\tilde{N} = N_{\text{th}}(\cosh^2(s) + \sinh^2(s)) + \sinh^2(s), \qquad (8.56)$$

$$\tilde{M} = -\frac{1}{2}\sinh(2s)e^{i\Phi}(2N_{\text{th}} + 1) \equiv Re^{i\Phi(\omega_0)}, \qquad (8.57)$$

with

$$\omega_0 = \frac{\omega_1 + \omega_2}{2}, \qquad (8.58)$$

and

$$N_{\text{th}} = \frac{1}{e^{\frac{\hbar\omega}{k_B T}} - 1}. \qquad (8.59)$$

N_{th} is the Planck distribution giving the number of thermal photons at the frequency ω and s, Φ are squeezing parameters. The analogous case of a thermal bath without squeezing can be obtained from the above expressions by setting these squeezing parameters to zero, while setting the temperature (T) to zero one recovers the case of the vacuum bath.

$$H_{\tilde{S}} = \hbar \sum_{n=1}^{2} \omega_n S_n^z + \hbar \sum_{\substack{i,j \\ (i \neq j)}}^{2} \Omega_{ij} S_i^+ S_j^-, \tag{8.60}$$

where

$$
\begin{aligned}
\Omega_{ij} &= \frac{3}{4} \sqrt{\Gamma_i \Gamma_j} \left[-[1 - (\hat{\mu}.\hat{r}_{ij})^2] \frac{\cos(k_0 r_{ij})}{k_0 r_{ij}} + [1 - 3(\hat{\mu}.\hat{r}_{ij})^2] \right. \\
&\quad \times \left[\frac{\sin(k_0 r_{ij})}{(k_0 r_{ij})^2} + \frac{\cos(k_0 r_{ij})}{(k_0 r_{ij})^3} \right].
\end{aligned}
$$

In the above expressions, $\hat{\mu} = \hat{\mu}_1 = \hat{\mu}_2$ and $\hat{r}_{ij}$ are unit vectors along the atomic transition dipole moments and $\vec{r}_{ij} = \vec{r}_i - \vec{r}_j$, respectively. Also, $k_0 = \omega_0/c$, $r_{ij} = |\vec{r}_{ij}|$. The wavevector $k_0 = 2\pi/\lambda_0$, λ_0 being the resonant wavelength, occurring in the term $k_0 r_{ij}$. It sets up a length scale into the problem depending upon the ratio r_{ij}/λ_0. This is thus the ratio between the interatomic distance and the resonant wavelength, allowing for a discussion of the dynamics in two regimes:

- *localized decoherence, where $k_0.r_{ij} \sim \frac{r_{ij}}{\lambda_0} \geq 1$ and*

- *collective decoherence, where $k_0.r_{ij} \sim \frac{r_{ij}}{\lambda_0} \to 0$.*

Localized decoherence implies that each qubit is interacting with its own reservoir. Hence, this regime could also be called the independent decoherence regime. Collective decoherence would arise when the qubits are close enough for them to feel the bath collectively or when the bath has a long correlation length (set by the resonant wavelength λ_0) in comparison to the interqubit separation r_{ij}. Ω_{ij}, Eq. (8.61), is a collective coherent effect due to the multiqubit interaction and is mediated via the bath through the terms

$$\Gamma_i = \frac{\omega_i^3 \mu_i^2}{3\pi\varepsilon\hbar c^3}. \tag{8.61}$$

The term Γ_i is present even in the case of single-qubit dissipative system bath interaction and is the spontaneous emission rate, while

$$\Gamma_{ij} = \Gamma_{ji} = \sqrt{\Gamma_i \Gamma_j} F(k_0 r_{ij}), \tag{8.62}$$

where $i \neq j$ with

$$
\begin{aligned}
F(k_0 r_{ij}) &= \frac{3}{2} \left[[1 - (\hat{\mu}.\hat{r}_{ij})^2] \frac{\sin(k_0 r_{ij})}{k_0 r_{ij}} + [1 - 3(\hat{\mu}.\hat{r}_{ij})^2] \right. \\
&\quad \times \left[\frac{\cos(k_0 r_{ij})}{(k_0 r_{ij})^2} - \frac{\sin(k_0 r_{ij})}{(k_0 r_{ij})^3} \right].
\end{aligned}
$$

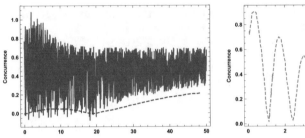
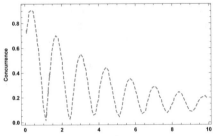

Figure 8.9: Concurrence $\mathcal{C}$ as a function of time of evolution t. The left figure deals with the case of vacuum bath ($T = s = 0$), while the figure in the right panel considers concurrence in the two-qubit system interacting with a squeezed thermal bath, for a temperature $T = 1$ and bath squeezing parameter s equal to 0.1. In the left figure the bold curve depicts the collective decoherence model ($k_0 r_{12} = 0.01$), while the dashed curve represents the independent decoherence model ($k_0 r_{12} = 1.1$). In the right figure, for the given settings, concurrence for the collective model is depicted by the dashed curve. Here, for the independent decoherence model, concurrence is negligible and is thus not seen. The figure is adapted from [275].

Γ_{ij} is a collective incoherent effect due to the dissipative multi-qubit interaction with the bath. For the case of identical qubits, as considered here, $\Omega_{12} = \Omega_{21}$, $\Gamma_{12} = \Gamma_{21}$ and $\Gamma_1 = \Gamma_2 = \Gamma$. Plots depicting the behavior of concurrence, entanglement, between the two qubits as a function of the evolution time t and inter-qubit separation r_{12} are depicted in figures (8.9) and (8.10), respectively. The different effects of the collective and localized regimes on the concurrence are clearly brought out.

Next, we made a comparative study, on states generated by the above discussed open system two-qubit model, of various features of quantum correlations like teleportation fidelity (F_{max}), violation of Bell's inequality $M(\rho)$ (violation takes place for $M(\rho) \geq 1$), concurrence $C(\rho)$ and discord with respect to various experimental parameters like, bath squeezing parameter s, inter-qubit spacing r_{12}, temperature T and time of evolution t [277]. A basic motivation for this is to have realistic open system models that generate entangled states which can be useful for teleportation, but at the same time, not violate Bell's inequality. We provide below some examples of such states. Interestingly, we also find examples of states with positive discord, but zero entanglement, reiterating the fact that entanglement is a subset of quantum correlations.

The Figs. (8.11) and (8.12) depict the evolution of various facets of quantum correlations in a two-qubit system undergoing a dissipative evolution. In particular, they represent the evolution of concurrence, maximum teleportation fidelity F_{max}, test of Bell's inequality $M(\rho)$, discord as a function of inter-qubit distance r_{12}. Here temperature $T = 300$, evolution time t is 0.1 and bath

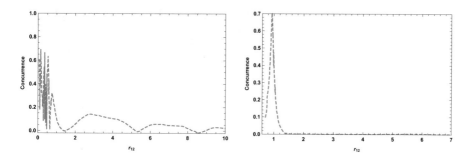

Figure 8.10: Concurrence $\mathcal{C}$ with respect to inter-qubit distance r_{12}. The figure on the left deals with the case of vacuum bath ($T = s = 0$), while the figure on the right considers concurrence in the two-qubit system interacting with a squeezed thermal bath, for $T = 1$, evolution time $t = 1$ and bath squeezing parameter s equal to 0.2. In the left figure the oscillatory behavior of concurrence is stronger in the collective decoherence regime, in comparison with the independent decoherence regime ($k_0 r_{12} \geq 1$); here k_0 is set equal to one. In the right figure, the effect of finite bath squeezing and T has the effect of diminishing the concurrence to a great extent in comparison to the vacuum bath case. Here the concurrence for the localized decoherence regime is negligible, in agreement with the previous figure. The figure is adapted from [275].

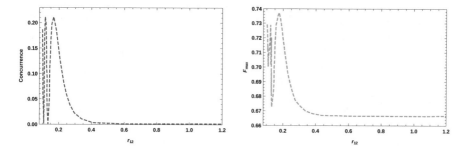

Figure 8.11: Dynamics of concurrence and maximum teleportation fidelity F_{max} as a function of the inter-qubit separation r_{12}. The figure is adapted from [277].

squeezing parameter $s = -1$. From the left Fig. (8.11), we find that the two qubit density matrix is entangled with a positive concurrence except at the point 0.133 (approx) and for $r_{12} \geq 0.4$. Figure (8.11), right panel, illustrates that $F_{max} > \frac{2}{3}$, for all values of r_{12} except where there is no entanglement. However, from the left Fig. (8.12) we find that $M(\rho) < 1$ for all values of r_{12}, clearly demonstrating that the states can be useful for teleportation despite the fact that they satisfy Bell's inequality. Moreover, from the right Fig. (8.12), a positive discord is seen for the complete range of r_{12}, even in the range where there is no entanglement. As a function of the inter-qubit distance, the various correlation measures exhibit oscillatory behavior, in the collective regime

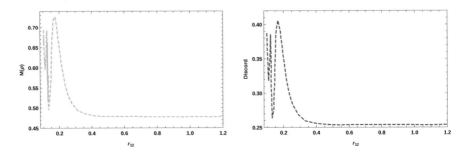

Figure 8.12: Dynamics of Bell's inequality factor $M(\rho)$ and discord as a function of the inter-qubit separation r_{12}. The figure is adapted from [277].

of the model, but flatten out subsequently to attain almost constant values in the independent regime of the model. This oscillatory behavior is due to the strong collective behaviour exhibited by the dynamics due to the relatively close proximity of the qubits in the collective regime.

8.3.6   Quantum Walk

Quantum walks (QWs) are the quantum analog of the classical random walks (CRWs) [278, 279, 280, 281, 282]. The quantum features of interference and superposition make the QW spread quadratically faster than the analogous CRW. Like their classical counterpart, QWs are also widely studied in two forms, *viz.*, continuous-time QW (CTQW) [282] and discrete-time QW (DTQW) [280, 281, 283, 284] and are found to be very useful from the perspective of quantum algorithms [285, 286, 287, 288]. Experimental implementation of QWs has been reported in nuclear magnetic resonance (NMR) systems [289, 290], in the continuous tunneling of light fields through waveguide lattices [291], in the phase space of trapped ions [292, 293] based on the scheme proposed by [294], with single neutral atoms in optically trapped atoms [295], and with single photons [296].

Decoherence, due to open system effects, in QW and the transition of QW to CRW is important from the perspective of practical implementations, and have been extensively studied [297, 298, 299, 300]. Here, keeping in line with the common theme in this chapter, we study the effect of the squeezed generalized amplitude damping (SGAD) channel [220], on the walk evolution.

Consider a particle (a qubit) which is executing a discrete time QW in one dimension, also called the discrete time quantum walk on the line, and its internal states $|0\rangle$ and $|1\rangle$ span $\mathcal{H}_c$, which is referred as the coin Hilbert space. The allowed position states of the particle are $|x\rangle$, which spans $\mathcal{H}_x$, where $x \in \mathbf{I}$, the set of integers. In an n-cycle walk, there are n allowed positions, and in addition the periodic boundary condition $|x\rangle = |x \bmod n\rangle$ is imposed. A t step coined QW is generated by iteratively applying a unitary operation W which acts on the Hilbert space $\mathcal{H}_c \otimes \mathcal{H}_x$:

$$|\psi_t\rangle = W^t|\psi_0\rangle , \tag{8.63}$$

where $|\psi_0\rangle = (\cos(\theta_0/2)|0\rangle + \sin(\theta_0/2)e^{i\phi_0}|1\rangle)|0\rangle$ is an arbitrary initial state of the particle and $W \equiv U\, \mathcal{C}(\xi, \theta, \zeta)$. The $\mathcal{C}(\xi, \theta, \zeta)$ is an arbitrary $SU(2)$ coin toss operation which acts on the coin space and is given by

$$\mathcal{C}(\xi, \theta, \zeta) = \begin{pmatrix} e^{i\xi}\cos(\theta) & e^{i\zeta}\sin(\theta) \\ e^{-i\zeta}\sin(\theta) & -e^{-i\xi}\cos(\theta) \end{pmatrix}. \tag{8.64}$$

The matrix $\mathcal{C}(\xi, \theta, \zeta)$, whose elements are written as $\mathcal{C}_{jk}$, controls the evolution of the walk, with the Hadamard walk corresponding to $\mathcal{C}(0, 45, 0)$. The U is a unitary controlled-shift operation:

$$U \equiv |0\rangle\langle 0| \otimes \sum_x |x-1\rangle\langle x| + |1\rangle\langle 1| \otimes \sum_x |x+1\rangle\langle x|. \tag{8.65}$$

The probability to find the particle at site x after t steps is given by

$$p(x, t) = \langle x|\mathrm{tr}_c(|\psi_t\rangle\langle\psi_t|)|x\rangle. \tag{8.66}$$

Variance of DTQW

Variance σ^2 is an important parameter of the quantum walk. It measures how much the walker has spread from the origin

$$\sigma^2 = \sum_{i=1}^{n} p_i(i - \mu)^2. \tag{8.67}$$

Here p_i is the probability of finding the walker at the i_{th} position and $\mu = \sum_i p_i x_i$. Figure (8.13) depicts variance as a function of time for 100 steps of Classical Random Walk (CRW) (black solid line) and Hadamard quantum walk (red line) $[U(H \otimes I)]^{100}$ on a particle starting from the initial state $\frac{1}{\sqrt{2}}(|\uparrow\rangle + i|\downarrow\rangle)$. It should be noted here that $H = \mathcal{C}(0, 45, 0)$.

Next, we consider the evolution of the quantum walk under the influence of the SGAD noise channel, modelled by the Kraus operators discussed earlier in this chapter. The noise is modeled to interact with the coin. The density operator ρ_c of the coin evolves according to $\rho_c \to \sum_j E_j \rho_c E_j^\dagger$, where E_j represent the relevant Kraus operators. The full evolution of the walker, described by density operator $\rho(t)$, is given by $\sum_j E_j(W\rho(t-1)W^t)E_j^\dagger$, where the E_j's are understood to act only in the coin space.

Figure (8.14) depicts the gradual classicalization of a QW on a line under the action of the SGAD noise for different channel parameters. It is seen that, as the QW turns into a CRW, correspondingly, the probability distribution becomes increasingly Gaussian, causing the quadratic functional dependence of variance on time, characteristic of quantum behavior, to become linear. This

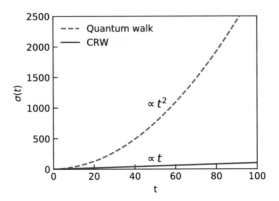

Figure 8.13: Variance of classical and quantum walk as a function of the time steps. The figure is adapted from [298].

pattern of quantum-to-classical transition has also been observed experimentally, in discrete-time quantum walks using a single-photon, subjected to decoherence of the pure dephasing type [296]. This behavior is quite generic, and can be shown, for example, to arise under arbitrary Markovian decoherence of a continuous-time quantum walk on a graph (cf. Fig. 3 of Ref. [301]). The Gaussianization is directly reflected in the fall of standard deviation.

8.3.7 Quasiprobability distributions in open quantum systems

A useful concept in the analysis of the dynamics of classical systems is the notion of phase space. A straightforward extension of this to quantum mechanics is foiled due to the uncertainty principle. Despite this, it is possible to construct quasiprobability distributions (QDs) for quantum mechanical systems in analogy with their classical counterparts [303, 304, 305, 306, 307, 308]. These QDs are very useful as they provide a quantum classical correspondence and facilitate the calculation of quantum mechanical averages in close analogy to classical phase space averages. Nevertheless, the QDs are not probability distributions as they can take negative values as well, a feature that could be used for the identification of quantumness in a system.

The first QD was developed by Wigner resulting in the epithet Wigner function (W) [309, 310, 311, 312, 313]. Other well known examples of QDs are the P and Q functions. The P function played a central role in the development of the field of quantum optics and was originally developed from the possibility of expressing any state of the radiation field in terms of a diagonal sum over coherent states [314, 315]. The P function can become singular for quantum states, a feature that lead to the development of other QDs such as the Q function [316, 317, 318]. These QDs are intimately related to the problem of operator orderings. The P and Q functions are related to the normal

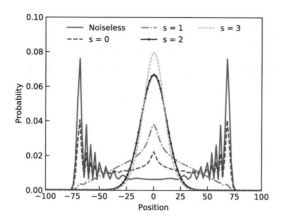

Figure 8.14: The effect of increasing the environmental squeezing parameter s of the SGAD channel, on the position probability distribution of a particle with the initial state $(1/\sqrt{2})(|0\rangle + i|1\rangle) \otimes |\psi_0\rangle$ and coin toss instruction given by $B(0°, 45°, 0°)$. Other channel parameters are fixed at $T = 3$, $\gamma_0 = 0.03$ and $\Delta = 0.2$. (a) The probability distribution on a line for different values of s, at time $t = 100$ steps. With increasing noise (parametrized by s), the distribution transforms from the characteristic QW twin-peaked distribution to the classical Gaussian. The figure is adapted from [302].

and antinormal orderings, respectively, while the W function is associated with symmetric operator ordering. There can be other QDs, apart from the above three, depending upon the operator ordering. However, among all the possible QDs the above three QDs are the most widely studied. There exist several reasons behind the intense interest in these QDs. Thus, for example, nonpositive values of P function define a nonclassical state. Nonpositivity of P is a necessary and sufficient criterion for nonclassicality, but other QDs provide only sufficient criteria.

A nonclassical state can be used to perform tasks that are classically impossible. This fact motivated many studies on nonclassical states, for example, studies on squeezed, antibunched and entangled states. Many of these applications make use of spin-qubit systems.

Quantum optics deals with atom-field interactions. The atoms, in their simplest forms, are modeled as qubits (two-level systems). These are also of immense practical importance as they can be the effective realizations of Rydberg atoms [319, 320]. This evokes the question whether one could have QDs for atomic systems as well. They are closely tied to the problem of development of QDs for $SU(2)$, spin-like (spin-j), systems. Such a development was made in [321], where a QD on the sphere, naturally related to the $SU(2)$ dynamical group [322, 323], was obtained. There are by now a number of constructions of spin QDs [324, 325, 326, 327, 114], among others.

Another approach, is to make use of the connection of $SU(2)$ geometry to that of a sphere. This approach is used here. The spherical harmonics provide a natural basis for functions on the sphere. This, along with the general theory of multipole operators [328, 329], can be made use of to construct QDs of spin (qubit) systems as functions of polar and azimuthal angles [127]. Other constructions, in the literature, of W functions for spin-1/2 systems can be found in [330, 331], among others. A concept that played an important role in the above developments, was the atomic coherent state [332], which lead to the definition of atomic P function in close analogy to their radiation field counterparts.

The fields of quantum optics and information have matured to the point where intense experimental investigations are being made. This motivates the use of open system ideas on the study of quasiprobability distributions. Here we concentrate on the spin systems, in particular, we will work out the QDs for a single two-level system under the influence of the SGAD channel.

The Wigner function

Exploiting the connection between spin-like, $SU(2)$, systems and the sphere, a QD can be expressed as a function of the polar and azimuthal angles. This expanded over a complete basis set, a convenient one being the spherical harmonics, the W function for a single spin-j state can be expressed as [127]

$$W(\theta, \phi) = \left(\frac{2j+1}{4\pi}\right)^{1/2} \sum_{K,Q} \rho_{KQ} Y_{KQ}(\theta, \phi), \qquad (8.68)$$

where $K = 0, 1, \ldots, 2j$, and $Q = -K, -K+1, \ldots, 0, \ldots, K-1, K$, and

$$\rho_{KQ} = \text{Tr}\left\{T_{KQ}^\dagger \rho\right\}. \qquad (8.69)$$

Here, Y_{KQ} are spherical harmonics and T_{KQ} are multipole operators given by

$$T_{KQ} = \sum_{m,m'} (-1)^{j-m} (2K+1)^{1/2} \begin{pmatrix} j & K & j \\ -m & Q & m' \end{pmatrix} |j,m\rangle\langle j,m'|, \qquad (8.70)$$

where $\begin{pmatrix} j_1 & j_2 & j \\ m_1 & m_2 & m \end{pmatrix} = \frac{(-1)^{j_1-j_2-m}}{\sqrt{2j+1}} \langle j_1 m_1 j_2 m_2 | j - m\rangle$ is the Wigner 3j symbol [333] and $\langle j_1 m_1 j_2 m_2 | j - m\rangle$ is the Clebsh-Gordon coefficient. The multipole operators T_{KQ} are orthogonal to each other and they form a complete set with property $T_{KQ}^\dagger = (-1)^Q T_{K,-Q}$. The W function is normalized as

$$\int W(\theta, \phi) \sin\theta d\theta d\phi = 1,$$

and $W^*(\theta, \phi) = W(\theta, \phi)$. Similarly, the W function of a two particle system, each with spin-j is [127]

$$W(\theta_1, \phi_1, \theta_2, \phi_2) = \left(\frac{2j+1}{4\pi}\right) \sum_{K_1,Q_1} \sum_{K_2,Q_2} \rho_{K_1 Q_1 K_2 Q_2} Y_{K_1 Q_1}(\theta_1, \phi_1) Y_{K_2 Q_2}(\theta_2, \phi_2), \qquad (8.71)$$

where $\rho_{K_1 Q_1 K_2 Q_2} = \text{Tr}\left\{\rho T_{K_1 Q_1}^\dagger T_{K_2 Q_2}^\dagger\right\}$. $W(\theta_1, \phi_1, \theta_2, \phi_2)$ is also normalized as

$$\int W(\theta_1, \phi_1, \theta_2, \phi_2) \sin\theta_1 \sin\theta_2 d\theta_1 d\phi_1 d\theta_2 d\phi_2 = 1.$$

Further, it is known that any arbitrary operator can be mapped into the W function or any other QD discussed here.

The P function

In analogy with the P function for continuous variable systems, the P function for a single spin-j state is defined as [127]

$$\rho = \int d\theta d\phi P(\theta, \phi) |\theta, \phi\rangle\langle\theta, \phi|, \tag{8.72}$$

and can be shown to be

$$P(\theta, \phi) = \sum_{K,Q} \rho_{KQ} Y_{KQ}(\theta, \phi) \left(\tfrac{1}{4\pi}\right)^{1/2} (-1)^{K-Q} \left(\frac{(2j-K)!(2j+K+1)!}{(2j)!(2j)!}\right)^{1/2}. \tag{8.73}$$

The P function for two spin-j particles is [127, 334]

$$P(\theta_1, \phi_1, \theta_2, \phi_2) = \sum_{K_1, Q_1 K_2, Q_2} \rho_{K_1 Q_1 K_2 Q_2} Y_{K_1 Q_1}(\theta_1, \phi_1) Y_{K_2 Q_2}(\theta_2, \phi_2)$$

$$\times (-1)^{K_1 - Q_1 + K_2 - Q_2} \left(\frac{1}{4\pi}\right)$$

$$\times \left(\frac{\sqrt{(2j - K_1)!(2j - K_2)!(2j + K_1 + 1)!(2j + K_2 + 1)!}}{(2j)!(2j)!}\right). \tag{8.74}$$

Here $|\theta, \phi\rangle$ is the atomic coherent state [332] and can be expressed in terms of the Wigner-Dicke states $|j, m\rangle$, the atomic analogues of the oscillator number states $|n\rangle$, as

$$|\theta, \phi\rangle = \sum_{m=-j}^{j} \binom{2j}{m+j}^{1/2} \sin^{j+m}\left(\frac{\theta}{2}\right) \cos^{j-m}\left(\frac{\theta}{2}\right) e^{-i(j+m)\phi} |j, m\rangle. \tag{8.75}$$

The Q function

Similarly, the Q function for a single spin-j state is

$$Q(\theta, \phi) = \frac{2j+1}{4\pi} \langle\theta, \phi|\rho|\theta, \phi\rangle, \tag{8.76}$$

and can be expressed as [127]

$$Q(\theta, \phi) = \sum_{K,Q} \rho_{KQ} Y_{KQ}(\theta, \phi) \left(\tfrac{1}{4\pi}\right)^{1/2} (-1)^{K-Q} (2j+1) \left(\frac{(2j)!(2j)!}{(2j-K)!(2j+K+1)!}\right)^{1/2}. \tag{8.77}$$

It is worth noting that from Eq. (8.76), the Q function being an expectation value is always positive. Further, the normalized Q function for a two particle system of spin-j [127, 334] particles is

$$Q\left(\theta_1, \phi_1, \theta_2, \phi_2\right) = \sum_{K_1, Q_1} \sum_{K_2, Q_2} \rho_{K_1 Q_1 K_2 Q_2} Y_{K_1 Q_1}\left(\theta_1, \phi_1\right) Y_{K_2 Q_2}\left(\theta_2, \phi_2\right) \left(\frac{(2j+1)^2}{4\pi}\right)$$

$$\times (-1)^{K_1 - Q_1 + K_2 - Q_2} \left(\frac{(2j)!\,(2j)!}{\sqrt{(2j - K_1)!\,(2j - K_2)!\,(2j + K_1 + 1)!\,(2j + K_2 + 1)!}}\right). \tag{8.78}$$

All the QDs discussed here are normalized to unity. They are also real functions as they correspond to probability density functions for classical states. The density matrix of a quantum state can be reconstructed from these QDs [308]. One can also calculate the expectation value of an operator from them [127].

For a single spin-$\frac{1}{2}$ starting in the atomic coherent state, the initial density matrix is

$$\rho\left(0\right) = |\alpha, \beta\rangle\langle\alpha, \beta|, \tag{8.79}$$

where the form of the atomic coherent state $|\alpha, \beta\rangle$ is as in Eq. (8.75). For evolution under an SGAD channel, making use of the appropriate Kraus operators as well as the multipole operators, the QDs can be computed, see [335] for details. Here we provide in Fig. (8.15), a visual depiction of the evolution of the various QDs with time. A comparison of the Figs. (8.15 a) and (8.15 b) brings out the effect of squeezing on the evolution of QDs. Further, it is easily observed that with the increase in T, the quantumness reduces.

8.3.8   Quantum error correction

At the heart of a quantum computational task lies quantum superposition and entanglement, which are fragile and decay due to noise arising from interactions with the surrounding, i.e., due to open system effects. The theory of quantum error correction and fault tolerant quantum computation deals with attempts to overcome such obstacles.

Quantum error correction was discovered independently by Shor and Steane. In [336] Shor introduced the 9-qubit quantum error correcting (QEC) code that encoded a qubit and could correct an arbitrary single-qubit error in the independent error limit, i.e., the errors on different qubits are not statistically correlated. Calderbank and Shor [337], and Steane [338] independently developed CSS class codes (named after the inventors) that encodes a qubit in 7-qubits, for correcting arbitrary independent single-qubit errors using ideas from the theory of classical error correction. The conditions for performing quantum error correction were introduced independently by Bennett, Divincenzo, Smolin and Wooters [339] and by Knill and Laflame [340], based on the work by Ekert and Macchiavello [341]. The 5-qubit QEC was discovered by Bennett, Divincenzo, Smolin and Wooters [339] and also independently by Laflame, Miquel,

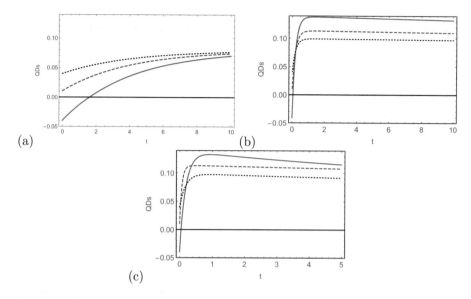

(a)   (b)

(c)

Figure 8.15:   The variation of all the QDs with time (t) are depicted for a single spin-$\frac{1}{2}$ atomic coherent state in the presence of the SGAD noise for, zero bath squeezing angle, in units of $\hbar = k_B = 1$, with $\omega = 1.0$, $\gamma_0 = 0.05$, and $\alpha = \frac{\pi}{2}, \beta = \frac{\pi}{3}, \theta = \frac{\pi}{2}, \phi = \frac{\pi}{3}$. In (a) the variation with time is shown for temperature $T = 5.0$ in the absence of squeezing parameter, i.e., $s = 0$. In (b) the effect of the change in squeezing parameter for same temperature, i.e., $T = 5.0$ is shown by using the squeezing parameter $s = 1.5$, keeping all the other values as same as that used in (a). Further, in (c) keeping $s = 1.5$ as in (b), the temperature is increased to $T = 15$ to show the effect of variation in T. In (c) time is varied only up to $t = 5$ to emphasize the effect of temperature. In all the three plots, dashed, solid and dotted lines correspond to the W, P and Q functions, respectively. The figure is adapted from [335].

Paz and Zurek [342]. This encodes one qubit in five qubits and corrects arbitrary independent single-qubit errors. A stabilizer description of QEC codes was introduced by Gottesman [343, 344]. In this method, attention is given to operators rather than on code words (the encoding states).

Correlated errors.   In [345], collective decoherence giving rise to correlated errors due to quantum non-demolition (QND) interaction on a n-qubit register was considered. In the noise model there, spatial correlation in the decoherence was controlled by varying the inter-qubit distance in the register and was described by the correlation matrix method. By applying the conventional single-error correcting schemes the performance of codes, evaluated via fidelity, against such noise was studied. It was found that such QEC schemes reduce the correlated decoherence. The action of correlated errors on CSS codes was considered in [346], where it was shown quantitatively how error correlations have adverse effect on the performance of CSS codes. Correlated 'semi-classical' (i.e., parametric) noise described by a set of Gaussian random variables was considered in Ref. [347].

Degenerate codes The idea of degenerate codes, where more than one error takes a QEC code to the same state and thus share the same recovery operation, was discovered by Shor and Smolin [348]. A heuristic method to construct degenerate codes for Pauli channels was provided in [349]. In [350], the Hamming bound in the presence of degeneracy was discussed for the codes. A degenerate version of Hamming bound was provided in [351] and showed that the codes work against correlated errors by violating the usual Hamming bound.

Approximate error correcting codes It has been seen that sometimes, constructing channel specific QEC codes would result in better performance against noise. Such codes were introduced in Ref. [352] for amplitude damping error. In [353] it was shown that it is possible to have approximate quantum error correcting codes for the channels that decrease the coherent information by a small ammount. The error correcting conditions for QEC codes were generalised to suit the approximate quantum error correction in [354, 355].

Basic Ideas

The basic idea behind the QEC is to embed a smaller dimensional space in a larger one, such that the added redundancy gives protection against noise. There is a suitable error operator basis, such that errors here shift the code space to mutually orthogonal subspaces in the larger dimensional space. This ensures that the errors can be detected and corrected by devising suitable measurements and recovery operations. The basic ideas of QEC can be illustrated by the following example.

3-qubit bit-flip code

The 3-qubit bit-flip code is similar to a repetition code used in classical error correction where the bits "0" and "1" are encoded as, $0 \to 000;\ 1 \to 111$. Classically, the only possible error is bit-flip. Recovery from error consists in outputting 0 or 1, depending on the majority of 0's and 1's. However, it is apparent that the scheme decodes wrongly if the flip happens on two or more bits. If the probability of a bit flip is p, the probability that two or more bits are flipped is $p_f = 3p^2(1 - p) + p^3 = 3p^2 - 2p^3$ (probability of failure). The encoding is useful if $p_f < p$, which happens when $p < 1/2$.

When trying to mimic the classical repetition to construct QEC codes for a qubit, one faces the following difficulties:

  i An arbitrary quantum state cannot be repeated, due to the *no cloning theorem*.

  ii Measurement performed on qubits destroys the state.

iii A continuum of errors is possible on qubits unlike just bit flip on classical states.

The above mentioned difficulties are surpassed by encoding a qubit $|\psi\rangle$ as

$$|\psi_L\rangle \equiv \alpha|000\rangle + \beta|111\rangle, \tag{8.80}$$

where the encodings are of the form

$$|0\rangle \longrightarrow |0_L\rangle \equiv |000\rangle; \quad |1\rangle \longrightarrow |1_L\rangle \equiv |111\rangle. \tag{8.81}$$

Note that the encoding, Eq. (8.80) is not a repetation of state $|\psi\rangle$. Also, error detection should not reveal any information about $|\psi_L\rangle$. If a bit-flip error happens on one of the three qubits of $|\psi_L\rangle$, it can be detected and corrected using the following two steps.

Error detection. To detect errors, a set of measurements are to be performed on the QECC. The measurement result is called *error syndrome*. For a bit-flip channel there are four error syndromes corresponding to the following four mutually commuting measurement operators:

$$
\begin{aligned}
P_1 &= |000\rangle\langle000| + |111\rangle\langle111|, \\
P_2 &= |100\rangle\langle100| + |011\rangle\langle011|, \\
P_3 &= |010\rangle\langle010| + |101\rangle\langle111|, \\
P_4 &= |001\rangle\langle001| + |110\rangle\langle110|.
\end{aligned} \tag{8.82}
$$

Since P_1, P_2, P_3, P_4 are mutually commuting, the operator $\hat{S} = \alpha P_1 + \beta P_2 + \gamma P_3 + \delta P_4$, where the coefficients are some real numbers, can be used for syndrome measurement. If the outcome of $\hat{S}$ is α, then no error occurred on qubits, while if the outcome is β, then error occurred on first qubit; if the outcome is γ, then error occurred on second qubit and finally, if the outcome is δ, then error occurred on the third qubit. Note that this *incomplete* measurement scheme would reveal nothing about the encoded state $|\psi_L\rangle$.

Recovery operation. Depending upon the outcome of syndrome measurement, the encoded state can be recovered by applying the X (Pauli σ_x) operator on the qubit identified as erroneous. The procedure works perfectly when error occurs at most on one qubit and the probability of not correcting the error is $3p^2 - 2p^3$ and is equivalent to the classical repetition code. Once $p < 1/2$, the encoding and decoding improves the reliability of storing the information. Equivalently, the errors can be determined by measuring the two commuting observables $Z_1 Z_2$ (short for $Z_1 \otimes Z_2 \otimes I$) and $Z_2 Z_3$ (short for $I \otimes Z_2 \otimes Z_3$), where Z is Pauli σ_z. Each of the observables Z_i has eigenvalues ± 1. If $Z_1 Z_2$ and $Z_2 Z_3$ have measurement outcome $+1$, then no bit flip occurred. If $Z_1 Z_2$ has measurement outcome $+1$ and $Z_2 Z_3$ has -1, then bit flip occurred on the third qubit. If $Z_1 Z_2$ has measurement outcome -1 and $Z_2 Z_3$ has $+1$, then bit flip occurred on the first qubit. If both have measurement outcomes -1, then bit flip occurred on second qubit. The recovery operation is performed as mentioned earlier.

To improve the error analysis, consider the *fidelity* which is a measure of closeness between two given states, that may be considered to quantify code performance. If ρ and σ are quantum states before and after the action of error respectively, the fidelity between them is defined as

$$F(\rho, \sigma) = \mathrm{Tr}\sqrt{\sqrt{\sigma}\rho\sqrt{\sigma}}. \qquad (8.83)$$

Consider the scenario where the unencoded state $|\psi\rangle$ is acted upon by bit-flip noise X. The corrupt state is $(1-p)|\psi\rangle\langle\psi| + pX|\psi\rangle\langle\psi|X$. The fidelity between the corrupt and uncorrupt states is $\sqrt{(1-p) + |\langle\psi|X|\psi\rangle|^2}$. The minimum fidelity is $\sqrt{1-p}$ (say when $|\psi\rangle = |0\rangle$). Upon encoding the state as $|\psi_L\rangle = |0\rangle_L + |1\rangle_L$, the state after noise and error correction is $((1-p)^3 + 3p(1-p)^2)|\psi\rangle\langle\psi|$. The fidelity in this case is $\sqrt{(1-p)^3 + 3p(1-p)^2}$, so that encoding of qubits is beneficial if $p \leq 1/2$.

Phase flip codes

Phase-flip error is special to quantum errors and has no classical analogue. In this error model, with probability p, the relative phase between the states $|0\rangle$ and $|1\rangle$ is flipped. This can be realized by applying Z (Pauli σ_z): $Z(\alpha|0\rangle + \beta|1\rangle) = \alpha|0\rangle - \beta|1\rangle$. To construct the QECC for the error, consider the states $|+\rangle = \frac{1}{\sqrt{2}}(|0\rangle + |1\rangle)$ and $|-\rangle = \frac{1}{\sqrt{2}}(|0\rangle - |1\rangle)$ which are eigenstates of X (σ_x). Notice that the action of Z takes $|+\rangle$ to $|-\rangle$ and vice versa. Thus, by changing the basis from $\{|0\rangle, |1\rangle\}$ to $\{|+\rangle, |-\rangle\}$, the action of Z changes from phase flipping to bit flipping. Hence, the action of bit flip on $\{|+\rangle, |-\rangle\}$ is equivalent to phase flip on $\{|0\rangle, |1\rangle\}$. Therefore the encoding

$$|\psi_L\rangle \to \alpha|0_L\rangle + \beta|1_L\rangle \equiv \alpha|+++\rangle + \beta|---\rangle, \qquad (8.84)$$

can protect the state $|\psi\rangle$ against phase-flip error.

For error detection, projectors $P'_j = H^{\otimes 3} P_j H^{\otimes 3}$, rotated by the Hadamard gate $H = \frac{1}{2}(|0\rangle\langle 0| - |0\rangle\langle 1| + |1\rangle\langle 0| + |1\rangle\langle 1|)$, are employed. Equivalently, one uses observables $X_1 X_2$ and $X_2 X_3$ for error detection. If $X_1 X_2$ and $X_2 X_3$ have measurement outcome $+1$, then no phase-flip occurred. If $X_1 X_2$ has measurement outcome $+1$ and $X_2 X_3$ has -1, then phase-flip occurred on third qubit. If $X_1 X_2$ has measurement outcome -1 and $X_2 X_3$ has $+1$, then phase-flip occurred on first qubit. If both have measurement outcomes -1, then phase-flip occurred on second qubit. For recovery, one performs Z operation on the erroneous bit.

Illustrating quantum error correction via the Shor code

The Shor code is a QEC code that encodes one qubit in 9 qubits and provides protection against an arbitrary single-qubit error. The code is a concatenation of bit-flip and phase-flip codes. First, a state is encoded using the phase-flip code: $|0\rangle \to |+++\rangle$, $|1\rangle \to |---\rangle$. Then each of these states are encoded using bit-flip codes: $|+\rangle \to \frac{1}{\sqrt{2}}(|000\rangle + |111\rangle)$, $|-\rangle \to \frac{1}{\sqrt{2}}(|000\rangle - |111\rangle)$.

Thus, the logical states for the Shor code are

$$|0\rangle \to |0_L\rangle \equiv \frac{(|000\rangle + |111\rangle)(|000\rangle + |111\rangle)(|000\rangle + |111\rangle)}{2\sqrt{2}},$$

$$|1\rangle \to |1_L\rangle \equiv \frac{(|000\rangle - |111\rangle)(|000\rangle - |111\rangle)(|000\rangle - |111\rangle)}{2\sqrt{2}}. \qquad (8.85)$$

Error detection and recovery

On the first set of the three qubits, error detection is carried out similar to the 3-qubit bit-flip code. The same procedure is followed for the second and the third set of qubits. For detecting the phase-errors, consider the two observables $\mathcal{X}_1 = X_1 X_2 X_3 X_4 X_5 X_6$ and $\mathcal{X}_2 = X_4 X_5 X_6 X_7 X_8 X_9$. If the measurement outcomes of both $\mathcal{X}_1$ and $\mathcal{X}_2$ is $+1$, then no phase-flip occurred. If the measurement outcome of $\mathcal{X}_1$ is $+1$ and that of $\mathcal{X}_2$ is -1, then phase-flip happened on the third set of qubits. If the measurement outcome of $\mathcal{X}_1$ is -1 and that of $\mathcal{X}_2$ is $+1$, then the phase of the first set of qubits flipped. If the measurement outcomes of both $\mathcal{X}_1$ and $\mathcal{X}_2$ is -1, then phase-flip occurred on the second set of qubits. The recovery procedure is same as the 3-qubit bit flip code for each set of three qubits. For recovering the QECC from phase-flip errors, we follow the 3-qubit phase-flip code.

Discretizing quantum errors

In classical communication theory, a continuum of errors is handled by *digitizing* the signal which carries the data. In the case of quantum information errors are not only bit and phase flip errors but a continuum of arbitrary errors, for e.g., the application of a phase gate instead of a phase flip, which might at first seem to be uncorrectable. But quantum error correction is possible essentially because measurement helps discretize noise. For the analogous classical systems, no such descretization exists.

As explained in Sec. 8.2.1, noise $\mathcal{E}$ on qubits has operator-sum representation with elements $\{K_i\}$ which is also most convenient for QEC. The action of noise on the QEC codes $|\psi_L\rangle$ is

$$\mathcal{E}(|\psi_L\rangle\langle\psi_L|) = \sum_i K_i |\psi_L\rangle\langle\psi_L| K_i^\dagger. \qquad (8.86)$$

Discretization of the continuum of errors is achieved by decomposing K_i in the error basis $E_j = \{I, X, Y, Z\}$ as

$$K_i = e_{i0}I + e_{i1}X + e_{i2}Y + e_{i3}Z = \sum_j e_{i,j} E_j. \qquad (8.87)$$

Note that $\mathcal{E}$ could be a multi-qubit error acting on m-qubits of the QEC code. In such cases the E_i is expanded in $\{I, X, Y, Z\}^{\otimes m}$. Due to discretizition, any erroneous QEC code can be written as (apart from normalization)

$$K_i |\psi_L\rangle = e_{i0} |\psi_L\rangle + e_{i1} X |\psi_L\rangle + e_{i2} Y |\psi_L\rangle + e_{i3} Z |\psi\rangle = \sum_j e_{i,j} E_j |\psi_L\rangle, \qquad (8.88)$$

due to which Eq. (8.86) can be written as

$$\mathcal{E}(|\psi_L\rangle\langle\psi_L|) = \sum_{i,j} e_{i,j} E_j |\psi_L\rangle\langle\psi_L| E_k^\dagger e_{i,k}^* = \sum_{j,k} \chi_{j,k} E_j |\psi_L\rangle\langle\psi_L| E_k^\dagger, \quad (8.89)$$

where $\chi_{i,j} = \sum_i e_{i,j} e_{i,k}^*$ is a Hermitian matrix known as the *process matrix*. The subject of quantum process tomography which deals with characterizing the quantum processes is about devising methods to determine the process matrix $\chi_{j,k}$. This suggests that there exists an overlap between QEC and quantum process tomography.

By performing syndrome measurement the state is collapsed to one of the mutually orthogonal states $|\psi_L\rangle$, $X|\psi_L\rangle$, $Y|\psi_L\rangle$, $Z|\psi_L\rangle$ which can be distinguished. For this reason, discretizing the continuum of errors works and is central to quantum error correction. Then by performing recovery operations, the QECC $|\psi_L\rangle$ is recovered.

The Theory of Quantum Error Correction

The theory of error correction generalizes the ideas introduced by the Shor code. In quantum information processing, a QECC protects quantum information from noise, provided the initial state $|\Psi\rangle$ is prepared within the code space $\mathcal{C}$, which satisfies suitable properties [38, 343, 340]. Let $\{|J\rangle\}$ be a n-qubit basis for $\mathcal{C}$, encoding k-qubit states $|j\rangle$ with $0 \leq j \leq 2^k - 1$. Such a code is a $[[n, k]]$ QECC, where k is the code rate. In this work, we will assume that the error basis elements are elements of the Pauli group $\mathcal{P}_n$, the set of all possible tensor products of n Pauli operators, with and without factors $\pm 1, \pm i$. Hence, $E_k^\dagger = E_k$ and $(E_j)^2 = I_n$, the identity operator over n qubits. The necessary and sufficient conditions for quantum error correction are:

$$\langle J|E_m^\dagger E_n|K\rangle = 0, \quad (8.90a)$$
$$\langle J|E_m^\dagger E_n|J\rangle = \langle K|E_m^\dagger E_n|K\rangle \in \{0, 1\}, \quad (8.90b)$$

where $|J\rangle \neq |K\rangle$, and E_m, E_n are two (possibly identical) basis elements of an operator basis for the space $\mathcal{E}$ of allowed errors. In Eq. (8.90b), the choice 0 corresponds to the non-degenerate case.

To see why Eqs. (8.90) are necessary, suppose that the total recovery operation is denoted by a unitary operation $\mathcal{R}$. Recovery involves preparing an ancilla in an initial state $|\alpha\rangle$ and applying $\mathcal{R}$ on the joint system. We thus have $\mathcal{R}|\alpha\rangle E_m|J\rangle = |\alpha_m\rangle|J\rangle$ and $\mathcal{R}|\alpha\rangle E_n|K\rangle = |\alpha_n\rangle|K\rangle$, or

$$\langle J|E_m^\dagger\langle\alpha|\mathcal{R}^\dagger\mathcal{R}|\alpha\rangle E_n|K\rangle = \langle\alpha_m|\alpha_n\rangle\langle J|K\rangle$$
$$\Rightarrow \langle J|E_m^\dagger E_n|K\rangle = 0,$$

from which Eq. (8.90a) follows. This ensures that two distinct code words are not confused even in the presence of noise, and has an obvious counterpart in classical error correction. We also have

$$\langle J|E_m^\dagger\langle\alpha|\mathcal{R}^\dagger\mathcal{R}|\alpha\rangle E_n|J\rangle = \langle\alpha_m|\alpha_n\rangle$$
$$\Rightarrow \langle J|E_m^\dagger E_n|J\rangle = \langle\alpha_m|\alpha_n\rangle,$$

as also for $|K\rangle$, from which Eq. (8.90b) follows. Note that we only require for equality between the left- and right-hand sides of Eq. (8.90b). If in addition the *LHS* and *RHS* vanish, this would correspond to the classical requirement that distinct errors on the same code word produce orthogonal erroneous words. In the quantum case, however, the *LHS* and *RHS* need not vanish, and we obtain *degenerate* codes, which have no classical counterpart.

To prove the sufficiency of (8.90) for quantum error correction, let the system be in an arbitrary logical state $|\Psi\rangle = \sum_J \alpha_J|J\rangle$, which encodes the state $|\psi\rangle = \sum_j \alpha_j|j\rangle$, where $\sum_j |\alpha_j|^2 = 1$. Let the error be an incoherent sum of Kraus operators of the form $F = \sum_k \beta_k E_k$, with $\sum_k \beta_k = 1$. This maps the initial state to $F|\Psi\rangle$. In the non-degenerate case, $F|\Psi\rangle = \beta_0 E_0|\Psi\rangle + \cdots \beta_{4^n-1}E_{4^n-1}|\Psi\rangle$. Each of the terms $E_j|\Psi\rangle$ must be in an orthogonal space given by (8.90). Thus, a projection to $E_j\mathcal{C}$, followed by an application of E_j constitutes the required recovery $\mathcal{R}$. In the degenerate case, suppose that E_m and E_n are degenerate. Then $E_m|J\rangle = E_n|J\rangle$, and a projection on to $E_m\mathcal{C} = E_n\mathcal{C}$ followed by either E_m or E_n constitutes the required recovery $\mathcal{R}$.

The conditions (8.90) can be equivalently stated as [340]:

$$\langle\Psi|G^\dagger G|\Psi\rangle = c(G), \tag{8.91}$$

where the function c depends only on the error G and not the encoded state $|\Psi\rangle$. By expanding $|\Psi\rangle$ in terms of $|J\rangle$ and G in terms of the basis elements E_j, we rewrite condition (8.91) as:

$$\langle J|E_m^\dagger E_n|K\rangle = c_{m,n}\delta_{JK}, \tag{8.92}$$

where $c_{m,n}$ is a Hermitian matrix of numbers that is independent of J, K.

Bounds on quantum error correcting codes

The quantum versions of the Hamming and Gilbert-Varshamov bounds were introduced by Ekert and Macchiavello [341]. Consider a k-qubit state being encoded into a n-qubit QEC code and at most independent single qubit errors happen on t-qubits. The possible errors are X, Y, Z. Also there are 2^k such logical states. The total number of possible errors is $\sum_{j=0}^t {}^nC_j 3^j$. For non-degenerate codes the total number of errors must be lesser than or equal to 2^n. Thus, we have the inequality

$$2^k \sum_{j=0}^t {}^nC_j 3^j \le 2^n. \tag{8.93}$$

For $k = 1$ and $t = 1$ the quantum Hamming bound reduces to $2(1 + 3n) \le 2^n$. The inequality is not satisfied for $n \le 4$, while for $n \ge 5$ it is. Therefore, it follows that there is no code with less than five qubits which can protect a qubit against single qubit errors.

For large n and some k there exists $[n, k]$ code correcting errors on at most t-qubits such that

$$\frac{k}{n} \geq 1 - 2H\left(\frac{2t}{n}\right), \tag{8.94}$$

where $H(x)$ is the binary Shannon entropy and the bound is known as Gilbert-Varshamov bound.

Stabilizer codes

A stabilizer description of error correction [344, 356] focusses attention on operators, which can be compact, rather than on code words, which can be large. A state $|\psi_L\rangle$ is said to be stabilized by an operator S if $S|\psi_L\rangle = |\psi_L\rangle$. Let $\mathcal{G}$ be a subset of $n - k$ independent, commuting elements from $\mathcal{P}_n$. A $[[n, k]]$ QECC is the 2^k-dimensional $+1$-eigenspace $\mathcal{C}$ of the elements of $\mathcal{G}$. The simultaneous eigenbasis of the elements of $\mathcal{G}$ are the code words $|j_L\rangle$. The set of 2^{n-k} operators generated by $\mathcal{G}$ constitute the stabilizer $\mathcal{S}$. The centralizer of $\mathcal{S}$ is the set of all elements of $\mathcal{P}_n$ that commute with each member of $\mathcal{S}$:

$$\mathcal{Z} = \{P \in \mathcal{P}_n \mid \forall S \in \mathcal{S}, [P, S] = 0\}, \tag{8.95}$$

while the normalizer of $\mathcal{S}$ is the set of all elements of $\mathcal{P}_n$ that conjugate the stabilizer to itself:

$$\mathcal{N} = \{P \in \mathcal{P}_n \mid PSP^\dagger = \mathcal{S}\}. \tag{8.96}$$

We note that $\mathcal{S} \subseteq \mathcal{N}$ because the elements of $\mathcal{S}$ are unitary and mutually commute. Similarly, $\mathcal{Z} \subseteq \mathcal{N}$ because elements of the centralizer are unitary and commute with all elements of the stabilizer. To see that the converse is true, we note that if $N \in \mathcal{N}$ then $NSN^\dagger = S'$, or $NS = S'N$. For Pauli operators, $NS = \pm SN$, meaning $S' = \pm S$. But if $S' = -S$, then $NSN^\dagger = -S$, which would require that both S and $-S$ are in $\mathcal{S}$. However if $S \in \mathcal{S}$, then $-S$ is not in the stabilizer, so the only possibility is $S' = S$, and we obtain $[N, S] = 0$, i.e., $\mathcal{N} \subseteq \mathcal{Z}$. It thus follows that here $\mathcal{Z} = \mathcal{N}$. We have $SN|j_L\rangle = NS|j_L\rangle = N|j_L\rangle$, which implies that the action of N is that of a logical Pauli operation on code words.

A set of operators $E_j \in \mathcal{P}_n$ constitutes a basis for correctable errors if one of the following conditions hold:

$$E_j E_k \ \in \ \mathcal{S} \tag{8.97a}$$
$$\exists G \ \in \ \mathcal{G} : [E_j E_k, G] \neq 0. \tag{8.97b}$$

The case (8.97a) corresponds to *degeneracy*. Here $\langle \psi_L | E_j E_k | \psi_L \rangle = \langle \psi_L | \psi_L \rangle = 1$, meaning that both errors produce the same effect, and the code space is indifferent as to which of them happened. Thus, either error can be applied as a recovery operation when one of them occurs. The case (8.97b) corresponds to $E_j^\dagger E_k \notin \mathcal{N}$. In that case, $\exists G \in \mathcal{G} : E_j E_k G = -G E_j E_k$, which ensures that G anti-commutes with precisely one of the operators E_j and E_k. Hence, the noisy logical states $E_j|\psi_L\rangle$ and $E_k|\psi_L\rangle$ will yield distinct eigenvalues (one being $+1$ and the other -1) when G is measured. The set of $n - k$ eigenvalues ± 1 obtained by measuring the generators G forms the error syndrome. The consolidated error correcting condition (8.97) can be stated as the requirement $E_j E_k^\dagger \notin \mathcal{N} - \mathcal{S}$.

Noise characterization

Characterizing the quantum dynamics forms a vital part in implementing quantum computation and information physically. This provides a comparison of the implemented quantum operations against the desired ones on the qubits and thus helps in benchmarking the quality of gates. Principal difficulty in the realization of quantum processing tasks is environmental-induced noise, which decoheres the quantum system, resulting in the loss of quantum superposition and entanglement. In this situation, complete or partial characterization of noise is essential to fight against it, say by constructing appropriate quantum error correcting codes.

Action of noise $\mathcal{E}$, described by a CP map, on a quantum state ρ of dimension d can be expressed in an error basis $\{E_i\}$ of $d \times d$ matrices, as described by Eq. (8.89)

$$\mathcal{E}(\rho) = \sum_{m,n}^{d^2} \chi_{m,n} E_m \rho E_n^\dagger. \tag{8.98}$$

The error basis $\{E_i\}$ satisfies the orthogonality condition $\mathrm{Tr}(E_i E_j^\dagger) = d\delta_{i,j}$, where $\delta_{i,j}$ is the Kronecker delta. $\chi_{m,n}$ is a Hermitian matrix, also known as the "process matrix", in the d^2-dimensional Hilbert-Schmidt space of linear operators acting on the system of dimension d. From the tracing preserving property of $\mathcal{E}$, we have $\sum_{m,n}^{d^2} \chi_{m,n} E_m^\dagger E_n = \mathbb{I}$, which imposes d^2 conditions, so that the matrix χ has $d^4 - d^2$ independent real elements. This forces the condition $\sum_j \chi_{j,j} = 1$, the (positive) diagonal elements of which can be interpreted as probabilities. Here, E_j are multi-qubit Pauli operators, which is appropriate for employing the QEC formalism.

Standard quantum process tomography

Characterization of a quantum noise is determining the elements of process matrix $\chi_{m,n}$. The first technique to address this was standard quantum process tomography (SQPT) [38, 357] where a set of suitably prepared states $\{\rho_i\}$ is input to unknown noisy dynamics $\mathcal{E}$ to be characterized. The action of $\mathcal{E}$ on each input state can be determined experimentally by state tomographic techniques as

$$\mathcal{E}(\rho_i) = \sum_k C_{i,k} \rho_k, \tag{8.99}$$

where $\{\rho_k\}$ is a basis for measurements on output states of $\mathcal{E}$ and $C_{i,k} = \mathrm{Tr}(\mathcal{E}(\rho_i)\rho_k)$ are measurement outcomes. Also, $E_m \rho_i E_n^\dagger = \sum_k \beta_{i,k}^{m,n} \rho_k$, where

$\beta_{i,k}^{m,n}$ is the matrix determined by the choice of $\{\rho_i\}$, $\{\rho_k\}$ and $\{E_i\}$. Substituting the expression in Eq. (8.98) and comparing with Eq. (8.99) we have

$$\sum_k \sum_{m,n} \chi_{m,n} \beta_{i,k}^{m,n} \rho_k = \sum_k C_{i,k} \rho_k. \qquad (8.100)$$

Obtaining the values of $C_{i,k}$ from state tomographic measurements one can determine the elements of the process matrix by $\chi_{m,n} = \sum_{i,k} (\beta_{i,k}^{m,n})^{-1} C_{i,k}$. By SQPT, one needs to perform $d^4 - 1$ measurements which grows exponentially with the number of qubits.

Ancilla-assisted process tomography

Another method to determine the $\chi_{m,n}$ is ancilla-assisted process tomography (AAPT) where the principal system **P** and an ancillary system **A** are prepared in suitable initial states. Noise to be characterized $\mathcal{E}$ is made to act on **P** while **A** is required to be clean. The initial state considered in [358] was an entangled state, not essentially maximally entangled, and later it was shown in [359] that even a non-entangled Werner state can be used. The information about the dynamics on **P** is extracted via quantum state tomography on the joint system using separable or non-separable basis measurements. By having the ancilla of dimension atleast equal to that of principal system it is guaranteed that the joint state after being subjected to noise will bear one to one correspondence with the noise [359]. $d^4 - 1$ measurements are needed using separable measurements, whereas the same can be achieved by d^2 with non-separable measurements [360].

Direct methods

SQPT and AAPT are not direct methods for QPT in the sense that they first obtain full state tomographic data of the output states of the channel $\mathcal{E}$, and then use this exponentially large (grows with number of qubits) data to derive χ. Subsequently, a method which bypasses the state tomography known as direct characterization of quantum dynamics (DCQD) was introduced in [361, 362]. DCQD uses the maximally entangled state $(|00\rangle + |11\rangle)/\sqrt{2}$ stabilized by ZZ and XX for determining the diagonal terms of χ. The detection of error, i.e., syndrome measurement is implemented via Bell state measurements. For determining the off-diagonal terms, stabilizer measurement of non-maximally entangled states $\alpha|00\rangle + \beta|11\rangle$ (stabilized by ZZ), where $\alpha \neq \beta$, is performed. The statistics of measurement outcomes can determine an offdiagonal term. Then on the stabilizer-measurement-collapsed states normalizer (XX) measurement (which commutes with stabilizer) is done to get the other offdiagonal term. The complete process matrix can be determined by using d^2 different input states and single measurement on each state. Totally, $d^2 - 1$ configurations are needed. Other recent developments include a characterization of noise using an efficient method for transforming a channel into a symmetrized (i.e., having only diagonal elements in the process matrix) channel via twirling [363], which is suitable for identifying QECCs [364]. A method similar to [363], but extended to estimate any given off-diagonal term, was introduced in Ref. [365].

Characterization of quantum dynamics using quantum error correction

The methods discussed above, of characterization of quantum dynamics (CQD), are *offline*, i.e., QEC and CQD are not concurrent, as they require distinct state preparations. Recently, "Quantum error correction based characterization of dynamics" (QECCD) was introduced [366], in which the initial state is any element from the code space of a quantum error correcting code (QECC) that can protect the state from arbitrary errors acting on the subsystem subjected to the unknown dynamics. The statistics of stabilizer measurements, with possible unitary pre-processing operations, are used to characterize the noise, while the observed syndrome can be used to correct the noisy state. This requires at most $2(d^2 - 1)$ configurations to characterize arbitrary noise acting on n qubits. The QECCD technique was used to characterize the dissipative 2-qubit noise due to interaction with a vacuum bath in [367].

8.4 Guide to Further Literature

We do hope that this chapter wets the appetite of at least some of the readers, encouraging them to venture into research in this field. What is presented here is just the tip of the proverbial iceberg! There have been extensive developments, both on the theoretical as well as experimental fronts, in attempts to control decoherence dynamically, see for e.g., [226] and references therein. The study of quasiprobability distributions is naturally followed by tomography, the art and craft of state reconstruction [357, 368, 369, 370]. Error correction is now used for characterizing noisy channels [366, 367]. A topic conspicuous by its absence here is quantum information in continuous variables, a veritable field with numerous developments [371, 372]. This field bears witness to a number of fundamental developments, such as the formulation of entanglement [373, 374], in analogy with Wootters notion of concurrence. Further, the developments in this chapter essentially use Markovian noise processes. The stage is now set for a foray into non-Markovian phenomena, an endeavour we shall attempt in the next chapter.

Chapter 9

Recent Trends

This chapter is devoted to some of the recent trends in open quantum systems and includes a foray into relativistic phenomena such as the Unruh effect as well as sub-atomic physics, including neutrinos and mesons. Unruh effect is the sobriquet given to the thermal like effect due to accelerated motion and is the flat space analogue of the celebrated Hawking effect. Unruh effect as a thermal like effect suggests an open quantum system, in the language of quantum information processing, a quantum channel, characterization of which should elucidate the general properties of the effect. Quantum correlations have been predominantly studied on stable electronic and photonic systems. However, with the advent of modern experimental progress there is no reason why one should not venture into the subatomic domains. This is taken up here in the form of studies of open system effects on neutrinos and mesons. Such studies, apart from their importance from the fundamental perspective could also shed light into our understanding of the universe. Also discussed are some of the developments into non-Markovian phenomena and quantum thermodynamics. Non-Markovian phenomena come to the forefront when the evolution has memory and has been the torchbearer of a lot of research in recent times. Further, with experimental progress in the quantum domain making steady inroads into the device sector, quantum technology is expected to make its indelible mark on society soon. In that case, the engines would be operational in the quantum regime and the working principle behind them would be quantum thermodynamics, the infusion of quantum mechanics into the unshakeable, solid bedrock of thermodynamics! These are all areas of current research and their discussion here would hopefully excite the reader.

9.1 Application of Open Quantum System to Unruh Effect and Sub-Atomic Physics

Here we will discuss, briefly, about the interface of open system ideas to the Unruh effect and some facets of sub-atomic physics, in particular, neutrino and

© Hindustan Book Agency 2018 and Springer Nature Singapore Pte Ltd. 2018   221
S. Banerjee, *Open Quantum Systems*, Texts and Readings in Physical Sciences 20,
https://doi.org/10.1007/978-981-13-3182-4_9

neutral meson physics. These are modern developments, but can be followed easily by the reader who has been patient enough to reach this stage of the book.

9.1.1   Unruh Effect

The Unruh effect is the name coined to the *surprising* phenomena of an accelerated observer, undergoing uniform acceleration, perceiving the Minkowski vacuum to be endowed with a thermal spectrum. That this thermal behavior, usually associated with a statistical phenomena, should have its origin, here, in uniform acceleration, is what comes as a surprise [375]. It is the flat space analog of the well known Hawking radiation [376, 377, 378], related to the quantum thermodynamics of a black hole [379]. This opened up the field of application of quantum field theory ideas to curved spacetime and is hoped to be the precursor to a genuine quantum theory of gravitation [380, 381].

The thermal spectrum observed by an accelerated observer could be codified into a thermalization theorem which states that the pure state which is the vacuum from the point of view of an inertial observer is a canonical ensemble, hence the temperature, from the perspective of a uniformly accelerated observer. The temperature characterizing the ensemble is proportional to the magnitude of the observer's acceleration. The basic idea involves quantization in different vacua and the associated particles. Uniform acceleration is very suitably described in the Rindler spacetime [382, 383] which divides spacetime into two parts that are separated from each other by an event horizon. On the other hand, in the Minkowski spacetime all parts of the spacetime are accessible. The difference in the two spacetimes lead to a difference in their respective vacuums and hence the nature of the particle spectrum. The statistics of the thermal distribution is bosonic or fermionic, corresponding to the vacuum of a scalar or Fermi particle, respectively. The transformation between the Rindler $\{\theta, \phi\}$, and Minkowski coordinates $x^{(0)}, x^{(n)}$ is

$$x^{(0)} = \theta \sinh(\phi), \quad x^{(1)} = \theta \cosh(\phi). \tag{9.1}$$

The remaining coordinates $x^{(n)}$, $n > 1$, are common to both the coordinate systems. The above transformation is the Minkowski version of transformation from Cartesian to cylindrical coordinates, θ and ϕ characterize the cylindrical coordinates, in Euclidean space. Although the Unruh channel could be envisaged, formally, as a class of quantum noise channel, it is pertinent to point out that it does not describe conventional quantum noise.

We make use of the tools of quantum information theory, introduced in the previous chapter, to shed light on the Unruh effect. The Unruh effect experienced by a mode of a free Dirac field, as seen by a relativistically accelerated observer, is treated as a noise channel, and is called the *Unruh channel*. We characterize this channel by providing its operator-sum representation. We compare and contrast this channel from conventional noise due to environmental decoherence. We also discuss, briefly, the effect of various facets of quantum correlations under the influence of the Unruh channel. Let us begin with two observers, Alice (A) and Rob (R) sharing a maximally entangled state of two Dirac field modes, and thus a qubit fermionic Unruh channel, at a point in Minkowski spacetime, of the form

$$|\psi\rangle_{A,R} = \frac{|00\rangle_{A,R} + |11\rangle_{A,R}}{\sqrt{2}}, \tag{9.2}$$

where $|i\rangle$ denote Fock states. We consider the scenario where Alice is stationary and Rob moves away with a uniform proper acceleration a. The effect of constant proper acceleration is described, as discussed above, by a Rindler spacetime, which manifests two causally disconnected regions I and II, where region I is accessible to Rob, and separated from region II by an event horizon.

It can be shown that from Rob's frame the Minkowski vacuum state is seen as a two-mode squeezed state, while the excited state appears as a product state [384]

$$|0\rangle_M \equiv \cos r |0\rangle_I |0\rangle_{II} + \sin r |1\rangle_I |1\rangle_{II},$$
$$|1\rangle_M \equiv |1\rangle_I |0\rangle_{II}, \tag{9.3}$$

where ω is a Dirac particle frequency while $\cos r = \dfrac{1}{\sqrt{e^{-\frac{2\pi\omega c}{a}} + 1}}$ is one of the Bogoliubov coefficients, connecting the Minkowski, indicated by the subscript M, and Rindler, subscripts I, II, vacua. It follows that $\cos r \in [\frac{1}{\sqrt{2}}, 1]$ as a ranges from ∞ to 0. Observe that the states in the left hand side of Eq. (9.3) are single-mode states, while those in the right hand side are not!

Under the representation (9.3), the state represented in Eq. (9.2) becomes

$$|\psi\rangle_{A,I,II} = \frac{1}{\sqrt{2}} \left(|0\rangle_A (\cos r |0\rangle_I |0\rangle_{II} + \sin r |1\rangle_I |1\rangle_{II}) + |1\rangle_A |1\rangle_I |1\rangle_{II} \right). \tag{9.4}$$

Tracing out mode II, which is not accessible to Rob, we obtain the following density matrix

$$\rho'_{A,R} = \frac{1}{2} \left[\cos^2(r)|00\rangle\langle 00| + \cos r(|00\rangle\langle 11| + |11\rangle\langle 00|) + \sin^2(r)|01\rangle\langle 01| + |11\rangle\langle 11| \right], \tag{9.5}$$

where the subscript I has been replaced with subscript R, for Rindler. The evolution of Rob's qubit to a mixed state under the transformation $\mathcal{E}_U : \rho_R \to \rho'_R$ constitutes what we call the Unruh channel for a fermionic qubit.

Next, we make use of the Choi-Jamiolkowski isomorphism, introduced in the previous chapter, to develop the Kraus operators characterizing the Unruh channel. Consider the maximally entangled two-mode state, Eq. (9.2), in which the second mode is Unruh accelerated. The resulting state, Eq. (9.5), is

$$\rho_U = \frac{1}{2} \begin{pmatrix} \cos^2 r & 0 & 0 & \cos r \\ 0 & \sin^2 r & 0 & 0 \\ 0 & 0 & 0 & 0 \\ \cos r & 0 & 0 & 1 \end{pmatrix}. \tag{9.6}$$

ρ_U is the Choi matrix $\sum_{j,k} |j\rangle\langle k| \otimes \mathcal{E}_U(|j\rangle\langle k|)$, modulo the factor $1/2$, corresponding to the Unruh channel $\mathcal{E}_U$. Spectral decomposition yields

$$\rho_U = \sum_{j=0}^{3} |\xi_j\rangle\langle\xi_j|, \tag{9.7}$$

where $|\xi_j\rangle$ are the eigenvectors normalized to the value of the eigenvalue. By Choi's theorem [221, 385], each $|\xi_j\rangle$ yields a Kraus operator obtained by folding the d^2 (here, 4) entries of the eigenvector in to $d \times d$ (2×2) matrix, essentially by taking each sequential d-element segment of $|\xi_j\rangle$, writing it as a column, and then juxtaposing these columns to form the matrix [385].

Corresponding to the two non-vanishing eigenvalues, the two eigenvectors are

$$\begin{aligned} |\xi_0\rangle &= (\cos r, 0, 0, 1), \\ |\xi_1\rangle &= (0, \sin r, 0, 0). \end{aligned} \tag{9.8}$$

The Kraus representation for $\mathcal{E}_U$ is now easily seen to be

$$K_1^U = \begin{pmatrix} \cos r & 0 \\ 0 & 1 \end{pmatrix}; \quad K_2^U = \begin{pmatrix} 0 & 0 \\ \sin r & 0 \end{pmatrix}. \tag{9.9}$$

Hence,

$$\mathcal{E}_U(\rho) = \sum_{j=1,2} K_j^U \rho \left(K_j^U\right)^\dagger, \tag{9.10}$$

with the completeness condition

$$\sum_{j=1,2} \left(K_j^U\right)^\dagger K_j^U = \mathbb{I}. \tag{9.11}$$

This is formally similar to the operator elements in the Kraus representation of an amplitude damping (AD) channel [38], which models the effect of a zero temperature thermal bath. This is surprising as the Unruh effect corresponds to a finite temperature and would naively be expected to correspond to the *generalized* AD or squeezed generalized amplitude damping (SGAD) channels, which are finite temperature channels. This is a pointer towards a fundamental difference between the Unruh and the AD channel. This can be seen by studying the behavior of the maximally mixed state under the Unruh channel. By virtue of linearity of the map, it follows that the maximally mixed state maps to the Bloch vector

$$\hat{n}^\infty(\mathbb{I}) = (0, 0, -\frac{1}{2}). \tag{9.12}$$

Thus the Unruh channel is non-unital and the Bloch sphere subjected to it does not converge to a point, as it does for the AD channel [213], but contracts by a finite factor. In fact, it can be shown [386] that the volume contraction factor of the Bloch sphere under the relativistic channel is $\mathcal{K} \equiv \frac{1}{4}$.

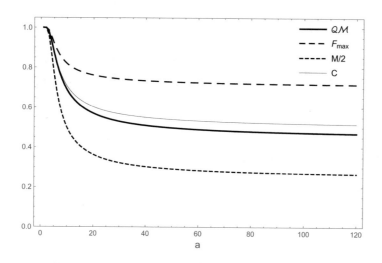

Figure 9.1: Degradation of QMID $\mathcal{QM}$ (dark bold line), maximum teleportation fidelity $F_{\max}$ (large dashed curve), Bell quantity $M/2$ (small dashed curve)) and concurrence (light bold line) C as a function of Unruh acceleration a, for $\omega = 0.1$ (in units where $\hbar \equiv c \equiv 1$). Figure adapted from [386].

In Fig. (9.1), are depicted various facets of quantum correlations, see Chapter 8, Sec. 3.5, under the influence of the Unruh channel. It is seen that quantum measurement induced disturbance (QMID) $\mathcal{QM} > 0$ throughout the range considered, implying that the system remains nonclassical. In particular, the system becomes local, i.e., the Bell quantity satisfies $M < 1$ at $a \approx 7$, but stays nonclassical with respect to the other parameters, such as concurrence $C > 0$, maximum teleportation fidelity $F_{\max} > \frac{2}{3}$, and QMID $\mathcal{QM} > 0$.

The Bloch vector formalism is very useful for understanding the Unruh channel as well as its behavior under the influence of other noisy channels, such as phase damping, AD, SGAD [387, 388]. Any two level system can be represented in the Bloch vector formalism as

$$\rho = \frac{1}{2}\left(\mathbb{I} + \vec{\chi}\cdot\sigma\right),\tag{9.13}$$

where σ are the standard Pauli matrices and $\vec{\chi}$ is the Bloch vector.

For the initial state $\rho = |0\rangle\langle 0|\cos^2\frac{\theta}{2} + |0\rangle\langle 1|e^{i\phi}\cos\frac{\theta}{2}\sin\frac{\theta}{2} + |1\rangle\langle 0|e^{-i\phi}\cos\frac{\theta}{2}\sin\frac{\theta}{2} + |1\rangle\langle 1|\sin^2\frac{\theta}{2}$, the Bloch vector is easily seen to be $\chi_0 = (\cos\phi\sin\theta, -\sin\phi\sin\theta, \cos\theta)$. Evolving this state under the Unruh channel, characterized by the above Kraus operators leads to a state, which could be called the Unruh-Dirac (UD) qubit state, whose Bloch vector is

$$\vec{\chi} = \begin{pmatrix} \cos r\cos\phi\sin\theta \\ -\cos r\sin\phi\sin\theta \\ \cos^2 r\cos\theta - \sin^2 r \end{pmatrix} = A\vec{\chi}_0 + C.\tag{9.14}$$

From this A and C can be found to be

$$A = \begin{pmatrix} \cos r & 0 & 0 \\ 0 & \cos r & 0 \\ 0 & 0 & \cos^2 r \end{pmatrix}, \quad C = \begin{pmatrix} 0 \\ 0 \\ -\sin^2 r \end{pmatrix}. \quad (9.15)$$

Problem 1: Show that for the state $|\psi\rangle = \cos(\theta/2)|0\rangle + e^{-i\phi}\sin(\theta/2)|1\rangle$, the Bloch vector is $\chi_0 = (\cos\phi\sin\theta, -\sin\phi\sin\theta, \cos\theta)$.

Problem 2: Sketch the steps leading to the Eqs. (9.14), (9.15).

The A and C matrices completely characterize the Unruh channel. Starting from the UD qubit state, Eq. (9.14), application of the external noise channel results in

$$\rho_{\text{in}} \xrightarrow{\mathcal{E}(\text{external noise})} \rho_{\text{new}}. \quad (9.16)$$

From ρ_{new}, we get the new Bloch vector $\vec{\chi}_{\text{new}}$ which is related to the original state Bloch vector as

$$\begin{aligned} \vec{\chi}_{\text{new}} &= A'\vec{\zeta} + C' = A'(A\vec{\chi}_0 + C) + C' \\ &= AA'\vec{\chi}_0 + (A'C + C') \equiv A_{\text{new}}\vec{\chi}_0 + C_{\text{new}}. \end{aligned} \quad (9.17)$$

Here $\vec{\chi}$ and $\vec{\chi}_0$ are as in Eq. (9.14). From the above equation, it can be seen that the effect of the external noise channel on the Unruh channel is encoded in $A_{\text{new}} = A'A$ and $C_{\text{new}} = (A'C + C')$. Hence, the effect of the external noise channel on the Unruh channel is reduced to the computation of A' and C' for the desired channels. This can be done in a straightforward fashion once the Kraus operators characterizing the external channels are known [387, 388].

9.1.2 Neutrinos

Neutrinos were first postulated by Wolfgang Pauli to explain how beta decay could conserve energy, momentum and angular momentum (spin) in the decay of neutron n into a proton p and electron e

$$n \rightarrow p + e + \bar{\nu}_e.$$

Here $\bar{\nu}_e$ is the electron anti-neutrino. Neutrinos, as we now understand, come in three varieties, called flavors, ν_e, ν_μ, ν_τ, i.e., the electron, muon and tau neutrinos. Due to their non-zero mass, they oscillate from one flavor to another. Neutrino oscillations are experimentally well established [389, 390, 391, 392]. Such oscillations are possible under the premise that the following conditions are satisfied:

- The neutrino flavour state is a linear superposition of non-degenerate mass eigenstates.

- The time evolution of a flavour state is a coherent superposition of the time evolution of the corresponding mass eigenstates.

Neutrinos are *Left-handed*, while the anti-neutrinos are *Right-handed*. It may be of some interest to note that around 65 billion (6.5×10^{10}) neutrinos comming from Sun's interior pass through 1 square centimeter of area on earth per second! Having said that, neutrinos are notoriously difficult to detect as they interact only via weak interaction.

It is assumed that neutrinos mix via a 3×3 unitary matrix to form the three mass eigenstates ν_1, ν_2 and ν_3. Neutrino oscillations occur only if the three corresponding masses, m_1, m_2 and m_3, are non-degenerate. Of the three mass-squared differences $\Delta_{kj} = m_k^2 - m_j^2$ (where $j, k = 1, 2, 3$ with $k > j$), only two are independent. From the oscillation data it is seen that $\Delta_{21} \approx 0.03 \times \Delta_{32}$, implying that $\Delta_{31} \approx \Delta_{32}$. One of the three mixing angles parametrizing the mixing matrix, θ_{13}, is measured to be small (about 0.14 radians) [393, 394, 395]. Neutrino oscillations are fundamentally three flavor oscillations. However, in a number of cases, the three flavor formula reduces to an effective two flavor formula, if one or both of the small parameters, Δ_{21}/Δ_{32} and θ_{13}, are set equal to zero. Some issues, related to neutrinos, that are still not clearly understood are the neutrino mass hierarchy, i.e., whether $m_1^2 \leq m_2^2 \leq m_3^2$ or $m_3^2 \leq m_1^2 \leq m_2^2$. What is the absolute neutrino mass scale? What is the origin of neutrino mass and flavor mixing? Is there CP (charge-parity) violation and what is the value of the CP violating phase δ?

Two flavor neutrino oscillations: In the case of two flavor mixing, the relation between the flavor and the mass eigenstates is described by a 2×2 rotation matrix, $U(\theta)$, where θ is the mixing angle, for example, θ_{23}

$$\begin{pmatrix} \nu_\alpha \\ \nu_\beta \end{pmatrix} = \begin{pmatrix} \cos\theta & \sin\theta \\ -\sin\theta & \cos\theta \end{pmatrix} \begin{pmatrix} \nu_j \\ \nu_k \end{pmatrix}. \tag{9.18}$$

Each flavor state can be expressed as a superposition of mass eigenstates,

$$|\nu_\alpha\rangle = \sum_j U_{\alpha j} |\nu_j\rangle, \tag{9.19}$$

where $\alpha = \mu$ or τ and $j = 2, 3$. The time evolution of the mass eigenstates $|\nu_j\rangle$ is given by

$$|\nu_j(t)\rangle = e^{-iE_j t} |\nu_j\rangle, \tag{9.20}$$

where $|\nu_j\rangle$ are the mass states at time $t = 0$. Thus, we can write

$$|\nu_\alpha(t)\rangle = \sum_j U_{\alpha j} e^{-iE_j t} |\nu_j\rangle. \tag{9.21}$$

The evolving flavor neutrino state $|\nu_\alpha\rangle$ can also be projected on to the flavor basis in the form

$$|\nu_\alpha(t)\rangle = \tilde{U}_{\alpha\alpha}(t) |\nu_\alpha\rangle + \tilde{U}_{\alpha\beta}(t) |\nu_\beta\rangle, \tag{9.22}$$

where $|\nu_\alpha\rangle$ is the flavor state at time $t = 0$ and $|\tilde{U}_{\alpha\alpha}(t)|^2 + |\tilde{U}_{\alpha\beta}(t)|^2 = 1$.

Problem 3: Find the explicit form of $\tilde{U}_{\alpha\alpha}(t)$ and $\tilde{U}_{\alpha\beta}(t)$ in Eq. (9.22).

Two flavor neutrino oscillation with matter effect: The above calculation corresponds to the case when neutrinos travel through vacuum. But the oscillation patterns can be significantly affected if neutrinos travel through a material medium. Therefore matter effect should also be taken care of. ν_e interacts with electrons (e^-) present in the matter via neutral and charged current interactions, while ν_μ and ν_τ interact only by neutral current interaction. The amplitude corresponding to neutral current interactions are identical for all of the three flavors. Therefore the amplitude corresponding to charged current interaction of ν_e with e^- only is considered. The equation of motion in mass eigenstate basis is

$$i\frac{d}{dt}\begin{bmatrix} \nu_1 \\ \nu_2 \end{bmatrix} = H \begin{bmatrix} \nu_1 \\ \nu_2 \end{bmatrix}, \tag{9.23}$$

where

$$H = \begin{bmatrix} E_1 & 0 \\ 0 & E_2 \end{bmatrix}. \tag{9.24}$$

We assume that ν is emitted in plane wave state with definite momentum, i.e., $E_i^2 = p^2 + m_i^2$ with ultra high relativistic approximation ($p^2 >>> m_i^2$). Then the Hamiltonian becomes

$$H = \begin{bmatrix} p + \frac{m_1^2}{2p} & 0 \\ 0 & p + \frac{m_2^2}{2p} \end{bmatrix}. \tag{9.25}$$

The Hamiltonian can also be expressed in terms of mass square difference $\Delta = m_2^2 - m_1^2$ as,

$$H = \begin{bmatrix} p + \frac{m_1^2+m_2^2}{4p} - \frac{\Delta}{4p} & 0 \\ 0 & p + \frac{m_1^2+m_2^2}{4p} + \frac{\Delta}{4p} \end{bmatrix}. \tag{9.26}$$

Thus, the equation of motion in flavor state basis is given by

$$i\frac{d}{dt}\begin{bmatrix} \nu_e(t) \\ \nu_\mu(t) \end{bmatrix} = \left[p + \frac{m_1^2 + m_2^2}{4p} I + \frac{\Delta}{4p} O^T \begin{pmatrix} -1 & 0 \\ 0 & 1 \end{pmatrix} O \right] \begin{bmatrix} \nu_e \\ \nu_\mu \end{bmatrix}. \tag{9.27}$$

where O is the mixing matrix. Neglecting the first term and putting $p \approx E$, the above equation becomes,

$$i\frac{d}{dt}\begin{bmatrix} \nu_e(t) \\ \nu_\mu(t) \end{bmatrix} = \frac{\Delta}{4E} \begin{bmatrix} -\cos 2\theta & \sin 2\theta \\ \sin 2\theta & \cos 2\theta \end{bmatrix} \begin{bmatrix} \nu_e \\ \nu_\mu \end{bmatrix}. \tag{9.28}$$

The survival P_{ee} and oscillation $P_{e\mu}$ probabilities take the form

$$P_{ee} = 1 - \sin^2 2\theta \, \sin^2 \frac{\Delta L}{4E\hbar c}, \tag{9.29}$$

$$P_{e\mu} = \sin^2 2\theta \, \sin^2 \frac{\Delta L}{4E\hbar c}.$$

Since ν_e only interacts with matter via charged current interaction, an extra term V, to account for the matter density potential, is added to this equation such that,

$$i\frac{d}{dt}\begin{bmatrix}\nu_e(t)\\ \nu_\mu(t)\end{bmatrix}=\begin{bmatrix}-\frac{\Delta\cos2\theta}{4E}+V & \frac{\Delta\sin2\theta}{4E}\\ \frac{\Delta\sin2\theta}{4E} & \frac{\Delta\cos2\theta}{4E}\end{bmatrix}\begin{bmatrix}\nu_e\\ \nu_\mu\end{bmatrix}. \tag{9.30}$$

Here $V=\sqrt{2}G_F N_e$ with $G_F\to$ Fermi constant, $N_e\to$ electron density. As a consequence of this, for constant matter density, survival and oscillation probabilities can be seen to be

$$P_{ee}=1-\sin^2 2\theta_m\ \sin^2\frac{\Delta_m L}{4E\hbar c}, \tag{9.31}$$

$$P_{e\mu}=\sin^2 2\theta_m\ \sin^2\frac{\Delta_m L}{4E\hbar c},$$

where θ_m and Δ_m are effective mixing angle and mass square difference, respectively, and can be expressed in the form of mixing angle θ and vacuum mass square difference Δ as

$$\theta_m = \frac{1}{2}\tan^{-1}\left(\frac{\tan2\theta}{1-\frac{2EV}{\Delta\cos2\theta}}\right), \tag{9.32}$$

$$\Delta_m = \sqrt{(\Delta\cos2\theta-2EV)^2+\Delta^2\sin^2 2\theta}.$$

The resonance condition, i.e., $2EV=\Delta\cos2\theta$, will cause maximal mixing. This is the $Mikheyev-Smirnov-Wolfenstein\ (MSW)$ effect [396].

   Three flavor neutrino Oscillations: To study the effect of CP violation in neutrino oscillations (for Dirac neutrinos), one has to go through the calculation of three flavor neutrino oscillations. Applying some appropriate approximations, the mathematical picture of two flavor oscillation can be reproduced for the three flavor case. In three flavor neutrino oscillation, the propagation states are $\{|\nu_1\rangle,\ |\nu_2\rangle,\ |\nu_3\rangle\}$ and the flavor states are $\{|\nu_e\rangle,\ |\nu_\mu\rangle,\ |\nu_\tau\rangle\}$. The general state of a neutrino can be expressed in flavor basis as

$$|\Psi(t)\rangle = \nu_e(t)\,|\nu_e\rangle + \nu_\mu(t)\,|\nu_\mu\rangle + \nu_\tau(t)\,|\nu_\tau\rangle. \tag{9.33}$$

The same state in propagation basis looks like

$$|\Psi(t)\rangle = \nu_1(t)\,|\nu_1\rangle + \nu_2(t)\,|\nu_2\rangle + \nu_3(t)\,|\nu_3\rangle. \tag{9.34}$$

Coefficients in the two representations are connected by a *unitary* matrix [396]

$$\begin{pmatrix}\nu_e(t)\\ \nu_\mu(t)\\ \nu_\tau(t)\end{pmatrix}=\begin{pmatrix}U_{e1} & U_{e2} & U_{e3}\\ U_{\mu1} & U_{\mu2} & U_{\mu3}\\ U_{\tau1} & U_{\tau2} & U_{\tau3}\end{pmatrix}\begin{pmatrix}\nu_1(t)\\ \nu_2(t)\\ \nu_3(t)\end{pmatrix}. \tag{9.35}$$

In short notation, this can be written as

$$\nu_\alpha(t)=\mathbf{U}\nu_i(t). \tag{9.36}$$

As in the two flavor case, the mass basis can be reexpressed in terms of the flavor basis and the evolution, in flavor basis, would have the form

$$|\Psi(t)\rangle = a(t)\,|\nu_e\rangle + b(t)\,|\nu_\mu\rangle + c(t)\,|\nu_\tau\rangle. \tag{9.37}$$

Assuming that the initial state was $|\nu_e\rangle$, the survival probability is $|\langle\nu_e|\Psi(t)\rangle|^2 = |a(t)|^2$, while the transition probability for $|\nu_e\rangle$ oscillating to $|\nu_\mu\rangle$ is $|\langle\nu_\mu|\Psi(t)\rangle|^2 = |b(t)|^2$.

As the neutrino propagates in matter and interacts with its environment, albeit very weakly, the interaction could lead to decoherence and dissipation; hence the evolution, in such scenarios, need to be treated as an open quantum system, which in the Markovian regime can be described by completely positive linear maps acting on the system density matrices. The general form of the evolution is

$$\frac{d\rho^\alpha(t)}{dt} = -i\,[H, \rho^\alpha(t)] + \mathcal{L}[\rho^\alpha(t)]. \tag{9.38}$$

Here $\alpha = \{e, \mu\}$, i.e., we are considering, for simplicity, the two flavor scenario. Needless to say, this is the well known Lindblad form of evolution, discussed a number of times before in the previous chapters. The first term on the RHS of the above equation is responsible for the coherent evolution, while the second term is one that causes dissipation, an incoherent process. The dissipator has the general form

$$\mathcal{L}[\rho^\alpha(t)] = \sum_{m,n=0}^{3} d_{mn}\left(\sigma_n \rho^\alpha \sigma_m - \frac{1}{2}\{\sigma_m \sigma_n, \rho^\alpha\}\right). \tag{9.39}$$

Here σ's are the Pauli matrices and d_{mn} are the coefficients that ensure complete positivity of the evolution. This form of the evolution has been used to study the geometric phase of neutrino propagating through dissipative matter [397]. It has also been used to fit available neutrino data [398, 399].

9.1.3 Mesons

Here we consider the open system dynamics of unstable massive systems such as correlated $B\bar{B}$ and $K\bar{K}$ meson systems [400]. B factories, electron-positron colliders tailor-made to study the production and decay of B mesons, and ϕ factories, which perform the same function for K mesons, provide an ideal testing ground. After production, the B (or K) mesons fly apart and decay on a much longer time scale. An important feature of these systems for the study of correlations is the oscillations of the bottom and strangeness flavors $b \leftrightarrow s$, giving rise to $B\bar{B}$ oscillations. A decaying system is intrinsically an open system, even without explicitly invoking an external environment, and as a result it can have surprises not seen in its stable counterpart [387].

We make use of the probability-preserving formalism of decaying systems [401, 402] to study various measures of quantum correlations in $B\bar{B}$ and $K\bar{K}$

systems. We employ the methods of open quantum systems [2, 403], which asserts that any real system interacts with its environment. In this context, the environment could be fluctuations of the quantum mechanical vacuum, resulting in loss of quantum coherence and the transformation from pure to mixed states [404]. This thus brings focus to the fundamental aspects of correlated neutral meson systems, and more generally of unstable quantum systems.

The flavor-space wave function of the correlated $M\bar{M}$ meson systems $(M = K, B_d, B_s)$ at the initial time $t = 0$ is

$$|\psi(0)\rangle = \frac{1}{\sqrt{2}} \left[|M\bar{M}\rangle - |\bar{M}M\rangle \right], \qquad (9.40)$$

where the first (second) particle in each ket is the one flying off in the left (right) direction and $|M\rangle$ and $|\bar{M}\rangle$ are flavor eigenstates. As seen from (9.40), the initial state of the neutral meson system is a maximally entangled, singlet state. The Hilbert space of a system of two correlated neutral mesons, as in (9.40), is

$$\mathcal{H} = (\mathcal{H}_L \oplus \mathcal{H}_0) \otimes (\mathcal{H}_R \oplus \mathcal{H}_0), \qquad (9.41)$$

where $\mathcal{H}_{L,R}$ are the Hilbert spaces of the left-moving and right-moving decay products, each of which can be either a meson or an anti-meson, and $\mathcal{H}_0$ is that of the zero-particle (vacuum) state. Thus, the total Hilbert space can be seen to be the tensor sum of a two-particle space, two one-particle spaces, and one zero-particle state. In order to compute quantum correlations in the resulting system, one needs to project the evolution from the full Hilbert space $\mathcal{H}$ down to the two-particle sector $\mathcal{H}_L \otimes \mathcal{H}_R$. This is facilitated by the operator-sum (Kraus) representation of the evolution, which can be show to be [387, 402]

$$
\begin{aligned}
E_0 &= |0\rangle \langle 0|, \\
E_1 &= \mathcal{E}_{1+} \left(|B^0\rangle \langle B^0| + |\bar{B}^0\rangle \langle \bar{B}^0 o| \right) + \mathcal{E}_{1-} \left(\frac{p}{q} |B^0\rangle \langle \bar{B}^0| + \frac{q}{p} |\bar{B}^0\rangle \langle B^0| \right), \\
E_2 &= \mathcal{E}_2 \left(\frac{p+q}{2p} |0\rangle \langle B^0| + \frac{p+q}{2q} |0\rangle \langle \bar{B}^0| \right), \\
E_3 &= \mathcal{E}_{3+} \frac{p+q}{2p} |0\rangle \langle B^0| + \mathcal{E}_{3-} \frac{p+q}{2q} |0\rangle \langle \bar{B}^0|, \\
E_4 &= \mathcal{E}_4 \left(|B^0\rangle \langle B^0| + |\bar{B}^0\rangle \langle \bar{B}^0| + \frac{p}{q} |B^0\rangle \langle \bar{B}^0| + \frac{q}{p} |\bar{B}^0\rangle \langle B^0| \right), \\
E_5 &= \mathcal{E}_5 \left(|B^0\rangle \langle B^0| + |\bar{B}^0\rangle \langle \bar{B}^0| - \frac{p}{q} |B^0\rangle \langle \bar{B}^0| - \frac{q}{p} |\bar{B}^0\rangle \langle B^0| \right).
\end{aligned}
$$

Here the coefficients are

$$\mathcal{E}_{1\pm} = \frac{1}{2}\left[e^{-(2im_L+\Gamma_L+\lambda)t/2} \pm e^{-(2im_H+\Gamma_H+\lambda)t/2}\right], \tag{9.42a}$$

$$\mathcal{E}_2 = \sqrt{\frac{Re[\frac{p-q}{p+q}]}{|p|^2-|q|^2}\left(1-e^{-\Gamma_L t}-(|p|^2-|q|^2)^2\frac{|1-e^{-(\Gamma+\lambda-i\Delta m)t}|^2}{1-e^{-\Gamma_H t}}\right)}, \tag{9.42b}$$

$$\mathcal{E}_{3\pm} = \sqrt{\frac{Re[\frac{p-q}{p+q}]}{(|p|^2-|q|^2)(1-e^{-\Gamma_H t})}}\left[1-e^{-\Gamma_H t}\pm(1-e^{-(\Gamma+\lambda-i\Delta m)t})(|p|^2-|q|^2)\right], \tag{9.42c}$$

$$\mathcal{E}_4 = \frac{e^{-\Gamma_L t/2}}{2}\sqrt{1-e^{-\lambda t}}, \tag{9.42d}$$

$$\mathcal{E}_5 = \frac{e^{-\Gamma_H t/2}}{2}\sqrt{1-e^{-\lambda t}}. \tag{9.42e}$$

A meson initially in state $\rho_{B^0}(0) = |B^0\rangle\langle B^0|$ or $\rho_{\bar{B}^0}(0) = |\bar{B}^0\rangle\langle\bar{B}^0|$, after time t, evolves to

$$\rho_{B^0}(t) = \frac{1}{2}e^{-\Gamma t}\begin{pmatrix} a_{ch}+e^{-\lambda t}a_c & (\frac{q}{p})^*(-a_{sh}-ie^{-\lambda t}a_s) & 0 \\ (\frac{q}{p})(-a_{sh}+ie^{-\lambda t}a_s) & |\frac{q}{p}|^2 a_{ch}-e^{-\lambda t}a_c & 0 \\ 0 & 0 & \rho_{33}(t) \end{pmatrix}, \tag{9.43}$$

and

$$\rho_{\bar{B}^0}(t) = \frac{1}{2}e^{-\Gamma t}\begin{pmatrix} |\frac{p}{q}|^2(a_{ch}-e^{-\lambda t}a_c) & (\frac{p}{q})(-a_{sh}+ie^{-\lambda t}a_s) & 0 \\ (\frac{p}{q})^*(-a_{sh}-ie^{-\lambda t}a_s) & a_{ch}+e^{-\lambda t}a_c & 0 \\ 0 & 0 & \tilde{\rho}_{33}(t) \end{pmatrix}. \tag{9.44}$$

Here, a_{ch} (a_{sh}) and a_c (a_s) denote the hyperbolic functions $\cosh[\frac{\Delta\Gamma t}{2}]$ ($\sinh[\frac{\Delta\Gamma t}{2}]$) and the trigonometric functions $\cos[\Delta mt]$ ($\sin[\Delta mt]$), respectively. Also, p and q are the CP (charge-parity) violating parameters and satisfy the relation $|p^2|+|q|^2 = 1$. $\Delta\Gamma = \Gamma_L - \Gamma_H$ is the difference of the decay width Γ_L (for B_L^o) and Γ_H (for B_H^o). $\Gamma = \frac{1}{2}(\Gamma_L+\Gamma_H)$ is the average decay width. The mass difference $\Delta m = m_H - m_L$, where m_H and m_L are the masses of B_H^o and B_L^o states, respectively. The strength of the interaction between the one particle system and its environment is quantified by λ, the *decoherence* parameter. The elements $\rho_{33}(t)$ and $\tilde{\rho}_{33}(t)$ are known functions of B physics parameters, which do not feature in what follows below. A similar analysis holds for the K mesons with appropriate change in notations. The approach used here can also be effectively applied to study observables of central importance in particle physics [405].

Using the above constructed density matrices, Eqs. (9.43) and (9.44), we can study the interplay of quantum correlations in meson systems. The nonclassicality of quantum correlations, in the neutral mesons, can be characterized in terms of nonlocality (which is the strongest condition), entanglement, teleportation fidelity or weaker nonclassicality measures like quantum discord, see

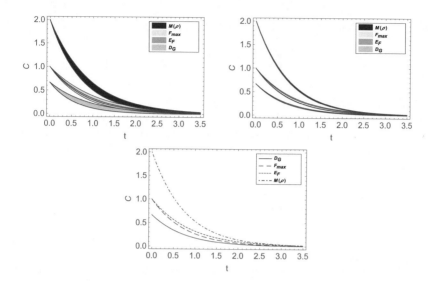

Figure 9.2: Average correlation measures c, i.e., the various measures modulated by the exponential factor $e^{-2\Gamma t}$, as a function of time t. The upper row corresponds to the correlations of a $K\bar{K}$ and $B_d\bar{B}_d$ pair, respectively, while the correlations of $B_s\bar{B}_s$ pair is depicted in the bottom row figure; here these pairs are created at $t = 0$. The four correlation measures are (top to bottom): $M(\rho)$ (Bell's inequality; blue band), $F_{\max}$ (teleportation fidelity; red band), E_F (entanglement of formation; grey band) and D_G (geometric discord; green band). For $K\bar{K}$ pairs, left panel, time is in units of 10^{-10} seconds whereas for the $B_d\bar{B}_d$ and $B_s\bar{B}_s$ pairs, time is in units of 10^{-12} seconds (in all cases, the approximate lifetime of the particles). In the figures in the upper row, the bands represent the effect of decoherence corresponding to a 3σ upper bound on the decoherence parameter λ. The bottom row has no such bands because there is currently no experimental evidence for decoherence in the case of B_s mesons. Figure adapted from [387].

Chapter 8, section 3.5. The fall in the pattern of the average value of these correlations, i.e., the various measures modulated by the exponential factor $e^{-2\Gamma t}$, as displayed in Fig. (9.2), are in accord with the fact that here we are dealing with unstable particles, which decay with time. From the left panel of Fig. (9.2), one can see that until about 50% of the average life time of K_S meson in the presence of decoherence and about 60% in its absence, $M(\rho) > 1$. This means that, in the conventional sense, until this time, the time evolution cannot be simulated by any local realistic theory. However, we find that even for some cases where $M(\rho)$ exceeds one, the teleportation fidelity $F_{\max}$ could be below the classical value of $2/3$. For example, from the left panel of Fig. (9.2), it is seen that, in the absence of decoherence, $F_{\max}$ drops below $2/3$ as $M(\rho)$ drops below 1.3, in violation of the inequality [267],

$$F_{max} \geq \frac{1}{2} \left(1 + \frac{1}{3} M(\rho) \right) \geq \frac{2}{3} \text{ if } M(\rho) > 1. \qquad (9.45)$$

according to which the cutoff is $M(\rho) = 1$. This violation is slightly reduced, but nonetheless still occurs, even in the presence of decoherence, starting at $M(\rho) \simeq 1.2$. This is consistent with the degradation of correlations with decoherence. Hence we see that the study of quantum correlations in unstable systems is nontrivially different from their stable counterparts.

9.2 Non-Markovian Phenomena

Though interest in non-Markovian phenomena has been there for a long time, tracing its roots to the development of (quantum) Brownian motion [10, 406, 407, 98], it has witnessed an upsurge of interest from the perspective of quantum information over the last decade, leading to a number of useful concepts, some of which will be sketched here. The basic idea behind Markovain approximation is the clean separation of the environmental time-scales from the system time-scales [408, 174]. Current advancement in experimental techniques allows for the possibility of getting into regimes where the reservoir (environment) effectuates memory effects in the system dynamics, blurring the above clean separation of system-reservoir time scales, and would be the so called *non-Markovian* regime.

Recent work, from a quantum information perspective, allows us to reach the following consensus related to the information theoretic witnesses of non-Markoviantity: (a) Information backflow, by which is meant the increase of distinguishability with time between any two given states, as witnessed by measures like trace distance [409, 410]; (b) The time-evolution generated by the dynamical maps cannot be divided into intermediate maps that are completely positive (CP-Divisiblity) [408]. In general not all given dynamics strictly satisfy both the above conditions for it to be termed non-Markovian. Condition (a) implies an increase in the distinguishability D, causing a recurrence or "backflow" of information back from the environment into the system, while from condition (b) one would infer that the intermediate map is non-CP (NCP), essentially because the system-bath interaction generates system-bath entanglement.

Our strategy in this section would be to briefly discuss a few prominent notions of non-Markovianity and illustrate them with concrete examples, from the perspective of open quantum systems.

9.2.1 Non-Markovian Master Equations

The dynamics of the system of interest s (Markovian or non-Markovian), represented by, for example, the density matrix $\rho_s(t)$, can be obtained from its dynamical map generated by the Kraus representation, see Section 8.2.1, provided the map is invertible and differentiable. Let us start with the time derivative of the dynamical evolution of the state via the Kraus operators E_i

$$\frac{d\rho_s(t)}{dt} = \sum_i \left(\frac{dE_i(t)}{dt} \rho_s(0) E_i^\dagger + E_i(t)\rho_s(0) \frac{d^\dagger E_i(t)}{dt} \right). \tag{9.46}$$

Reversibility of the dynamical map allows us to express the initial state as

$$\rho_s(0) = \sum_n G_n(t)\rho_s(t)K_n(t). \tag{9.47}$$

Substituting Eq. (9.47) into Eq. (9.46), we can repackage the evolution equation as

$$\frac{d\rho_s(t)}{dt} = \sum_m M_m(t)\rho_s(t)N_m^\dagger(t). \tag{9.48}$$

Note that the *RHS* of the above equation has $\rho_s(t)$ and not $\rho_s(0)$. The label m in the above equation is a collective symbol for $\{\phi, i, n\}$, where $\phi = 1, 2$ and i, n are as they appear in the above equations. Thus, we can see that $M_{1,i,n} = \{dE_i(t)/dt\}G_n(t)$, $M_{2,i,n} = E_i(t)G_n(t)$, $N_{1,i,n}^\dagger = K_n(t)E_i^\dagger(t)$ and $N_{2,i,n}^\dagger = K_n(t)\{dE_i^\dagger(t)/dt\}$.

Following [411], the system operators $M_m(t)$ and $N_m(t)$ are expanded in the basis of $N = d^2$ operators $\{F_i, \ i = 0, \cdots N - 1\}$. Here d is the dimension of the system and $F_0 = I/\sqrt{d}$, $F_i = F_i^\dagger$, $\mathrm{Tr}F_i = \delta_{i0}$ and $\mathrm{Tr}\{F_iF_j\} = \delta_{ij}$. For a two-level system, $F_i \equiv \sigma_i$, the three Pauli matrices. We have

$$M_m(t) = \sum_i \alpha_{im}(t)F_i,$$

$$N_m(t) = \sum_j \beta_{jm}(t)F_j. \tag{9.49}$$

Here $\alpha_{im}(t) = \mathrm{Tr}\{M_m(t)F_i\}$ and $\beta_{im}(t) = \mathrm{Tr}\{N_m(t)F_i\}$. In terms of these, the master equation (9.48) can be expressed as

$$\frac{d\rho_s(t)}{dt} = \sum_{i,j=0}^{N-1} \gamma_{ij}F_i\rho_s(t)F_j, \tag{9.50}$$

where $\gamma_{ij} = \sum_m \alpha_{im}(t)\beta_{jm}^*(t)$ are elements of an $N \times N$ Hermitian matrix. This follows from the hermiticity of $\rho_s(t)$ and F_i. Separating the $i, j = 0$ terms, the master equation can be written as

$$\frac{d\rho_s(t)}{dt} = -i[H_s(t), \rho_s(t)] + \Gamma\rho_s(t) + \rho_s(t)\Gamma^\dagger + \sum_{i,j=0}^{N-1} \gamma_{ij}F_i\rho_s(t)F_j. \tag{9.51}$$

Here

$$\Gamma = \frac{I_s}{2d}\gamma_{00} + \sum_i \frac{\gamma_{i0}}{\sqrt{d}}F_i. \tag{9.52}$$

Using conservation of trace, $\Gamma + \Gamma^\dagger = -\sum_{i,j=0}^{N-1} \gamma_{ij} F_j F_i$. The Eq. (9.51) can be expressed in terms of combinations of $\Gamma - \Gamma^\dagger$ and $\Gamma + \Gamma^\dagger$ to yield

$$\frac{d\rho_s(t)}{dt} = -i[H_s(t), \rho_s(t)] + \sum_{ij=1}^{d^2-1} \gamma_{ij}(t) \left(F_i \rho_s(t) F_j - \frac{1}{2}\{F_j F_i, \rho_s(t)\} \right). \quad (9.53)$$

Note that $\{A, B\}$ denotes the anticommutation of operators A and B, $H_s(t) = (i/2)(\Gamma - \Gamma^\dagger)$ and γ_{ij} is the decoherence matrix. Being Hermitian it can be diagonalized as $\gamma_{ij}(t) = \sum_n U_{in}(t)\Theta_n(t)U_{jn}^*(t)$, where $\Theta_n(t)$ and $U_{in}(t)$ are its eigenvalues and eigenvectors, respectively. Defining

$$\Gamma_n(t) = \sum_{i=1}^{N-1} U_{in}(t) F_i, \quad (9.54)$$

Eq. (9.53) can be rewritten as

$$\frac{d\rho_s(t)}{dt} = \mathcal{K}(t)\rho_s(t)$$

$$= -i[H_s(t), \rho_s(t)] + \sum_{n=1}^{d^2-1} \Theta_n(t) \left(2\Gamma_n(t)\rho_s(t)\Gamma_n^\dagger(t) - \{\Gamma_n^\dagger(t)\Gamma_n(t), \rho_s(t)\} \right).$$

$$(9.55)$$

This is the non-Markovian generalization of the Lindblad equation. Complete positivity of the resultant dynamics can be ensured only when all the $\Theta_n(t) \geq 0$, which holds for a Markovian evolution. Note that in the standard Lindbladian evolution, $\Theta_n \geq 0$ and is time independent. This lead to the defination of a measure of non-Markovianity as a sum of all intervals where $\Theta_n(t)$ are negative [411]. See also [412], where non-Markovian evolution was studied for noninteracting bosons (fermions) linearly coupled to thermal environments of noninteracting bosons (fermions).

9.2.2 Information backflow and breakdown of CP divisibility of the intermediate map

Information Backflow: As we remarked above, in non-Markovian evolutions it is possible to have situations where distance measures like the trace distance, which quantify the closeness of two states as they evolve under the given evolution and hence are connected to their distinguishability, increase with time as compared to the monotonic fall experienced under a Markovian, for example, Lindbladian evolution. This leads to a *backflow* of information from the environment to the system, which updates the status of the system and is thus a powerful diagnostic of non-Markovian behavior. This, in turn manifests in the form of oscillations in correlation measures such as quantum mutual information [413], which are otherwise monotonic functions if the dynamics is Markovian. The distance between any two quantum states defined on the space of density matrices is given by a metric called the trace distance D, which is defined as

$$D(\rho_1, \rho_2) = \frac{1}{2}\text{Tr}||\rho_1 - \rho_2||_1, \tag{9.56}$$

where $||O||_1$ is the operator norm given by $\sqrt{O^\dagger O}$. The use of trace distance is rooted in the idea of distinguishability of a pair of quantum states, which is monotonically decreasing under completely positive (CP) maps Λ, i.e., the CP maps are contractions for this metric,

$$D(\Lambda \, \rho_1, \Lambda \, \rho_2) \le D(\rho_1, \rho_2). \tag{9.57}$$

For non-Markovian processes, due to the backflow of information from the environment to the system, there is a temporary increase in the distinguishability of quantum states and hence the above inequality may be violated (this being the characteristic of backflow). This idea has been exploited in an effort to quantify non-Markovianity [409].

The connection of the problem to distinguishability can be clarified by taking up the unbiased two state discrimination problem. Consider two parties, Alice and Bob. Alice prepares a quantum system in one of two states ρ^1 or ρ^2 with probability $\frac{1}{2}$ each, and then sends the system to Bob. It is Bob's task to find out by a single measurement on the system whether the system state was ρ^1 or ρ^2. It turns out that Bob cannot always distinguish the states with certainty, but there is an optimal strategy which allows him to achieve the maximal possible success probability given by

$$P_{\text{max}} = \frac{1}{2}\left[1 + D(\rho^1, \rho^2)\right]. \tag{9.58}$$

The trace distance $D(\rho^1, \rho^2) = \frac{1}{2}||\rho^1 - \rho^2||_1 = \frac{1}{2}\text{Tr}|\rho^1 - \rho^2|$ can therefore be interpreted as a measure for the distinguishability of the quantum states ρ^1 and ρ^2. Here $\text{Tr}|A| = \text{Tr}\sqrt{A^\dagger A}$.

The trace distance between any pair of states satisfies the following properties:

(a). $0 \le D(\rho^1, \rho^2) \le 1$.

(b). The trace distance is sub-additive with respect to tensor products of states

$$D(\rho^1 \otimes \sigma^1, \rho^2 \otimes \sigma^2) \le D(\rho^1, \rho^2) + D(\sigma^1, \sigma^2). \tag{9.59}$$

(c). The trace distance is invariant under unitary transformations U,

$$D(U\rho^1 U^\dagger, U\rho^2 U^\dagger) = D(\rho^1, \rho^2). \tag{9.60}$$

(d). More generally, all trace preserving and completely positive maps, i.e., all trace preserving quantum operations Λ are contractions of the trace distance,

$$D(\Lambda\rho^1, \Lambda\rho^2) \le D(\rho^1, \rho^2). \tag{9.61}$$

No quantum process that can be described by a family of completely positive, trace preserving (CPT) dynamical maps can ever increase the distinguishability of a pair of states over its initial value. Thus, when a quantum process reduces the distinguishability of states, information is flowing from the system to the environment. Correspondingly, an increase of the distinguishability signifies that information flows from the environment back to the system, i.e., information backflow is taking place. The definition for quantum non-Markovianity, discussed here, is based on the idea that for Markovian processes any two quantum states become less distinguishable under the dynamics, leading to a perpetual loss of information into the environment. Quantum memory effect thus arise if there is a temporal flow of information from the environment to the system. The information flowing back from the environment allows the earlier open system states to have an effect on the later dynamics of the system, which implies the emergence of memory effects [409]. As a corollary, the class of quantum dynamical semigroups, generated by the Lindbladian evolution, which are divisible families of dynamical maps, are Markovian.

A quantum process described in terms of a family of quantum dynamical maps $\Phi(t,0)$ is non-Markovian if there is a pair of initial states $\rho_S^{1,2}(0)$ such that the trace

$$\sigma(t, \rho_S^{1,2}(0)) \equiv \frac{d}{dt} D(\rho_S^1(t), \rho_S^2(t)) > 0, \tag{9.62}$$

where $\sigma(t, \rho_S^{1,2}(0))$ denotes the rate of change of the trace distance at time t corresponding to the initial pair of states.

This suggests defining a measure $\mathcal{N}(\Phi)$ for the non-Markovianity of a quantum process through [409]

$$\mathcal{N}(\Phi) = \max_{\rho_S^{1,2}(0)} \int_{\sigma>0} dt\, \sigma(t, \rho_S^{1,2}(0)). \tag{9.63}$$

The time integration is extended over all time intervals (a_i, b_i) in which σ is positive and the maximum is taken over all pairs of initial states. The measure can be written as

$$\mathcal{N}(\Phi) = \max_{\rho^{1,2}(0)} \sum_i \left[D(\rho_S^1(b_i), \rho_S^2(b_i)) - D(\rho_S^1(a_i), \rho_S^2(a_i)) \right]. \tag{9.64}$$

To calculate this quantity one first determines for any pair of initial states the total growth of the trace distance over each time interval (a_i, b_i) and sums up the contribution of all intervals. $\mathcal{N}(\Phi)$ is then obtained by determining the maximum over all pairs of initial states.

CP Divisibility: A family of dynamical maps $\Phi(t,0)$ is defined to be divisible if for all $t_2 \geq t_1 \geq 0$ there exists a CPT map $\Phi(t_2, t_1)$ such that the relation $\Phi(t_2, 0) = \Phi(t_2, t_1)\Phi(t_1, 0)$ holds. Here, $\Phi(t_2, t_1) = \Phi(t_2, 0)\Phi^{-1}(t_1, 0)$ is the intermediate map whose CP behavior is an indication of whether the underlying dynamics is Markovian, wherein the intermediate map would be CP, or not. The simplest example of a divisible quantum process is given by a dynamical semigroup. For a semigroup $\Phi(t,0) = \exp[\mathcal{L}t]$ and divisibility is satisfied with the CPT map $\Phi(t_2,t_1) = \exp[\mathcal{L}(t_2 - t_1)]$.

Consider now a quantum process given by the time-local master equation with a time dependent generator. The dynamical maps can then be represented in terms of a time-ordered exponential,

$$\Phi(t,0) = \mathrm{T} \exp\left[\int_0^t dt'\mathcal{K}(t')\right], \quad t \geq 0, \tag{9.65}$$

where T denotes the chronological time-ordering operator. We can also define the maps

$$\Phi(t_2,t_1) = \mathrm{T} \exp\left[\int_{t_1}^{t_2} dt'\mathcal{K}(t')\right], \quad t_2 \geq t_1 \geq 0, \tag{9.66}$$

such that the composition law $\Phi(t_2,0) = \Phi(t_2,t_1)\Phi(t_1,0)$ holds by construction. The maps $\Phi(t_2,t_1)$ are completely positive, as is required by the divisibility condition, if and only if the decay rates $\Theta_i(t)$, Eq. (9.55), of the generator are positive functions. Thus divisibility is equivalent to positive rates in the time-local master equation [409]. It follows that non-Markovian quantum processes could be described by time-local master equations whose generator involves at least one temporarily negative rate $\Theta_i(t)$ [411], as discussed above.

In this context a characterization of non-Markovianity was given by [408] using the Choi-Jamiolkowski isomorphism, see Chapter 8.2.2, to quantify the degree of non-complete positiveness of the intermediate map $\Phi(t + \epsilon, t)$

$$f(t) = \lim_{\epsilon\to 0^+} \frac{||\left[\Phi(t + \epsilon, t) \otimes \mathcal{I}\right]\left(|\Psi\rangle\langle\Psi|\right)||_1 - 1}{\epsilon}. \tag{9.67}$$

The central quantity in the above equation is $\left[\Phi(t + \epsilon, t) \otimes \mathcal{I}\right]\left(|\Psi\rangle\langle\Psi|\right)$, where $|\Psi\rangle = \frac{1}{\sqrt{d}}\sum_{i=0}^{d-1}|i\rangle|i\rangle$ is the maximally entangled state of two copies of the system and d is the dimension. It follows from Choi's theorem that $\Phi(t+\epsilon, t)$ is CP if the matrix $\left[\Phi(t+\epsilon, t)\otimes\mathcal{I}\right]\left(|\Psi\rangle\langle\Psi|\right) \geq 0$. Since $\Phi(t+\epsilon, t)$ is trace preserving, it follows that $||\left[\Phi(t + \epsilon, t) \otimes \mathcal{I}\right]\left(|\Psi\rangle\langle\Psi|\right)||_1$ is equal to one for CP $\Phi(t + \epsilon, t)$ and greater than one otherwise, which would be an indicator of non-Markovian behavior. This implies that $f(t) > 0$, Eq. (9.67), for non-Markovian evolution. This lead to the formulation of the following measure of non-Markovianity $\mathcal{M} = \int_I dt f(t)$, for $t \in I$, where I is a time interval.

9.2.3 Illustrative Examples

We will now illustrate the above discussions on two open system models.

(A). *Garraway Model*: We make use of a model, introduced by Garraway [414], of a two-level system decaying spontaneously into a vacuum bath. The model is worked out under the assumption of a single excitation in the system-bath Hilbert space.

The system Hamiltonian is

$$H_S = \omega_0 \sigma_+ \sigma_-, \tag{9.68}$$

describing a two-state system (qubit) with ground state $|0\rangle$, excited state $|1\rangle$ and transition frequency ω_0; $\sigma_+ = |1\rangle\langle 0|$ and $\sigma_- = |0\rangle\langle 1|$ are the raising and lowering operators of the qubit. The Hamiltonian of the environment is

$$H_R = \sum_k \omega_k b_k^\dagger b_k, \tag{9.69}$$

and represents a reservoir of harmonic oscillators with creation and annihilation operators $b_k^\dagger$ and b_k, respectively. The interaction Hamiltonian takes the form

$$H_{SR} = \sum_k \left(g_k \sigma_+ \otimes b_k + g_k^* \sigma_- \otimes b_k^\dagger \right). \tag{9.70}$$

Due to the RWA, as well as the fact that the number of excitations are restricted to one, the total number of excitations in the system,

$$N = \sigma_+ \sigma_- + \sum_k b_k^\dagger b_k, \tag{9.71}$$

is a conserved quantity. Assuming the environment to be in the vacuum state $|0\rangle$, it can be shown that [2]

$$\begin{aligned}
\rho_{11}(t) &= |c(t)|^2 \rho_{11}(0), \\
\rho_{00}(t) &= \rho_{00}(0) + (1 - |c(t)|^2)\rho_{11}(0), \\
\rho_{10}(t) &= c(t)\rho_{10}(0), \\
\rho_{01}(t) &= c^*(t)\rho_{01}(0),
\end{aligned} \tag{9.72}$$

where the $\rho_{ij}(t) = \langle i|\rho_S(t)|j\rangle$ denote the matrix elements of $\rho_S(t)$.

The function $c(t)$ is the solution of the integro-differential equation

$$\frac{d}{dt}c(t) = -\int_0^t dt_1 f(t - t_1)c(t_1), \tag{9.73}$$

corresponding to the initial condition $c(0) = 1$, where the kernel $f(t - t_1)$ represents a reservoir two-point correlation function,

$$\begin{aligned}
f(t - t_1) &= \langle 0|\mathcal{O}(t)\mathcal{O}^\dagger(t_1)|0\rangle e^{i\omega_0(t-t_1)} \\
&= \sum_k |g_k|^2 e^{i(\omega_0 - \omega_k)(t-t_1)} \\
&= \int d\omega I(\omega)e^{i(\omega_0 - \omega)(t-t_1)},
\end{aligned} \tag{9.74}$$

of the environmental (reservoir) operators

$$\mathcal{O}(t) = \sum_k g_k b_k e^{-i\omega_k t}. \tag{9.75}$$

Here $I(\omega)$ is the reservoir spectral density. These results hold for a generic environmental spectral density and the corresponding two-point correlation function. To make things specific, we use this on a well known model of damped atom-photon interaction, i.e., the *damped Jaynes-Cummings* model. This models the coupling of a two-level atom to a single cavity mode which in turn is coupled to a reservoir of harmonic oscillators in vacuum. Considering a single excitation in the atom-cavity system, the cavity mode can be eliminated to give an effective Lorentzian spectral density

$$I(\omega) = \frac{1}{2\pi} \frac{\Theta_0 \chi^2}{(\omega_0 - \omega)^2 + \chi^2}. \tag{9.76}$$

Using Eq. (9.76) in Eq. (9.74), we find an exponential two-point correlation function

$$f(\tau) = \frac{1}{2}\Theta_0 \lambda e^{-\chi|\tau|}, \tag{9.77}$$

where Θ_0 describes the strength of the system-environment coupling and χ the spectral width which is related to the environmental correlation time by $\tau_R = \chi^{-1}$. Using this we find

$$c(t) = e^{-\chi t/2}\left[\cosh\left(\frac{dt}{2}\right) + \frac{\chi}{d}\sinh\left(\frac{dt}{2}\right)\right], \tag{9.78}$$

where $d = \sqrt{\chi^2 - 2\Theta_0\chi}$.

For the dissipative Jaynes-Cummings model, studied here, the generator $\mathcal{K}(t)$ of the time local master equation, Eq. (9.55), is

$$\begin{aligned}\mathcal{K}(t)\rho_S &= -\frac{i}{2}S(t)[\sigma_+\sigma_-, \rho_S] \\ &\quad + \Theta(t)\left[\sigma_-\rho_S\sigma_+ - \frac{1}{2}\{\sigma_+\sigma_-, \rho_S\}\right],\end{aligned} \tag{9.79}$$

where $\Theta(t) = -2\Re\left(\frac{\dot{c}(t)}{c(t)}\right)$, $S(t) = -2\Im\left(\frac{\dot{c}(t)}{c(t)}\right)$. The quantity $S(t)$ plays the role of a time-dependent frequency shift, and $\Theta(t)$ can be interpreted as a time-dependent decay rate. Due to the time dependence of these quantities the process does not generally represent a dynamical semigroup.

Non-Markovian behavior in the Garraway model: In the limit of small $\alpha = \Theta_0/\chi$ we may approximate $c(t) \approx e^{-\Theta_0 t/2}$. $S(t) = 0$ and $\Theta(t) = \Theta_0$, i.e., the generator $\mathcal{K}(t)$ assumes the form of a Lindblad generator of a quantum dynamical semigroup. Note that α can also be written as the ratio of the environmental correlations time $\tau_R = \chi^{-1}$ and the relaxation time $\tau_{rel} = \Theta_0^{-1}$ of the system, i.e., $\alpha = \frac{\tau_R}{\tau_{rel}}$. Thus we see that the standard Markov condition $\Theta_0 \ll \chi$ indeed leads to a Markovian semigroup here.

For the Garraway model, the necessary and sufficient condition for the complete positivity of the intermediate map $\Phi(t_2, t_1)$, Eq. (9.66), is given by $|c(t_2)| \leq |c(t_1)|$. Thus the dynamical map of the model is divisible if and only if $|c(t)|$ is a monotonically decreasing function of time. The rate $\Theta(t)$ can be written as

$$\Theta(t) = -\frac{2}{|c(t)|}\frac{d}{dt}|c(t)|. \tag{9.80}$$

This brings out the point that any increase of $|c(t)|$ leads to a negative decay rate in the corresponding generator, and illustrates the equivalence of the non-divisibility of the dynamical map and the occurrence of a temporarily negative rate in the time-local master equation.

Further, the time evolution of the trace distance corresponding to any pair of initial states $\rho_S^1(0)$ and $\rho_S^2(0)$ is given by

$$D(\rho_S^1(t), \rho_S^2(t)) = |c(t)|\sqrt{|c(t)|^2 a^2 + |b|^2}, \tag{9.81}$$

where $a = \rho_{11}^1(0) - \rho_{11}^2(0)$ and $b = \rho_{10}^1(0) - \rho_{10}^2(0)$. The time derivative of this expression yields

$$\sigma(t, \rho_S^{1,2}(0)) = \frac{2|c(t)|^2 a^2 + |b|^2}{\sqrt{|c(t)|^2 a^2 + |b|^2}}\frac{d}{dt}|c(t)|. \tag{9.82}$$

From this we conclude that the trace distance increases at time t if and only if the function $|c(t)|$ increases at this point of time. It follows that the process is non-Markovian, $\mathcal{N}(\Phi) > 0$, Eq. (9.64), if and only if the dynamical map is non-divisible, which in turn is equivalent to a temporarily negative rate $\gamma(t)$.

(B). *One Dimesional Quantum Walk subjected to the Random Telegraph Noise*: Next, we consider the one dimensional quantum walk, see Chapter 8.3.6, with the coin degree of freedom being subjected to the Random Telegraph noise (RTN) [415, 416]. Consider the time dependent stochastic Hamiltonian describing the evolution of a single qubit [417]

$$H(t) = \hbar \sum_{i=1}^{3} V_i(t)\sigma_i, \tag{9.83}$$

where σ_i denote Pauli matrices and $V_i(t) = a_i(-1)^{n_i(t)}$ is the representation of the RTN signal. RTN is a non-Gaussian stationary stochastic process that fluctuates randomly between binary amplitude values $\pm a$, following the Poisson probability distribution realized by the random variable $n_i(t)$. An example of this would be a two-level atom driven by a laser source with rapidly varying phase noise. Autocorrelation function for RTN $V_i(t)$ is

$$\langle V_i(t)V_j(t') \rangle = \delta_{ij} a^2 e^{-|t-t'|/\tau}. \tag{9.84}$$

Here a has the significance of the strength of the system-environment coupling and $1/\tau$ is proportional to the fluctuation rate of the RTN. The Fourier transform of the correlation function results in a Lorentzian power spectral density with peak value given by $2a^2\tau$. The Kraus operators representing the process are

$$
\begin{aligned}
K_1 &= \sqrt{1 + \Lambda(\nu)/2} I, \\
K_2 &= \sqrt{1 - \Lambda(\nu)/2} \sigma_3,
\end{aligned}
\tag{9.85}
$$

satisfying the completeness relation $\sum_{n=1}^{2} K_n^\dagger K_n = I$. $\Lambda(\nu)$ represents the damped harmonic function which encodes both the Markovian and non-Markovian behavior of the qubit,

$$
\Lambda(\nu) = e^{-\nu}[\cos(\nu\mu) + \sin(\nu\mu)/\mu],
\tag{9.86}
$$

where $\mu = \sqrt{(\frac{2a}{\gamma})^2 - 1}$ is the frequency of the harmonic oscillators and $\nu = \gamma t$ is the dimensionless time. γ is the fluctuation rate and is equal to $1/2\tau$. The function $\Lambda(\nu)$ corresponds to two regimes; the purely damping regime, where $a\tau < 0.25$, and damped oscillations for $a\tau > 0.25$. Corresponding to these two regimes of $\Lambda(\nu)$, we observe Markovian and non-Markovian behavior, respectively.

Non-Markovian behavior in the model: First we look at the model from the perspective of CP divisibility; any violations from which, as indicated above, would be signatures of non-Markovian behavior. We begin with a dynamical map $\mathcal{E}(t_2, t_0)$ connecting the system's density operator at times t_0 and $t_2 > t_0$. The intermediate map $\mathcal{E}^{\text{IM}}(t_2, t_1)$ for some intermediate time t_1 such that $t_2 > t_1 > t_0$, is given by:

$$
\mathcal{E}^{\text{IM}}(t_2, t_1) = \mathcal{E}(t_2, t_0)\mathcal{E}^{-1}(t_1, t_0),
\tag{9.87}
$$

provided that the inverse map $\mathcal{E}^{-1}(t_1, t_0)$ exists. The Choi matrix for the intermediate map can be obtained as:

$$
M_{\text{Choi}} = (\mathcal{E}^{\text{IM}}(t_2, t_1) \otimes \mathbb{I})|\Phi^+\rangle\langle\Phi^+|,
\tag{9.88}
$$

where $|\Phi^+\rangle \equiv |00\rangle + |11\rangle$. This is found to be

$$
M_{\text{Choi}} =
\begin{bmatrix}
1 & 0 & 0 & \frac{\Lambda(t_2)}{\Lambda(t_1)} \\
0 & 0 & 0 & 0 \\
0 & 0 & 0 & 0 \\
\frac{\Lambda(t_2)}{\Lambda(t_1)} & 0 & 0 & 1
\end{bmatrix}.
\tag{9.89}
$$

The non-vanishing eigenvalues of M_{Choi} in Eq. (9.89) are,

$$
\lambda_3 = \left[1 - \frac{\Lambda(t_2)}{\Lambda(t_1)}\right]; \lambda_4 = \left[1 + \frac{\Lambda(t_2)}{\Lambda(t_1)}\right].
\tag{9.90}
$$

It may be checked that if $2a \ll \gamma$, then $\Lambda(t)$ is a monotonically decreasing function and all eigenvalues are positive at all times, Fig. (9.3), consistent with

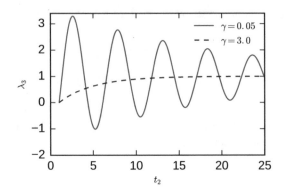

Figure 9.3: The eigenvalues of the Choi matrix obtained from intermediate dynamical map in both the Markovian and non-Markovian regime for RTN. We can observe that the eigenvalue (λ_3) of RTN become negative only in the non-Markovian (solid line) regime indicative of NCP character, whereas in the Markovian regime (dashed line) the eigenvalue is always positive. Note that both the intermediate time $t_1 = 1$ and the noise amplitude $a = 0.6$ is common for both the curves. Figure adapted from [415].

Markovian behavior. On the other hand, if $2a \gg \gamma$, then $\Lambda(t)$ can have regions of increase, and correspondingly, some eigenvalues can be negative, Fig. (9.3), a clear diagnostic of non-Markovian pattern.

The Kraus operators for the intermediate map are obtained by folding the eigenvectors of M_{Choi} [210].

$$K_{\pm}^{\text{IM}} = \sqrt{\frac{1}{2}\left|1 \pm \frac{\Lambda(t_2)}{\Lambda(t_1)}\right|} \begin{pmatrix} 1 & 0 \\ 0 & \pm 1 \end{pmatrix}. \tag{9.91}$$

As discussed above, trace distance (TD) [409] is a measure of distinguishability between two states. It has been used as a measure of non-Markovianity to quantify the amount of backflow from the environment to the system. For the RTN noise channels and initial states $|\pm\rangle = \frac{|0\rangle \pm |1\rangle}{\sqrt{2}}$, it is straightforward to compute the evolution of TD, which is $D(|+\rangle, |-\rangle) = \Lambda(t)$. TD measure between the reduced coin states, obtained by tracing over the position degrees of freedom of the quantum walk, undergo high frequency oscillation, as shown in Fig. (9.4) (a). Such TD oscillations are a signature of non-Markovian backflow behavior leading to non-zero value of the non-Markovian measures [418] and could be attributed to the *position environment* due to tracing over the position degrees of freedom of the walk. In addition to the oscillations present in the noiseless evolution of quantum walk, a further oscillatory feature arises when the coin is exposed to an external noise such as RTN in the non-Markovian regime, as depicted in Fig. (9.4) (c). This is a signature of non-Markovian backflow of information from the RTN environment, updating the coin dynamics.

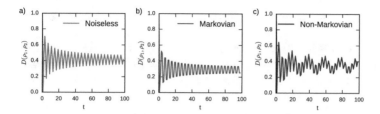

Figure 9.4: Plot of TD (Trace Distance) evolution under the influence of RTN, between the initial states $|\psi(\pm\frac{\pi}{4},0)\rangle_c$. (a) Noiseless quantum walk: Oscillations in TD are observed due to the interaction with position environment. Effect of RTN on the QW: (b). in the Markovian regime (middle curve) damping of the position induced oscillations is observed, while (c). in the non-Markovian regime (right) an additional oscillatory frequency component is present due to RTN-induced decoherence. Figure adapted from [415].

9.3   Quantum Thermodynamics

As a result of impressive progress on the experimental front, it is now considered within the reach of experimental feasibility to envisage the existence of engines on the nanonscale that operate within the realms of quantum mechanics. The pertinent question to ask then would be that do the laws of thermodynamics, as developed phenomenologically more than a century back, still hold in the quantum regime? This is the question that is addressed by the newly developed field of quantum thermodynamics [419, 420] and will be discussed here briefly.

Quantum mechanics infused dynamics into thermodynamics. The theory of open quantum systems, in particular, the Lindblad, Gorini-Kossakowski-Sudarshan (LGKS) master equation, discussed extensively in this book, plays a key role in quantum thermodynamics. Two types of devices have been studied: (a). reciprocating engines utilizing the Otto and Carnot cycle and (b). continuous engines resembling lasers and laser cooling devices. A reciprocating cycle is partitioned typically into four segments; two heat transfer segments, either isotherms, constant temperature, for the Carnot cycle or isochores, constant volume, for the Otto cycle, achieved with the help of interaction of the system with a bath using the tools of open quantum systems and two adiabats, where the working system is isolated from the environment. The same cycles can also be used as models for refrigerators. A well known example of a continuous engine would be a three-level laser. An example of a continuous refrigerator is laser cooling, obtained by reversing the operation of a three-level laser. In all these examples, a thermodynamic description is appropriate up to the level of a single open quantum system. Developments in the field of quantum thermodynamics are closely related to those in quantum information processing.

A generic model of a microscopic quantum engine consists of the following elements:

(a). A *working fluid*, which is a microscopic quantum system S;

(b). Hot and cold reservoirs modelled by infinite quantum systems with states at thermal equilibrium, for example, bath of harmonic oscillators;

(c). External periodic driving.

This can be treated as a model of open quantum systems. Using the first law of thermodynamics, which is basically the conservation of energy, we have

$$\frac{dU}{dt} = \sum_i \mathcal{J}_i + \mathcal{P}. \tag{9.92}$$

This formulation envisages a quantum network, which is a collection of interconnected quantum systems and baths at different temperatures. Here U is the internal energy of the system, which in the present thermodynamic context would be the working fluid S. In a simple scenario of a thermodynamic substance operating between a hot and a cold bath, $\sum_i^2 \mathcal{J}_i = \mathcal{J}_h + \mathcal{J}_c$, where $\mathcal{J}_h$ and $\mathcal{J}_c$ would be the heat currents entering the system S from the hot and cold baths, respectively. Also, $\mathcal{P}$ is the power provided by external sources. Recalling our discussion on the LGKS form of master equation, see Eq. (97) of Chapter 3, the equation of motion of an operator or a thermodynamic observable O can be obtained from the LGKS equation in the Heisenberg picture, i.e.,

$$\frac{dO}{dt} = i[H_S, O(t)] + \sum_{j=1} \gamma_j \left(A_j O A_j^\dagger - \frac{1}{2}\{A_j A_j^\dagger, O\} \right) + \frac{\partial}{\partial t} O. \tag{9.93}$$

The operators A_j are, as usual, the Lindblad operators. By substituting H_S for the observable O in Eq. (9.93) and keeping in mind that thermodynamic observations are modulo averages, a comparison with Eq. (9.92) yields

$$\mathcal{J}_i = \langle A_j H_S A_j^\dagger - \frac{1}{2}\{A_j A_j^\dagger, H_S\}\rangle, \tag{9.94}$$

$$\mathcal{P} = \langle \frac{\partial}{\partial t} H_S \rangle. \tag{9.95}$$

Here the angular brackets indicate the operation of an average. The second law of thermodynamics implies

$$\frac{d\mathcal{S}}{dt} - \sum_i \frac{\mathcal{J}_i}{T_i} \geq 0. \tag{9.96}$$

Here $\mathcal{S}$ is the entropy of the system S and would be the von-Neumann entropy in the quantum mechanical case. As before, in the simple case of the engine operating between the hot and cold baths, $\sum_i \frac{\mathcal{J}_i}{T_i} = \frac{\mathcal{J}_h}{T_h} + \frac{\mathcal{J}_c}{T_c}$, where T_h and T_c are the temperatures of the hot and cold baths, respectively. The terms $\sum_i \frac{\mathcal{J}_i}{T_i}$ would be the entropy flows from the baths and the *LHS* of Eq. (9.96) indicates positive entropy production. In the steady state regime, the entropy and energy averaged over a cycle, i.e., $\langle S \rangle$ and $\langle U \rangle$, respectively, are constant and the first and the second laws assume the following forms

$$\sum_i \langle \mathcal{J}_i \rangle + \langle \mathcal{P} \rangle = 0, \tag{9.97}$$

$$\sum_i \frac{\langle \mathcal{J} \rangle_i}{T_i} \leq 0. \tag{9.98}$$

Getting back to the case of the working fluid operating between a hot and a cold bath, for it to operate as an engine,

$$\langle \mathcal{J}_c \rangle < 0, \quad \langle \mathcal{P} \rangle < 0. \tag{9.99}$$

This implies that work, which is $-\langle \mathcal{P} \rangle$, performed by the engine on the external surroundings in positive and a part of heat extracted from the hot bath must be dumped into a cold bath. The efficiency of the engine η satisfies the bounds set by the Carnot cyle

$$\eta = \frac{-\langle \mathcal{P} \rangle}{\langle \mathcal{J}_h \rangle} \leq \frac{T_h - T_c}{T_h}. \tag{9.100}$$

In the refrigeration regime,

$$\langle \mathcal{J}_c \rangle > 0, \quad \langle \mathcal{P} \rangle > 0. \tag{9.101}$$

Here heat is extracted from the cold bath, cooling, at the cost of positive work $\langle \mathcal{P} \rangle$ supplied by the external driving. Here, the performance parameter ζ would be

$$\zeta = \frac{\langle \mathcal{J}_c \rangle}{\langle \mathcal{P} \rangle} \leq \frac{T_c}{T_h - T_c}. \tag{9.102}$$

Similarly, it is possible to show that the third law of thermodynamics implies that at absolute zero temperature, the entropy production at the cold bath is zero. This requirement leads to a scaling condition of the heat current. Hence, no refrigerator can cool a system to absolute zero temperature in finite time. Further, consistency with the third law ensures the existence of the ground state of the model open quantum system [419].

9.4 What next?

Here we have covered, albeit briefly, some of the modern lines of research where use is made of ideas related to open quantum systems. The readers who have had the patience to reach upto this chapter should have no difficulty in following the contents presented here. According to their personal interests, they can then follow up further developments from the scientific literature. Of course, there is a lot more that could be done but we need to draw the line somewhere and this would be a resonable place to do so. We thus bid the readers adieu.

References

[1] Ulrich Weiss. *Quantum dissipative systems*, volume 13. World scientific, 2012.

[2] Heinz-Peter Breuer and Francesco Petruccione. *The theory of open quantum systems*. Oxford University Press on Demand, 2002.

[3] Sushanta Dattagupta and Sanjay Puri. *Dissipative phenomena in condensed matter: some applications*, volume 71. Springer Science & Business Media, 2013.

[4] G.S. Agarwal and S. Dattagupta. *Stochastic processes–formalism and applications: proceedings of the winter school held at the University of Hyderabad, India, December 15-24, 1982*. Lecture notes in physics. Springer-Verlag, 1983.

[5] Richard Phillips Feynman and Frank Lee Vernon. The theory of a general quantum system interacting with a linear dissipative system. *Annals of physics*, 24:118–173, 1963.

[6] William Henry Louisell and William H Louisell. *Quantum statistical properties of radiation*, volume 7. Wiley New York, 1973.

[7] Girish S Agarwal. Quantum statistical theories of spontaneous emission and their relation to other approaches. In *Quantum Optics*, pages 1–128. Springer, 1974.

[8] AO Caldeira and AJ Leggett. Physica (utrecht) a, 121, 587 (1983). *MathSciNet ADS MATH Google Scholar*, 1985.

[9] Vincent Hakim and Vinay Ambegaokar. Quantum theory of a free particle interacting with a linearly dissipative environment. *Physical Review A*, 32(1):423, 1985.

[10] Hermann Grabert, Peter Schramm, and Gert-Ludwig Ingold. Quantum brownian motion: the functional integral approach. *Physics reports*, 168(3):115–207, 1988.

[11] Wojciech H Zurek. Decoherence and the transition from quantum to classical. *Physics today*, 44(10):36–44, 1991.

© Hindustan Book Agency 2018 and Springer Nature Singapore Pte Ltd. 2018
S. Banerjee, *Open Quantum Systems*, Texts and Readings in Physical Sciences 20,
https://doi.org/10.1007/978-981-13-3182-4

[12] QA Turchette, CJ Myatt, BE King, CA Sackett, David Kielpinski, WM Itano, C Monroe, and DJ Wineland. Decoherence and decay of motional quantum states of a trapped atom coupled to engineered reservoirs. *Physical Review A*, 62(5):053807, 2000.

[13] Christopher J Myatt, Brian E King, Quentin A Turchette, Cass A Sackett, David Kielpinski, Wayne M Itano, CWDJ Monroe, and David J Wineland. Decoherence of quantum superpositions through coupling to engineered reservoirs. *Nature*, 403(6767):269–273, 2000.

[14] Michel Brune, E Hagley, J Dreyer, X Maitre, A Maali, C Wunderlich, JM Raimond, and S Haroche. Observing the progressive decoherence of the meter in a quantum measurement. *Physical Review Letters*, 77(24):4887, 1996.

[15] Subrahmanyan Chandrasekhar. Stochastic problems in physics and astronomy. *Reviews of modern physics*, 15(1):1, 1943.

[16] MD Kostin. On the schrödinger-langevin equation. *The Journal of Chemical Physics*, 57(9):3589–3591, 1972.

[17] Kunio Yasue. Quantum mechanics of nonconservative systems. *Annals of Physics*, 114(1-2):479–496, 1978.

[18] Edward Nelson. Derivation of the schrödinger equation from newtonian mechanics. *Physical review*, 150(4):1079, 1966.

[19] H Dekker. Quantization of the linearly damped harmonic oscillator. *Physical Review A*, 16(5):2126, 1977.

[20] HJ Carmichael. Quantum trajectory theory for cascaded open systems. *Physical review letters*, 70(15):2273, 1993.

[21] Crispin Gardiner and Peter Zoller. *Quantum Noise (Springer- Verlag, Berlin, 2000)*. Springer Science & Business Media, 2004.

[22] Eugen Merzbacher. *Quantum Mechanics*. Wiley, 1970.

[23] J. J. Sakurai, San-Fu Tuan, and Eugene D Commins. *Modern quantum mechanics, revised edition*. Addison-Wesley, 1995.

[24] Paul Adrien Maurice Dirac. *The principles of quantum mechanics*. Number 27. Oxford university press, 1981.

[25] Ramamurti Shankar. *Principles of quantum mechanics*. Springer Science & Business Media, 2012.

[26] C Cohen-Tannoudji, B Diu, and F Laloe. *Quantum Mechanics, vols. I and II*. John Wiley and Sons, 1977.

[27] William Henry Louisell and William H Louisell. *Quantum statistical properties of radiation*, volume 7. Wiley New York, 1973.

[28] FT Arecchi, Eric Courtens, Robert Gilmore, and Harry Thomas. Atomic coherent states in quantum optics. *Physical Review A*, 6(6):2211, 1972.

[29] Marlan O. Scully and M. Suhail Zubairy. *Quantum Optics*. Cambridge University Press, 1997.

[30] Ludwig D Faddeev and Oleg Aleksandrovich IAkubovskiĭ. *Lectures on quantum mechanics for mathematics students*, volume 47. American Mathematical Soc., 2009.

[31] Max Born and Kun Huang. *Dynamical theory of crystal lattices*. Clarendon press, 1954.

[32] LD Landau. *Statistical Physics, Part I, Revised by Lifshitz, EM and Pitaevskii, LP, Translated by Sykes JB and Kearsley, MJ*. Pergamon Press, 1980.

[33] David Chandler. Introduction to modern statistical mechanics. *Introduction to Modern Statistical Mechanics, by David Chandler, pp. 288. Foreword by David Chandler. Oxford University Press, Sep 1987. ISBN-10: 0195042778. ISBN-13: 9780195042771*, page 288, 1987.

[34] Linda E Reichl. *A modern course in statistical physics*. John Wiley & Sons, 2016.

[35] Thomas M Cover and Joy A Thomas. *Elements of information theory*. John Wiley & Sons, 2012.

[36] Claude Elwood Shannon. Communication in the presence of noise. *Proceedings of the IRE*, 37(1):10–21, 1949.

[37] Richard Chace Tolman. *Relativity, thermodynamics, and cosmology*. Courier Corporation, 1987.

[38] M.A. Nielsen and I.L. Chuang. *Quantum Computation and Quantum Information*. Cambridge Series on Information and the Natural Sciences. Cambridge University Press, 2000.

[39] Edwin T Jaynes. Information theory and statistical mechanics. *Physical review*, 106(4):620, 1957.

[40] Richard P Feynman, Albert R Hibbs, and Daniel F Styer. *Quantum mechanics and path integrals*. Courier Corporation, 2010.

[41] Lawrence S Schulman. *Techniques and applications of path integration*. Courier Corporation, 2012.

[42] Hagen Kleinert. *Path integrals in quantum mechanics, statistics, polymer physics, and financial markets*. World scientific, 2009.

[43] Ashok Das. *Field theory: a path integral approach*, volume 52. World Scientific, 1993.

[44] Walter Greiner and Joachim Reinhardt. *Field quantization.* Springer Science & Business Media, 2013.

[45] P. A. M. Dirac. On the analogy between classical and quantum mechanics. *Rev. Mod. Phys.*, 17:195–199, Apr 1945.

[46] Cécile DeWitt-Morette, Amar Maheshwari, and Bruce Nelson. Path integration in non-relativistic quantum mechanics. *Physics Reports*, 50(5):255–372, 1979.

[47] Richard Phillips Feynman and Frank Lee Vernon. The theory of a general quantum system interacting with a linear dissipative system. *Annals of physics*, 24:118–173, 1963.

[48] Amir O Caldeira and Anthony J Leggett. Path integral approach to quantum brownian motion. *Physica A: Statistical mechanics and its Applications*, 121(3):587–616, 1983.

[49] M. Toda, R. Kubo, and N. Saito. *Statistical Physics I: Equilibrium Statistical Mechanics.* Springer Series in Solid-State Sciences. Springer Berlin Heidelberg, 2012.

[50] Morikazu Toda, Ryogo Kubo, Nobuhiko Saito, and Natsuki Hashitsume. *Statistical Physics II: Nonequilibrium Statistical Mechanics*, volume 30. Springer Science & Business Media, 1992.

[51] M. Chaichian and A. Demichev. *Path Integrals in Physics: Volume I Stochastic Processes and Quantum Mechanics.* CRC Press, 2001.

[52] Masud Chaichian and Andrei Demichev. *Path Integrals in Physics: Volume II Quantum Field Theory, Statistical Physics and other Modern Applications*, volume 2. CRC Press, 2001.

[53] Jean Zinn-Justin. *Path integrals in quantum mechanics.* Oxford University Press, 2010.

[54] Dinkar C Khandekar, SV Lawande, and KV Bhagwat. *Path-integral methods and their applications.* Allied Publishers, 2002.

[55] Dmitrii Nikolaevich Zubarev, Vladimir Morozov, and Gerd Röpke. *Statistical mechanics of nonequilibrium processes*, volume 1. Akademie Verlag Berlin, 1996.

[56] Robert Zwanzig. *Nonequilibrium statistical mechanics.* Oxford University Press, 2001.

[57] Göran Lindblad. Completely positive maps and entropy inequalities. *Communications in Mathematical Physics*, 40(2):147–151, 1975.

[58] Vittorio Gorini, Andrzej Kossakowski, and Ennackal Chandy George Sudarshan. Completely positive dynamical semigroups of n-level systems. *Journal of Mathematical Physics*, 17(5):821–825, 1976.

[59] Sadao Nakajima. On quantum theory of transport phenomena steady diffusion. *Progress of Theoretical Physics*, 20(6):948–959, 1958.

[60] Robert Zwanzig. Ensemble method in the theory of irreversibility. *The Journal of Chemical Physics*, 33(5):1338–1341, 1960.

[61] Hermann Grabert. *Projection operator techniques in nonequilibrium statistical mechanics*, volume 95. Springer, 2006.

[62] Fumiaki Shibata, Yoshinori Takahashi, and Natsuki Hashitsume. A generalized stochastic liouville equation. non-markovian versus memoryless master equations. *Journal of Statistical Physics*, 17(4):171–187, 1977.

[63] S Chaturvedi and F Shibata. Time-convolutionless projection operator formalism for elimination of fast variables. applications to brownian motion. *Zeitschrift für Physik B Condensed Matter*, 35(3):297–308, 1979.

[64] Fumiaki Shibata and Toshihico Arimitsu. Expansion formulas in nonequilibrium statistical mechanics. *Journal of the Physical Society of Japan*, 49(3):891–897, 1980.

[65] Herbert Goldstein, Charles P Poole, and John L Safko. *Classical Mechanics*. Pearson Higher Ed, 2014.

[66] R Brown. Philosophical magazine ns 4, 161-173,(1828); see also r. brown. *Philosophical Magazine NS*, 6:161–166, 1829.

[67] Albert Einstein. On the movement of small particles suspended in statiunary liquids required by the molecular-kinetic theory 0f heat. *Ann. d. Phys*, 17:549–560, 1905.

[68] A. J. Leggett, S. Chakravarty, A. T. Dorsey, Matthew P. A. Fisher, Anupam Garg, and W. Zwerger. Dynamics of the dissipative two-state system. *Rev. Mod. Phys.*, 59:1–85, Jan 1987.

[69] George W Ford, John T Lewis, and RF Oconnell. Quantum langevin equation. *Physical Review A*, 37(11):4419, 1988.

[70] L. D. Landau. *Zh. Eksp. Teor. Fiz.*, 7, 1937.

[71] Benjamin Svetitsky. Diffusion of charmed quarks in the quark-gluon plasma. *Physical Review D*, 37(9):2484, 1988.

[72] N. G. Van Kampen. *Stochastic processes in physics and chemistry*. Elsevier, 1992.

[73] C. W. Gardiner. *Handbook of Stochastic Methods*. Springer, 1985.

[74] Hannes Risken. *Fokker-Planck equation*. Springer, 1984.

[75] Fritz Haake. Statistical treatment of open systems by generalized master equations. In *Springer tracts in modern physics*, pages 98–168. Springer, 1973.

[76] Leon Van Hove. Quantum-mechanical perturbations giving rise to a statistical transport equation. *Physica*, 21(1-5):517–540, 1954.

[77] A Sndulescu and H Scutaru. Open quantum systems and the damping of collective modes in deep inelastic collisions. *Annals of Physics*, 173(2):277–317, 1987.

[78] Alfred G Redfield. On the theory of relaxation processes. *IBM Journal of Research and Development*, 1(1):19–31, 1957.

[79] E Brian Davies. Markovian master equations. *Communications in mathematical Physics*, 39(2):91–110, 1974.

[80] R Dümcke and H Spohn. The proper form of the generator in the weak coupling limit. *Zeitschrift für Physik B Condensed Matter*, 34(4):419–422, 1979.

[81] Subhashish Banerjee and R Srikanth. Geometric phase of a qubit interacting with a squeezed-thermal bath. *The European Physical Journal D-Atomic, Molecular, Optical and Plasma Physics*, 46(2):335–344, 2008.

[82] R Srikanth and Subhashish Banerjee. Squeezed generalized amplitude damping channel. *Physical Review A*, 77(1):012318, 2008.

[83] X Maître, E Hagley, J Dreyer, A Maali, C Wunderlich, M Brune, JM Raimond, and S Haroche. An experimental study of a schrödinger cat decoherence with atoms and cavities. *journal of modern optics*, 44(11-12):2023–2032, 1997.

[84] William G Unruh. Maintaining coherence in quantum computers. *Physical Review A*, 51(2):992, 1995.

[85] G Massimo Palma, Kalle-Antti Suominen, and Artur K Ekert. Quantum computers and dissipation. In *Proceedings of the Royal Society of London A: Mathematical, Physical and Engineering Sciences*, volume 452, pages 567–584. The Royal Society, 1996.

[86] QA Turchette, CJ Myatt, BE King, CA Sackett, David Kielpinski, WM Itano, Ch Monroe, and DJ Wineland. Decoherence and decay of motional quantum states of a trapped atom coupled to engineered reservoirs. *Physical Review A*, 62(5):053807, 2000.

[87] Braginskiĭ Vladimir Borisovich Braginsky, Vladimir B and Farid Ya Khalili. *Quantum measurement*. Cambridge University Press, 1995.

[88] R. W. P. Drever V. D. Sandberg C. M. Caves, K. D. Thorne and M. Zimmerman. Not uniquely identified, 1980.

[89] Subhashish Banerjee and R Ghosh. Dynamics of decoherence without dissipation in a squeezed thermal bath. *Journal of Physics A: Mathematical and Theoretical*, 40(45):13735, 2007.

[90] Carlton M Caves and Bonny L Schumaker. New formalism for two-photon quantum optics. i. quadrature phases and squeezed states. *Physical Review A*, 31(5):3068, 1985.

[91] Heinz-Peter Breuer, Bernd Kappler, and Francesco Petruccione. The time-convolutionless projection operator technique in the quantum theory of dissipation and decoherence. *Annals of Physics*, 291(1):36–70, 2001.

[92] Gerald D Mahan. *Many-particle physics*. Springer Science & Business Media, 2013.

[93] Robert Alicki. General theory and applications to unstable particles. *Quantum Dynamical Semigroups and Applications*, pages 1–94, 1987.

[94] Robert Alicki. General theory and applications to unstable particles. In *Quantum Dynamical Semigroups and Applications*, pages 1–46. Springer, 2007.

[95] Vincent Hakim and Vinay Ambegaokar. Quantum theory of a free particle interacting with a linearly dissipative environment. *Physical Review A*, 32(1):423, 1985.

[96] C Morais Smith and AO Caldeira. Generalized feynman-vernon approach to dissipative quantum systems. *Physical Review A*, 36(7):3509, 1987.

[97] Bei Lok Hu, Juan Pablo Paz, and Yuhong Zhang. Quantum brownian motion in a general environment: Exact master equation with nonlocal dissipation and colored noise. *Physical Review D*, 45(8):2843, 1992.

[98] BL Hu, Juan Pablo Paz, and Yuhong Zhang. Quantum brownian motion in a general environment. ii. nonlinear coupling and perturbative approach. *Physical Review D*, 47(4):1576, 1993.

[99] BL Hu and Andrew Matacz. Quantum brownian motion in a bath of parametric oscillators: A model for system-field interactions. *Physical Review D*, 49(12):6612, 1994.

[100] J. Audretsch B.L. Hu and V. de Sabbata. Quantum mechanics in curved spacetime, 1990. Not uniquely identified.

[101] Juan Pablo Paz and Sukanya Sinha. Decoherence and back reaction: The origin of the semiclassical einstein equations. *Physical Review D*, 44(4):1038, 1991.

[102] BL Hu. Dissipation in quantum fields and semiclassical gravity. *Physica A: Statistical Mechanics and its Applications*, 158(1):399–424, 1989.

[103] Arlen Anderson and Jonathan J Halliwell. Information-theoretic measure of uncertainty due to quantum and thermal fluctuations. *Physical Review D*, 48(6):2753, 1993.

[104] Subhashish Banerjee and R. Ghosh. Quantum theory of a stern-gerlach system in contact with a linearly dissipative environment. *Phys. Rev. A*, 62:042105, Sep 2000.

[105] Wojciech H Zurek. Pointer basis of quantum apparatus: Into what mixture does the wave packet collapse? *Physical Review D*, 24(6):1516, 1981.

[106] Murray Gell-Mann and James B Hartle. Classical equations for quantum systems. *Physical Review D*, 47(8):3345, 1993.

[107] Y-C Chen, JL Lebowitz, and C Liverani. Dissipative quantum dynamics in a boson bath. *Physical Review B*, 40(7):4664, 1989.

[108] E Brian Davies and John T Lewis. An operational approach to quantum probability. *Communications in Mathematical Physics*, 17(3):239–260, 1970.

[109] Dieter Forster. *Hydrodynamic fluctuations, broken symmetry, and correlation functions*. CRC Press, 2018.

[110] R Srikanth and Subhashish Banerjee. Squeezed generalized amplitude damping channel. *Physical Review A*, 77(1):012318, 2008.

[111] Francisco M Fernández. Time-evolution operator and lie algebras. *Physical Review A*, 40(1):41, 1989.

[112] Subhashish Banerjee and Joachim Kupsch. Applications of canonical transformations. *Journal of Physics A: Mathematical and General*, 38(23):5237, 2005.

[113] MS Kim and V Bužek. Photon statistics of superposition states in phase-sensitive reservoirs. *Physical Review A*, 47(1):610, 1993.

[114] J. J. Halliwell J. P. Paz. The physical origin of time asymmetry, 1994.

[115] JJ Halliwell and T Yu. Alternative derivation of the hu-paz-zhang master equation of quantum brownian motion. *Physical Review D*, 53(4):2012, 1996.

[116] Luciana Dávila Romero and Juan Pablo Paz. Decoherence and initial correlations in quantum brownian motion. *Physical Review A*, 55(6):4070, 1997.

[117] Robert Karrlein and Hermann Grabert. Exact time evolution and master equations for the damped harmonic oscillator. *Physical Review E*, 55(1):153, 1997.

[118] Subhashish Banerjee and R Ghosh. General quantum brownian motion with initially correlated and nonlinearly coupled environment. *Physical Review E*, 67(5):056120, 2003.

[119] Subhashish Banerjee and Abhishek Dhar. Classical limit of master equation for a harmonic oscillator coupled to an oscillator bath with separable initial conditions. *Physical Review E*, 73(6):067104, 2006.

[120] GG Emch. Algebraic methods in statistical mechanics and quantum field theory wiley. *New York*, 1972.

[121] Aurelian Isar, Aurel Sandulescu, and Werner Scheid. Density matrix for the damped harmonic oscillator within the lindblad theory. *Journal of mathematical physics*, 34(9):3887–3900, 1993.

[122] H Hofmann, C Gregoire, R Lucas, and Ch Ngô. A theoretical model for the charge equilibration in heavy ion collisions. *Zeitschrift für Physik A Hadrons and Nuclei*, 293(3):229–240, 1979.

[123] Rainer W Hasse. Microscopic derivation of quantum fluctuations in nuclear reactions. *Nuclear Physics A*, 318(3):480–506, 1979.

[124] EM Spina and HA Weidenmüller. Damping of collective modes in dic: Quantal versus statistical fluctuations. *Nuclear Physics A*, 425(2):354–372, 1984.

[125] CW Gardiner and MJ Collett. Input and output in damped quantum systems: Quantum stochastic differential equations and the master equation. *Physical Review A*, 31(6):3761, 1985.

[126] TAB Kennedy and DF Walls. Squeezed quantum fluctuations and macroscopic quantum coherence. *Physical Review A*, 37(1):152, 1988.

[127] GS Agarwal. Master equations in phase-space formulation of quantum optics. *Physical Review*, 178(5):2025, 1969.

[128] S Dattagupta. Brownian motion of a quantum system. *Physical Review A*, 30(3):1525, 1984.

[129] CM Savage and DF Walls. Damping of quantum coherence: The master-equation approach. *Physical Review A*, 32(4):2316, 1985.

[130] GS Agarwal. Brownian motion of a quantum oscillator. *Physical Review A*, 4(2):739, 1971.

[131] AO Caldeira and AJ Leggett. Influence of damping on quantum interference: An exactly soluble model. *Physical Review A*, 31(2):1059, 1985.

[132] Subhashish Banerjee and Joachim Kupsch. Applications of canonical transformations. *Journal of Physics A: Mathematical and General*, 38(23):5237, 2005.

[133] Paul AM Dirac. The quantum theory of the emission and absorption of radiation. In *Proceedings of the Royal Society of London A: Mathematical, Physical and Engineering Sciences*, volume 114, pages 243–265. The Royal Society, 1927.

[134] SM Barnett and DT Pegg. On the hermitian optical phase operator. *Journal of Modern Optics*, 36(1):7–19, 1989.

[135] Jeffrey H Shapiro, Scott R Shepard, and Ngai C Wong. Erratum:ultimate quantum limits on phase measurement[phys. rev. lett. 62, 2377 (1989)]. *Physical Review Letters*, 63(18):2002, 1989.

[136] M-J W Hall. The quantum description of optical phase. *Quantum Optics: Journal of the European Optical Society Part B*, 3(1):7, 1991.

[137] GS Agarwal, S Chaturvedi, K Tara, and V Srinivasan. Classical phase changes in nonlinear processes and their quantum counterparts. *Physical Review A*, 45(7):4904, 1992.

[138] L. Susskind and J. Glogower. Quantum mechanical phase and time operator. Technical report, Cornell Univ., Ithaca, NY, 1964.

[139] Huai-Xin Lu, Jie Yang, Yong-De Zhang, and Zeng-Bing Chen. Algebraic approch to master equations with superoperator generators of su (1, 1) and su (2) lie algebras. *Physical Review A*, 67(2):024101, 2003.

[140] Subhashish Banerjee and R Srikanth. Phase diffusion in quantum dissipative systems. *Physical Review A*, 76(6):062109, 2007.

[141] Artur K Ekert and Peter L Knight. Canonical transformation and decay into phase-sensitive reservoirs. *Physical Review A*, 42(1):487, 1990.

[142] SM Roy and Virendra Singh. Generalized coherent states and the uncertainty principle. *Physical Review D*, 25(12):3413, 1982.

[143] M Venkata Satyanarayana. Generalized coherent states and generalized squeezed coherent states. *Physical Review D*, 32(2):400, 1985.

[144] A Érdelyi, W Magnus, F Oberhettinger, and FG Tricomi. Higher transcendental functions (california institute of technology h. bateman ms project), vol. 2, 1953.

[145] Hans Maassen and Jos BM Uffink. Generalized entropic uncertainty relations. *Physical Review Letters*, 60(12):1103, 1988.

[146] Alberto Galindo and Miguel Angelo Martin-Delgado. Information and computation: Classical and quantum aspects. *Reviews of Modern Physics*, 74(2):347, 2002.

[147] Masanori Ohya and Dénes Petz. *Quantum entropy and its use*. Springer Science & Business Media, 2004.

[148] R Srikanth and Subhashish Banerjee. Complementarity in atomic (finite-level quantum) systems: an information-theoretic approach. *The European Physical Journal D-Atomic, Molecular, Optical and Plasma Physics*, 53(2):217–227, 2009.

[149] Solomon Kullback and Richard A Leibler. On information and sufficiency. *The annals of mathematical statistics*, 22(1):79–86, 1951.

[150] Iwo Białynicki-Birula and Jerzy Mycielski. Uncertainty relations for information entropy in wave mechanics. *Communications in Mathematical Physics*, 44(2):129–132, 1975.

[151] William Beckner. Inequalities in fourier analysis. *Annals of Mathematics*, pages 159–182, 1975.

[152] M Brune, S Haroche, JM Raimond, L Davidovich, and N Zagury. Manipulation of photons in a cavity by dispersive atom-field coupling: Quantum-nondemolition measurements and generation of schrödinger catstates. *Physical Review A*, 45(7):5193, 1992.

[153] V Bužek, Ts Gantsog, and MS Kim. Phase properties of schrödinger cat states of light decaying in phase-sensitive reservoirs. *Physica scripta*, 1993(T48):131, 1993. Not uniquely identified.

[154] P W Anderson, BI Halperin, and C M Varma. Anomalous low-temperature thermal properties of glasses and spin glasses. *Philosophical Magazine*, 25(1):1–9, 1972.

[155] WA Phillips. Tunneling states in amorphous solids. *Journal of Low Temperature Physics*, 7(3):351–360, 1972.

[156] Jun Kondo. Resistance minimum in dilute magnetic alloys. *Progress of theoretical physics*, 32(1):37–49, 1964.

[157] Rudolph A Marcus. On the theory of oxidation-reduction reactions involving electron transfer. i. *The Journal of Chemical Physics*, 24(5):966–978, 1956. Not uniquely identified.

[158] Anthony J Leggett, S Chakravarty, AT Dorsey, Matthew PA Fisher, Anupam Garg, and W Zwerger. Dynamics of the dissipative two-state system. *Reviews of Modern Physics*, 59(1):1, 1987.

[159] DJ Wallace. Solitons and instantons: An introduction to solitons and instantons in quantum field theory, 1983.

[160] Jean Zinn-Justin. The principles of instanton calculus: A few applications. Technical report, 1982. Les Houches, Session XXXIX, 1982.

[161] Joachim Kupsch and Subhashish Banerjee. Ultracoherence and canonical transformations. *Infinite Dimensional Analysis, Quantum Probability and Related Topics*, 9(03):413–434, 2006.

[162] Subhashish Banerjee and Joachim Kupsch. Applications of canonical transformations. *Journal of Physics A: Mathematical and General*, 38(23):5237, 2005.

[163] Alex I Braginski and Yi Zhang. Practical rf squids: configuration and performance. *The SQUID Handbook: Fundamentals and Technology of SQUIDs and SQUID Systems, Volume I*, pages 219–250, 2004.

[164] M Morillo, RI Cukier, and M Tij. A projection operator approach to a dissipative two-level system. *Physica A: Statistical Mechanics and its Applications*, 179(3):411–427, 1991.

[165] H Dekker. Noninteracting-blip approximation for a two-level system coupled to a heat bath. *Physical Review A*, 35(3):1436, 1987.

[166] Anupam Garg, José Nelson Onuchic, and Vinay Ambegaokar. Effect of friction on electron transfer in biomolecules. *The Journal of chemical physics*, 83(9):4491–4503, 1985.

[167] AJ Leggett. Quantum tunneling in the presence of an arbitrary linear dissipation mechanism. *Physical Review B*, 30(3):1208, 1984.

[168] M. Wakker. The dissipative two-state system. Masters Thesis, Institute for Theoretical Physics, Utrecht University.

[169] Milena Grifoni and Peter Hänggi. Driven quantum tunneling. *Physics Reports*, 304(5):229–354, 1998.

[170] Milena Grifoni, Maura Sassetti, Peter Hänggi, and Ulrich Weiss. Cooperative effects in the nonlinearly driven spin-boson system. *Physical Review E*, 52(4):3596, 1995.

[171] PW Anderson and G Yuval. Some numerical results on the kondo problem and the inverse square one-dimensional ising model. *Journal of Physics C: Solid State Physics*, 4(5):607, 1971.

[172] Manfred Winterstetter and Ulrich Weiss. Dynamical simulation of the driven spin-boson system: The influence of interblip correlations. *Chemical physics*, 217(2):155–166, 1997.

[173] Dara PS McCutcheon, Nikesh S Dattani, Erik M Gauger, Brendon W Lovett, and Ahsan Nazir. A general approach to quantum dynamics using a variational master equation: Application to phonon-damped rabi rotations in quantum dots. *Physical Review B*, 84(8):081305, 2011.

[174] Inés de Vega and Daniel Alonso. Dynamics of non-markovian open quantum systems. *Reviews of Modern Physics*, 89(1):015001, 2017.

[175] Mohsen Razavy. *Quantum theory of tunneling*, World Scientific, 2013.

[176] Joachim Ankerhold and LS Schulman. Quantum tunneling in complex systems: The semiclassical approach. *SIAM review*, 50(1):192, 2008.

[177] S Coleman. The uses of instantons, erice lectures, 1977, to be published; c. callan, r. dashen and d. gross. *Phys. Rev. D*, 17:2717, 1978.

[178] J. S. Langer. Theory of the condensation point. *Ann. Phys. (N.Y.)*, 41, 1967.

[179] J. S. Langer. Systems far from equilibrium. *Lecture Notes in Physics*, 132:12, 1980.

[180] George J Papadopoulos and Jozef Devreese. *Path Integrals: And Their Applications in Quantum, Statistical and Solid State Physics*, volume 34. Springer Science & Business Media, 2013.

[181] Roderick Wong. *Asymptotic approximations of integrals*. SIAM, 2001.

[182] Alan Jeffrey and Daniel Zwillinger. *Table of integrals, series, and products.* Academic press, 2007. Please check author's details.

[183] L. D. Landau and L. M. Lifshitz. *Quantum Mechanics.* Pergamon Press, 1977. 3-rd Edition.

[184] Mikhail Shifman. *Advanced topics in quantum field theory: A lecture course.* Cambridge University Press, 2012.

[185] Walter Greiner and Joachim Reinhardt. *Field quantization.* Springer Science & Business Media, 2013.

[186] Erick J Weinberg. *Classical solutions in quantum field theory: Solitons and Instantons in High Energy Physics.* Cambridge University Press, 2012.

[187] George Gamow. Zur quantentheorie des atomkernes. *Zeitschrift für Physik A Hadrons and Nuclei,* 51(3):204–212, 1928.

[188] A.O Caldeira and A.J Leggett. Quantum tunnelling in a dissipative system. *Annals of Physics,* 149(2):374 – 456, 1983.

[189] Peter Hänggi, Peter Talkner, and Michal Borkovec. Reaction-rate theory: fifty years after kramers. *Reviews of modern physics,* 62(2):251, 1990.

[190] Hendrik Anthony Kramers. Brownian motion in a field of force and the diffusion model of chemical reactions. *Physica,* 7(4):284–304, 1940.

[191] Amir O Caldeira. *An introduction to macroscopic quantum phenomena and quantum dissipation.* Cambridge University Press, 2014.

[192] Hermann Grabert, Peter Olschowski, and Ulrich Weiss. Quantum decay rates for dissipative systems at finite temperatures. *Physical Review B,* 36(4):1931, 1987.

[193] Friedrich Hund. On the interpretation of the molecular spectra-iii. *Zeitschriftfur Physik,* 43(11):805–826, 1927.

[194] J Robert Oppenheimer. Three notes on the quantum theory of aperiodic effects. *Physical review,* 31(1):66, 1928.

[195] Ronald W Gurney and Edward U Condon. Wave mechanics and radioactive disintegration. *Nature,* 122(3073):439, 1928.

[196] Eugene Wigner. Crossing of potential thresholds in chemical reactions. *journal for physical chemistry,* 19(1):203–216, 1932.

[197] Kishore Thapliyal, Subhashish Banerjee, Anirban Pathak, S Omkar, and V Ravishankar. Quasiprobability distributions in open quantum systems: spin-qubit systems. *Annals of Physics,* 362:261–286, 2015.

[198] William H Miller. Semiclassical limit of quantum mechanical transition state theory for nonseparable systems. *The Journal of chemical physics,* 62(5):1899–1906, 1975.

[199] S Levit and U Smilansky. A new approach to gaussian path integrals and the evaluation of the semiclassical propagator. *Annals of Physics*, 103(1):198–207, 1977.

[200] Victor Pavlovich Maslov. Stationary-phase method for feynman's continual integral. *Theoretical and Mathematical Physics*, 2(1):21–25, 1970.

[201] Martin C Gutzwiller. Phase-integral approximation in momentum space and the bound states of an atom. *Journal of mathematical Physics*, 8(10):1979–2000, 1967.

[202] Michael V Berry and KE Mount. Semiclassical approximations in wave mechanics. *Reports on Progress in Physics*, 35(1):315, 1972.

[203] Ian Affleck. Quantum-statistical metastability. *Physical Review Letters*, 46(6):388, 1981.

[204] Ulrich Weiss and Walter Haeffner. Complex-time path integrals beyond the stationary-phase approximation: Decay of metastable states and quantum statistical metastability. *Physical Review D*, 27(12):2916, 1983.

[205] U Eckern, A Schmid, A Leggett, and Yu Kagan. Quantum tunneling in condensed media. 1992.

[206] Hermann Grabert, Ulrich Weiss, and Peter Hanggi. Quantum tunneling in dissipative systems at finite temperatures. *Physical review letters*, 52(25):2193, 1984.

[207] Shin Takagi. *Macroscopic quantum tunneling*. Cambridge University Press, 2002.

[208] Joachim Ankerhold and Hermann Grabert. Semiclassical time evolution of the density matrix and tunneling. *Physical Review E*, 61(4):3450, 2000.

[209] Mark M Wilde. From classical to quantum shannon theory. *arXiv preprint arXiv:1106.1445 v5*, 2013.

[210] S Omkar, R Srikanth, and Subhashish Banerjee. Dissipative and non-dissipative single-qubit channels: dynamics and geometry. *Quantum information processing*, 12(12):3725–3744, 2013.

[211] A Buchleitner, M Tiersch, and C Viviescas. Entanglement and decoherence: Foundations and modern trends lect. *Notes Phys*, 768, 2009.

[212] Florian Mintert, André RR Carvalho, Marek Kuś, and Andreas Buchleitner. Measures and dynamics of entangled states. *Physics Reports*, 415(4):207–259, 2005.

[213] R Srikanth and Subhashish Banerjee. An environment-mediated quantum deleter. *Physics Letters A*, 367(4):295–299, 2007.

[214] Erich Joos, H Dieter Zeh, Claus Kiefer, Domenico JW Giulini, Joachim Kupsch, and Ion-Olimpiu Stamatescu. *Decoherence and the appearance of*

a classical world in quantum theory. Springer Science & Business Media, 2013.

[215] Amir O Caldeira and Anthony J Leggett. Influence of dissipation on quantum tunneling in macroscopic systems. *Physical Review Letters*, 46(4):211, 1981.

[216] WH Zurek. *Phys. Today*, 44:36, 1991.

[217] Wojciech H Zurek. Preferred states, predictability, classicality and the environment-induced decoherence. *Progress of Theoretical Physics*, 89(2):281–312, 1993.

[218] ECG Sudarshan, PM Mathews, and Jayaseetha Rau. Stochastic dynamics of quantum-mechanical systems. *Physical Review*, 121(3):920, 1961.

[219] Karl Kraus. States, effects and operations, vol. 190 of lecture notes in physics, 1983.

[220] R Srikanth and Subhashish Banerjee. Squeezed generalized amplitude damping channel. *Physical Review A*, 77(1):012318, 2008.

[221] Man-Duen Choi. Completely positive linear maps on complex matrices. *Linear algebra and its applications*, 10(3):285–290, 1975.

[222] Andrzej Jamiołkowski. Linear transformations which preserve trace and positive semidefiniteness of operators. *Reports on Mathematical Physics*, 3(4):275–278, 1972.

[223] Alexander S Holevo and Vittorio Giovannetti. Quantum channels and their entropic characteristics. *Reports on progress in physics*, 75(4):046001, 2012.

[224] Thomas Konrad, Fernando De Melo, Markus Tiersch, Christian Kasztelan, Adriano Aragao, and Andreas Buchleitner. Evolution equation for quantum entanglement. *Nature physics*, 4(2):99–102, 2008.

[225] William K Wootters. Entanglement of formation of an arbitrary state of two qubits. *Physical Review Letters*, 80(10):2245, 1998.

[226] Sandeep K Goyal, Subhashish Banerjee, and Sibasish Ghosh. Effect of control procedures on the evolution of entanglement in open quantum systems. *Physical Review A*, 85(1):012327, 2012.

[227] Subhashish Banerjee and R Srikanth. Geometric phase of a qubit interacting with a squeezed-thermal bath. *The European Physical Journal D-Atomic, Molecular, Optical and Plasma Physics*, 46(2):335–344, 2008.

[228] Debbie W Leung. Chois proof as a recipe for quantum process tomography. *Journal of Mathematical Physics*, 44(2):528–533, 2003.

[229] Rudolf Ahlswede, Lars Baumer, Ning Cai, Harout Aydinian, Vladimir Blinovsky, Christian Deppe, and Haik Mashurian. *General theory of information transfer and combinatorics*. Springer, 2006.

[230] K. Blum. *Density Matrix Theory and Applications (Physics of Atoms and Molecules)*. (Plenum Press, New York), 1996.

[231] Geetu Narang et al. Simulating a single-qubit channel using a mixed-state environment. *Physical Review A*, 75(3):032305, 2007.

[232] Peter W Shor. Polynomial-time algorithms for prime factorization and discrete logarithms on a quantum computer. *SIAM review*, 41(2):303–332, 1999.

[233] William K Wootters and Wojciech H Zurek. A single quantum cannot be cloned. *Nature*, 299(5886):802–803, 1982.

[234] Arun Kumar Pati and Samuel L Braunstein. Impossibility of deleting an unknown quantum state. *Nature*, 404(6774):164–165, 2000.

[235] A Robert Calderbank and Peter W Shor. Good quantum error-correcting codes exist. *Physical Review A*, 54(2):1098, 1996.

[236] Lorenza Viola and Seth Lloyd. Dynamical suppression of decoherence in two-state quantum systems. *Physical Review A*, 58(4):2733, 1998.

[237] Daniel A Lidar, Isaac L Chuang, and K Birgitta Whaley. Decoherence-free subspaces for quantum computation. *Physical Review Letters*, 81(12):2594, 1998.

[238] Shivaramakrishnan Pancharatnam. Generalized theory of interference, and its applications. In *Proceedings of the Indian Academy of Sciences-Section A*, volume 44, pages 247–262. Springer, 1956.

[239] Michael V Berry. Quantal phase factors accompanying adiabatic changes. In *Proceedings of the Royal Society of London A: Mathematical, Physical and Engineering Sciences*, volume 392, pages 45–57. The Royal Society, 1984.

[240] Barry Simon. Holonomy, the quantum adiabatic theorem, and berry's phase. *Physical Review Letters*, 51(24):2167, 1983.

[241] Yakir Aharonov and J Anandan. Phase change during a cyclic quantum evolution. *Physical Review Letters*, 58(16):1593, 1987.

[242] Joseph Samuel and Rajendra Bhandari. General setting for berry's phase. *Physical Review Letters*, 60(23):2339, 1988.

[243] N Mukunda and R Simon. Quantum kinematic approach to the geometric phase. i. general formalism. *Annals of Physics*, 228(2):205–268, 1993.

[244] Erik Sjöqvist, Arun K Pati, Artur Ekert, Jeeva S Anandan, Marie Ericsson, Daniel KL Oi, and Vlatko Vedral. Geometric phases for mixed states in interferometry. *Physical Review Letters*, 85(14):2845, 2000.

[245] DM Tong, Erik Sjöqvist, Leong Chuan Kwek, and Choo Hiap Oh. Kinematic approach to the mixed state geometric phase in nonunitary evolution. *Physical review letters*, 93(8):080405, 2004.

[246] Giuseppe Falci, Rosario Fazio, G Massimo Palma, Jens Siewert, and Vlatko Vedral. Detection of geometric phases in superconducting nanocircuits. *Nature*, 407(6802):355–358, 2000.

[247] Yasunobu Nakamura, Yu A Pashkin, and JS Tsai. Coherent control of macroscopic quantum states in a single-cooper-pair box. *nature*, 398(6730):786–788, 1999.

[248] Robert S Whitney, Yuriy Makhlin, Alexander Shnirman, and Yuval Gefen. Geometric nature of the environment-induced berry phase and geometric dephasing. *Physical review letters*, 94(7):070407, 2005.

[249] R Srikanth and Subhashish Banerjee. Squeezed generalized amplitude damping channel. *Physical Review A*, 77(1):012318, 2008.

[250] SN Sandhya and Subhashish Banerjee. Geometric phase: an indicator of entanglement. *The European Physical Journal D-Atomic, Molecular, Optical and Plasma Physics*, 66(6):1–6, 2012.

[251] Subhashish Banerjee, CM Chandrashekar, and Arun K Pati. Enhancement of geometric phase by frustration of decoherence: A parrondo-like effect. *Physical Review A*, 87(4):042119, 2013.

[252] K Singh, DM Tong, K Basu, JL Chen, and JF Du. Geometric phases for nondegenerate and degenerate mixed states. *Physical Review A*, 67(3):032106, 2003.

[253] Charles H Bennett and Gilles Brassard. Quantum cryptography: Public key distribution and coin tossing. *Theoretical computer science*, 560:7–11, 2014.

[254] Artur K Ekert. Quantum cryptography based on bells theorem. *Physical review letters*, 67(6):661, 1991.

[255] Charles H Bennett. Quantum cryptography using any two nonorthogonal states. *Physical review letters*, 68(21):3121, 1992.

[256] Anirban Pathak. *Elements of quantum computation and quantum communication*. Taylor & Francis, 2013.

[257] Kaoru Shimizu and Nobuyuki Imoto. Communication channels secured from eavesdropping via transmission of photonic bell states. *Physical Review A*, 60(1):157, 1999.

[258] Kim Boström and Timo Felbinger. Deterministic secure direct communication using entanglement. *Physical Review Letters*, 89(18):187902, 2002.

[259] Lior Goldenberg and Lev Vaidman. Quantum cryptography based on orthogonal states. *Physical Review Letters*, 75(7):1239, 1995.

[260] Marco Lucamarini and Stefano Mancini. Secure deterministic communication without entanglement. *Physical review letters*, 94(14):140501, 2005.

[261] Gui-lu Long, Fu-guo Deng, Chuan Wang, Xi-han Li, Kai Wen, and Wan-ying Wang. Quantum secure direct communication and deterministic secure quantum communication. *Frontiers of Physics in China*, 2(3):251–272, 2007.

[262] Charles H Bennett and Stephen J Wiesner. Communication via one-and two-particle operators on einstein-podolsky-rosen states. *Physical review letters*, 69(20):2881, 1992.

[263] R. Srikanth S. Banerjee N. Srinatha, S. Omkar and A. Pathak. The quantum cryptographic switch. *Quantum Information Processing*, 13, 2012.

[264] Reinhard F Werner. Quantum states with einstein-podolsky-rosen correlations admitting a hidden-variable model. *Physical Review A*, 40(8):4277, 1989.

[265] Vishal Sharma, Kishore Thapliyal, Anirban Pathak, and Subhashish Banerjee. A comparative study of protocols for secure quantum communication under noisy environment: single-qubit-based protocols versus entangled-state-based protocols. *Quantum Information Processing*, 15(11):4681–4710, 2016.

[266] Pathak A. Thapliyal, K. and S. Banerjee. Quantum cryptography over non-markovian channels. 16.

[267] Ryszard Horodecki, Michał Horodecki, and Paweł Horodecki. Teleportation, bell's inequalities and inseparability. *Physics Letters A*, 222(1-2):21–25, 1996.

[268] Micha Horodecki, Pawe Horodecki, and Ryszard Horodecki. Separability of mixed states: necessary and sufficient conditions. *Physics Letters A*, 223(1):1 – 8, 1996.

[269] Wojciech Hubert Zurek. Decoherence and the transition from quantum to classicalrevisited. In *Quantum Decoherence*, pages 1–31. Springer, 2006.

[270] Leah Henderson and Vlatko Vedral. Classical, quantum and total correlations. *Journal of physics A: mathematical and general*, 34(35):6899, 2001.

[271] Borivoje Dakić, Vlatko Vedral, and Časlav Brukner. Necessary and sufficient condition for nonzero quantum discord. *Physical review letters*, 105(19):190502, 2010.

[272] Satyabrata Adhikari and Subhashish Banerjee. Operational meaning of discord in terms of teleportation fidelity. *Physical Review A*, 86(6):062313, 2012.

[273] Shunlong Luo. Quantum discord for two-qubit systems. *Physical Review A*, 77(4):042303, 2008.

[274] Z Ficek and Ryszard Tanaś. Entangled states and collective nonclassical effects in two-atom systems. *Physics Reports*, 372(5):369–443, 2002.

[275] Subhashish Banerjee, V Ravishankar, and R Srikanth. Dynamics of entanglement in two-qubit open system interacting with a squeezed thermal bath via dissipative interaction. *Annals of Physics*, 325(4):816–834, 2010.

[276] Subhashish Banerjee, V Ravishankar, and R Srikanth. Entanglement dynamics in two-qubit open system interacting with a squeezed thermal bath via quantum nondemolition interaction. *The European Physical Journal D-Atomic, Molecular, Optical and Plasma Physics*, 56(2):277–290, 2010.

[277] S. Banerjee I. Chakrabarty and N. Siddharth. A study of quantum correlations in open quantum systems. *Quantum Information and Computation (QIC)*, 11:0541, 2011.

[278] GV Riazanov. The feynman path integral for the dirac equation. *Soviet Journal of Experimental and Theoretical Physics*, 6:1107, 1958.

[279] Richard Phillips Feynman and AR Hibbs. *Quantum mechanics and path integrals [by] RP Feynman [and] AR Hibbs.* McGraw-Hill, 1965.

[280] Yakir Aharonov, Luiz Davidovich, and Nicim Zagury. Quantum random walks. *Physical Review A*, 48(2):1687, 1993.

[281] David A Meyer. From quantum cellular automata to quantum lattice gases. *Journal of Statistical Physics*, 85(5-6):551–574, 1996.

[282] Edward Farhi and Sam Gutmann. Quantum computation and decision trees. *Physical Review A*, 58(2):915, 1998.

[283] A Ambainis, E Bach, A Nayak, A Vishwanath, and J Watrous. Proceeding of the 33rd acm symposium on theory of computing. 2001.

[284] Ashwin Nayak and Ashvin Vishwanath. Quantum walk on the line. *arXiv preprint quant-ph/0010117*, 2000.

[285] Andris Ambainis. Quantum walks and their algorithmic applications. *International Journal of Quantum Information*, 1(04):507–518, 2003.

[286] Andrew M Childs, Richard Cleve, Enrico Deotto, Edward Farhi, Sam Gutmann, and Daniel A Spielman. Proceedings of the 35th acm symposium on theory of computing. 2003.

[287] Neil Shenvi, Julia Kempe, and K Birgitta Whaley. Quantum random-walk search algorithm. *Physical Review A*, 67(5):052307, 2003.

[288] Andris Ambainis, Julia Kempe, and Alexander Rivosh. Coins make quantum walks faster. In *Proceedings of the sixteenth annual ACM-SIAM symposium on Discrete algorithms*, pages 1099–1108. Society for Industrial and Applied Mathematics, 2005.

[289] Jiangfeng Du, Hui Li, Xiaodong Xu, Mingjun Shi, Jihui Wu, Xianyi Zhou, and Rongdian Han. Experimental implementation of the quantum random-walk algorithm. *Physical Review A*, 67(4):042316, 2003.

[290] Colm A Ryan, Martin Laforest, Jean-Christian Boileau, and Raymond Laflamme. Experimental implementation of a discrete-time quantum random walk on an nmr quantum-information processor. *Physical Review A*, 72(6):062317, 2005.

[291] Hagai B Perets, Yoav Lahini, Francesca Pozzi, Marc Sorel, Roberto Morandotti, and Yaron Silberberg. Realization of quantum walks with negligible decoherence in waveguide lattices. *Physical review letters*, 100(17):170506, 2008.

[292] Hector Schmitz, Robert Matjeschk, Ch Schneider, Jan Glueckert, Martin Enderlein, Thomas Huber, and Tobias Schaetz. Quantum walk of a trapped ion in phase space. *Physical review letters*, 103(9):090504, 2009.

[293] F Zähringer, G Kirchmair, R Gerritsma, E Solano, R Blatt, and CF Roos. Realization of a quantum walk with one and two trapped ions. *Physical review letters*, 104(10):100503, 2010.

[294] Ben C Travaglione and Gerald J Milburn. Implementing the quantum random walk. *Physical Review A*, 65(3):032310, 2002.

[295] Michal Karski, Leonid Förster, Jai-Min Choi, Andreas Steffen, Wolfgang Alt, Dieter Meschede, and Artur Widera. Quantum walk in position space with single optically trapped atoms. *Science*, 325(5937):174–177, 2009.

[296] Matthew A Broome, Alessandro Fedrizzi, Benjimain P Lanyon, Ivan Kassal, Alan Aspuru-Guzik, and Andrew G White. Discrete single-photon quantum walks with tunable decoherence. *Physical review letters*, 104(15):153602, 2010.

[297] Viv Kendon. Decoherence in quantum walks–a review. *Mathematical Structures in Computer Science*, 17(6):1169–1220, 2007.

[298] CM Chandrashekar, R Srikanth, and Subhashish Banerjee. Symmetries and noise in quantum walk. *Physical Review A*, 76(2):022316, 2007.

[299] Subhashish Banerjee, R Srikanth, CM Chandrashekar, and Pranaw Rungta. Symmetry-noise interplay in a quantum walk on an n-cycle. *Physical Review A*, 78(5):052316, 2008.

[300] Chaobin Liu and Nelson Petulante. Quantum walks on the n-cycle subject to decoherence on the coin degree of freedom. *Physical Review E*, 81(3):031113, 2010.

[301] James D Whitfield, César A Rodríguez-Rosario, and Alán Aspuru-Guzik. Quantum stochastic walks: A generalization of classical random walks and quantum walks. *Physical Review A*, 81(2):022323, 2010.

[302] R Srikanth, Subhashish Banerjee, and CM Chandrashekar. Quantumness in a decoherent quantum walk using measurement-induced disturbance. *Physical Review A*, 81(6):062123, 2010.

[303] John R Klauder and Ennackel Chandy George Sudarshan. *Fundamentals of quantum optics*. Courier Corporation, 2006.

[304] Marlan O. Scully and M. Suhail Zubairy. *Quantum Optics*. Cambridge University Press, 1997.

[305] Wolfgang P Schleich. *Quantum optics in phase space*. John Wiley & Sons, 2011.

[306] G. S. Agarwal. *Quantum Optics*. Cambridge University Press, 2013.

[307] Ravinder R Puri. *Mathematical methods of quantum optics*, volume 79. Springer Science & Business Media, 2001.

[308] Andrei B Klimov and Sergei M Chumakov. *A group-theoretical approach to quantum optics*. John Wiley & Sons, 2009.

[309] E. P. Wigner. On the quantum correction for thermodynamic equilibrium. *Phys. Rev.*, 40:749, 1932.

[310] José E Moyal. Quantum mechanics as a statistical theory. In *Mathematical Proceedings of the Cambridge Philosophical Society*, volume 45, pages 99–124. Cambridge University Press, 1949.

[311] M. Hillery, Robert F O'Connell, Marlan O Scully, and Eugene P Wigner. Distribution functions in physics: fundamentals. *Physics reports*, 106(3):121–167, 1984.

[312] Young-Suk Kim and Marilyn E Noz. *Phase space picture of quantum mechanics: group theoretical approach*, volume 40. World Scientific, 1991.

[313] Adam Miranowicz, Wieslaw Leonski, and Nobuyuki Imoto. Quantum-optical states in finite-dimensional hilbert space. i. general formalism. *Modern nonlinear optics*, (Part I):155–193, 2003.

[314] Roy J Glauber. Coherent and incoherent states of the radiation field. *Physical Review*, 131(6):2766, 1963.

[315] ECG Sudarshan. Equivalence of semiclassical and quantum mechanical descriptions of statistical light beams. *Physical Review Letters*, 10(7):277, 1963.

[316] CL Mehta and ECG Sudarshan. Relation between quantum and semiclassical description of optical coherence. *Physical Review*, 138(1B):B274, 1965.

[317] Yutaka Kano. Probability distribution functions relating to blackbody radiation. *Journal of the Physical Society of Japan*, 19(9):1555–1560, 1964.

[318] Kôdi Husimi. Some formal properties of the density matrix. *Proceedings of the Physico-Mathematical Society of Japan. 3rd Series*, 22(4):264–314, 1940.

[319] Mark Saffman, Thad G Walker, and Klaus Mølmer. Quantum information with rydberg atoms. *Reviews of Modern Physics*, 82(3):2313, 2010.

[320] Yevhen Miroshnychenko, Tatjana Wilk, Amodsen Chotia, Matthieu Viteau, Daniel Comparat, Pierre Pillet, Antoine Browaeys, Philippe Grangier, et al. Observation of collective excitation of two individual atoms in the rydberg blockade regime. *Nature Physics*, 5(2):115–118, 2009.

[321] RL Stratonovich. On distributions in representation space. *SOVIET PHYSICS JETP-USSR*, 4(6):891–898, 1957.

[322] AB Klimov and SM Chumakov. On the su (2) wigner function dynamics. In *Quantum Theory and Symmetries*, pages 431–436. World Scientific, 2002.

[323] Sergey M Chumakov, Andrei B Klimov, and Kurt Bernardo Wolf. Connection between two wigner functions for spin systems. *Physical Review A*, 61(3):034101, 2000.

[324] William K Wootters. A wigner-function formulation of finite-state quantum mechanics. *Annals of Physics*, 176(1):1–21, 1987.

[325] Apostolos Vourdas. Factorization in finite quantum systems. *Journal of Physics A: Mathematical and General*, 36(20):5645, 2003.

[326] S Chaturvedi, E Ercolessi, G Marmo, G Morandi, N Mukunda, and R Simon. Wigner–weyl correspondence in quantum mechanics for continuous and discrete systemsa dirac-inspired view. *Journal of Physics A: Mathematical and General*, 39(6):1405, 2006.

[327] Ulf Leonhardt. Discrete wigner function and quantum-state tomography. *Physical Review A*, 53(5):2998, 1996.

[328] K. Blum. *Density Matrix Theory and Applications*. Plenum Press, New York, 1996.

[329] Richard N Zare. *Angular momentum: understanding spatial aspects in chemistry and physics*. New York, 1988.

[330] Leon Cohen and Marlan O Scully. Joint wigner distribution for spin-1/2 particles. *Foundations of physics*, 16(4):295–310, 1986.

[331] Joseph C Várilly and JoséM Gracia-Bondía. The moyal representation for spin. *Annals of physics*, 190(1):107–148, 1989.

[332] FT Arecchi, Eric Courtens, Robert Gilmore, and Harry Thomas. Atomic coherent states in quantum optics. *Physical Review A*, 6(6):2211, 1972.

[333] D. A. Varshalovich, A. N. Moskalev, and V. K. Khersonskii. *Quantum theory of angular momentum*. World Scientific, 1988.

[334] G Ramachandran, AR Usha Devi, P Devi, and Swarnamala Sirsi. Quasi-probability distributions for arbitrary spin-j particles. *Foundations of Physics*, 26(3):401–412, 1996.

[335] A. Pathak S. Omkar K. Thapliyal, S. Banerjee and V. Ravishankar. Quasiprobability distributions in open quantum systems: spin-qubit systems. *Ann. of Phys.*, 362:261286, 2015.

[336] Peter W Shor. Scheme for reducing decoherence in quantum computer memory. *Physical review A*, 52(4):R2493, 1995.

[337] A Robert Calderbank and Peter W Shor. Good quantum error-correcting codes exist. *Physical Review A*, 54(2):1098, 1996.

[338] Andrew M Steane. Error correcting codes in quantum theory. *Physical Review Letters*, 77(5):793, 1996.

[339] Charles H Bennett, David P DiVincenzo, John A Smolin, and William K Wootters. Mixed-state entanglement and quantum error correction. *Physical Review A*, 54(5):3824, 1996.

[340] Emanuel Knill and Raymond Laflamme. Theory of quantum error-correcting codes. *Physical Review A*, 55(2):900, 1997.

[341] Artur Ekert and Chiara Macchiavello. Quantum error correction for communication. *Physical Review Letters*, 77(12):2585, 1996.

[342] Raymond Laflamme, Cesar Miquel, Juan Pablo Paz, and Wojciech Hubert Zurek. Perfect quantum error correcting code. *Physical Review Letters*, 77(1):198, 1996.

[343] Daniel Gottesman. Class of quantum error-correcting codes saturating the quantum hamming bound. *Physical Review A*, 54(3):1862, 1996.

[344] Daniel Gottesman. Stabilizer codes and quantum error correction. *arXiv preprint quant-ph/9705052*, 1997.

[345] Lu-Ming Duan and Guang-Can Guo. Quantum error correction with spatially correlated decoherence. *Physical Review A*, 59(5):4058, 1999.

[346] Rochus Klesse and Sandra Frank. Quantum error correction in spatially correlated quantum noise. *Physical review letters*, 95(23):230503, 2005.

[347] James P Clemens, Shabnam Siddiqui, and Julio Gea-Banacloche. Quantum error correction against correlated noise. *Physical Review A*, 69(6):062313, 2004.

[348] Peter W Shor and John A Smolin. Quantum error-correcting codes need not completely reveal the error syndrome. *arXiv preprint quant-ph/9604006*, 1996.

[349] Graeme Smith and John A Smolin. Degenerate quantum codes for pauli channels. *Physical review letters*, 98(3):030501, 2007.

[350] Pradeep Sarvepalli and Andreas Klappenecker. Degenerate quantum codes and the quantum hamming bound. *Physical Review A*, 81(3):032318, 2010.

[351] Giulio Chiribella, Michele DallArno, Giacomo Mauro DAriano, Chiara Macchiavello, and Paolo Perinotti. Quantum error correction with degenerate codes for correlated noise. *Physical Review A*, 83(5):052305, 2011.

[352] Debbie W Leung, Michael A Nielsen, Isaac L Chuang, and Yoshihisa Yamamoto. Approximate quantum error correction can lead to better codes. *Physical Review A*, 56(4):2567, 1997.

[353] Benjamin Schumacher and Michael D Westmoreland. Approximate quantum error correction. *Quantum Information Processing*, 1(1):5–12, 2002.

[354] Hui Khoon Ng and Prabha Mandayam. Simple approach to approximate quantum error correction based on the transpose channel. *Physical Review A*, 81(6):062342, 2010.

[355] Cédric Bény and Ognyan Oreshkov. General conditions for approximate quantum error correction and near-optimal recovery channels. *Physical review letters*, 104(12):120501, 2010.

[356] Daniel Gottesman. An introduction to quantum error correction and fault-tolerant quantum computation. In *Quantum information science and its contributions to mathematics, Proceedings of Symposia in Applied Mathematics*, volume 68, pages 13–58, 2009.

[357] Giacomo Mauro D'Ariano. Quantum tomography: general theory and new experiments. *Fortschritte der Physik*, 48(5-7):579–588, 2000.

[358] GM D'Ariano and P Lo Presti. Quantum tomography for measuring experimentally the matrix elements of an arbitrary quantum operation. *Physical review letters*, 86(19):4195, 2001.

[359] Joseph B Altepeter, David Branning, Evan Jeffrey, TC Wei, Paul G Kwiat, Robert T Thew, Jeremy L OBrien, Michael A Nielsen, and Andrew G White. Ancilla-assisted quantum process tomography. *Physical Review Letters*, 90(19):193601, 2003.

[360] Giacomo Mauro D'Ariano. Universal quantum observables. *Physics Letters A*, 300(1):1–6, 2002.

[361] Masoud Mohseni and DA Lidar. Direct characterization of quantum dynamics. *Physical review letters*, 97(17):170501, 2006.

[362] Masoud Mohseni and Daniel A Lidar. Direct characterization of quantum dynamics: General theory. *Physical Review A*, 75(6):062331, 2007.

[363] Joseph Emerson, Marcus Silva, Osama Moussa, Colm Ryan, Martin Laforest, Jonathan Baugh, David G Cory, and Raymond Laflamme. Symmetrized characterization of noisy quantum processes. *Science*, 317(5846):1893–1896, 2007.

[364] Marcus Silva, Easwar Magesan, David W Kribs, and Joseph Emerson. Scalable protocol for identification of correctable codes. *Physical Review A*, 78(1):012347, 2008.

[365] Ariel Bendersky, Fernando Pastawski, and Juan Pablo Paz. Selective and efficient estimation of parameters for quantum process tomography. *Physical review letters*, 100(19):190403, 2008.

[366] S Omkar, R Srikanth, and Subhashish Banerjee. Characterization of quantum dynamics using quantum error correction. *Physical Review A*, 91(1):012324, 2015.

[367] S Omkar, R Srikanth, and Subhashish Banerjee. Quantum code for quantum error characterization. *Physical Review A*, 91(5):052309, 2015.

[368] S Mancini and VI Manko. Quantum semiclass. opt. 7 615 dariano gm, mancini s, manko vi and tombesi p 1996. *Quantum Semiclass. Opt*, 8:1017, 1995.

[369] V. I. Manko and O. V. Manko. Spin state tomography. *Jour. of Exp. and Theor. Phys.*, 85, 1997.

[370] Kishore Thapliyal, Subhashish Banerjee, and Anirban Pathak. Tomograms for open quantum systems: in (finite) dimensional optical and spin systems. *Annals of Physics*, 366:148–167, 2016.

[371] Alessandro Ferraro, Stefano Olivares, and Matteo GA Paris. Gaussian states in continuous variable quantum information. *arXiv preprint quant-ph/0503237*, 2005.

[372] Gerardo Adesso, Sammy Ragy, and Antony R Lee. Continuous variable quantum information: Gaussian states and beyond. *Open Systems & Information Dynamics*, 21(01n02):1440001, 2014.

[373] Rajiah Simon. Peres-horodecki separability criterion for continuous variable systems. *Physical Review Letters*, 84(12):2726, 2000.

[374] Lu-Ming Duan, Géza Giedke, Juan Ignacio Cirac, and Peter Zoller. Inseparability criterion for continuous variable systems. *Physical Review Letters*, 84(12):2722, 2000.

[375] S. Takagi. Progress of theoretical physics supplement. 88:1, 1986.

[376] Stephen W Hawking. Black hole explosions. *Nature*, 248(5443):30–31, 1974.

[377] Paul CW Davies. Scalar production in schwarzschild and rindler metrics. *Journal of Physics A: Mathematical and General*, 8(4):609, 1975.

[378] William G Unruh. Notes on black-hole evaporation. *Physical Review D*, 14(4):870, 1976.

[379] Jacob D Bekenstein. Black holes and entropy. *Physical Review D*, 7(8):2333, 1973.

[380] Nicholas David Birrell and Paul Charles William Davies. *Quantum fields in curved space*. Number 7. Cambridge university press, 1984.

[381] R. M. Wald. *General Relativity*. The University of Chicago Press, 1984.

[382] Wolfgang Rindler. Kruskal space and the uniformly accelerated frame. *American Journal of Physics*, 34(12):1174–1178, 1966.

[383] Stephen A Fulling. Nonuniqueness of canonical field quantization in riemannian space-time. *Physical Review D*, 7(10):2850, 1973.

[384] Paul M Alsing, Ivette Fuentes-Schuller, Robert B Mann, and Tracey E Tessier. Entanglement of dirac fields in noninertial frames. *Physical Review A*, 74(3):032326, 2006.

[385] Debbie W. Leung. Chois proof as a recipe for quantum process tomography. *Journal of Mathematical Physics*, 44(2):528–533, 2003.

[386] S. Omkar, Subhashish Banerjee, R. Srikanth, and Ashutosh Kumar Alok. The Unruh effect interpreted as a quantum noise channel. *Quant. Inf. Comput.*, 16:0757, 2016.

[387] Subhashish Banerjee, Ashutosh Kumar Alok, and S Omkar. Quantum fisher and skew information for unruh accelerated dirac qubit. *The European Physical Journal C*, 76(8):437, 2016.

[388] Subhashish Banerjee, Ashutosh Kumar Alok, S Omkar, and R Srikanth. Characterization of unruh channel in the context of open quantum systems. *Journal of High Energy Physics*, 2017(2):82, 2017.

[389] John N Bahcall, Maria C Gonzalez-Garcia, and Carlos Pena-Garay. Solar neutrinos before and after neutrino 2004. *Journal of High Energy Physics*, 2004(08):016, 2004.

[390] T Araki, K Eguchi, S Enomoto, K Furuno, K Ichimura, H Ikeda, K Inoue, K Ishihara, T Iwamoto, T Kawashima, et al. Measurement of neutrino oscillation with kamland: Evidence of spectral distortion. *Physical Review Letters*, 94(8):081801, 2005.

[391] K Abe, J Adam, H Aihara, T Akiri, C Andreopoulos, S Aoki, Akitaka Ariga, Tomoko Ariga, S Assylbekov, D Autiero, et al. Observation of electron neutrino appearance in a muon neutrino beam. *Physical review letters*, 112(6):061802, 2014.

[392] K Abe, J Adam, H Aihara, T Akiri, C Andreopoulos, S Aoki, A Ariga, T Ariga, S Assylbekov, D Autiero, et al. Measurement of neutrino oscillation parameters from muon neutrino disappearance with an off-axis beam. *Physical review letters*, 111(21):211803, 2013.

[393] FP An, JZ Bai, AB Balantekin, HR Band, D Beavis, W Beriguete, M Bishai, S Blyth, K Boddy, RL Brown, et al. Observation of electron-antineutrino disappearance at daya bay. *Physical Review Letters*, 108(17):171803, 2012.

[394] JK Ahn, S Chebotaryov, JH Choi, S Choi, W Choi, Y Choi, HI Jang, JS Jang, EJ Jeon, IS Jeong, et al. Observation of reactor electron antineutrinos disappearance in the reno experiment. *Physical Review Letters*, 108(19):191802, 2012.

[395] Daniel A Dwyer, Daya Bay Collaboration, et al. Improved measurement of electron-antineutrino disappearance at daya bay. *Nuclear Physics B-Proceedings Supplements*, 235:30–32, 2013.

[396] Carlo Giunti and Chung W Kim. *Fundamentals of neutrino physics and astrophysics*. Oxford university press, 2007.

[397] J Dajka, J Syska, and J Łuczka. Geometric phase of neutrino propagating through dissipative matter. *Physical Review D*, 83(9):097302, 2011.

[398] EM Spina and HA Weidenmüller. Damping of collective modes in dic: Quantal versus statistical fluctuations. *Nuclear Physics B*, 758(90111), 2006.

[399] Yasaman Farzan, Thomas Schwetz, and Alexei Yu Smirnov. Reconciling results of lsnd, miniboone and other experiments with soft decoherence. *Journal of High Energy Physics*, 2008(07):067, 2008.

[400] David Griffiths. *Introduction to elementary particles*. John Wiley & Sons, 2008.

[401] Aleksander Weron, AK Rajagopal, and Karina Weron. Quantum theory of decaying systems from the canonical decomposition of dynamical semigroups. *Physical Review A*, 31(3):1736, 1985.

[402] Paweł Caban, Jakub Rembieliński, Kordian A Smoliński, and Zbigniew Walczak. Unstable particles as open quantum systems. *Physical Review A*, 72(3):032106, 2005.

[403] Robert Alicki. General theory and applications to unstable particles. In *Quantum Dynamical Semigroups and Applications*, pages 1–46. Springer, 2007.

[404] Stephen W Hawking. Particle creation by black holes. *Communications in mathematical physics*, 43(3):199–220, 1975.

[405] Ashutosh Kumar Alok, Subhashish Banerjee, and S Uma Sankar. Re-examining sin 2β and δmd from evolution of mesons with decoherence. *Physics Letters B*, 749:94–97, 2015.

[406] Subhashish Banerjee and R Ghosh. Quantum theory of a stern-gerlach system in contact with a linearly dissipative environment. *Physical Review A*, 62(4):042105, 2000.

[407] Subhashish Banerjee and R Ghosh. General quantum brownian motion with initially correlated and nonlinearly coupled environment. *Physical Review E*, 67(5):056120, 2003.

[408] Angel Rivas, Susana F Huelga, and Martin B Plenio. Quantum non-markovianity: characterization, quantification and detection. *Reports on Progress in Physics*, 77(9):094001, 2014.

[409] Heinz-Peter Breuer, Elsi-Mari Laine, Jyrki Piilo, and Bassano Vacchini. Colloquium: Non-markovian dynamics in open quantum systems. *Reviews of Modern Physics*, 88(2):021002, 2016.

[410] AK Rajagopal, AR Usha Devi, and RW Rendell. Kraus representation of quantum evolution and fidelity as manifestations of markovian and non-markovian forms. *Physical Review A*, 82(4):042107, 2010.

[411] Michael JW Hall, James D Cresser, Li Li, and Erika Andersson. Canonical form of master equations and characterization of non-markovianity. *Physical Review A*, 89(4):042120, 2014.

[412] Wei-Min Zhang, Ping-Yuan Lo, Heng-Na Xiong, Matisse Wei-Yuan Tu, and Franco Nori. General non-markovian dynamics of open quantum systems. *Physical review letters*, 109(17):170402, 2012.

[413] Shunlong Luo, Shuangshuang Fu, and Hongting Song. Quantifying non-markovianity via correlations. *Physical Review A*, 86(4):044101, 2012.

[414] BM Garraway. Nonperturbative decay of an atomic system in a cavity. *Physical Review A*, 55(3):2290, 1997.

[415] Subhashish Banerjee, N Pradeep Kumar, R Srikanth, Vinayak Jagadish, and Francesco Petruccione. Non-markovian dynamics of discrete-time quantum walks. *arXiv preprint arXiv:1703.08004*, 2017.

[416] Pradeep Kumar, Subhashish Banerjee, R Srikanth, Vinayak Jagadish, and Francesco Petruccione. Non-markovian evolution: a quantum walk perspective. *arXiv preprint arXiv:1711.03267*, 2017.

[417] Sonja Daffer, Krzysztof Wódkiewicz, James D Cresser, and John K McIver. Depolarizing channel as a completely positive map with memory. *Physical Review A*, 70(1):010304, 2004.

[418] Margarida Hinarejos, Carlo Di Franco, Alejandro Romanelli, and Armando Pérez. Chirality asymptotic behavior and non-markovianity in quantum walks on a line. *Physical Review A*, 89(5):052330, 2014.

[419] Ronnie Kosloff. Quantum thermodynamics: A dynamical viewpoint. *Entropy*, 15(6):2100–2128, 2013.

[420] Robert Alicki. Quantum thermodynamics: An example of two-level quantum machine. *Open Systems and Information Dynamics*, 21(01n02):1440002, 2014.

Index

Printed in the United States
By Bookmasters